Student Solutions Manual for
Swokowski and Cole's
FUNDAMENTALS OF
TRIGONOMETRY

Ninth Edition

JEFFERY A. COLE
Anoka-Ramsey Community College

Brooks/Cole Publishing Company

I(T)P® *An International Thomson Publishing Company*

Pacific Grove • Albany • Belmont • Bonn • Boston • Cincinnati • Detroit • Johannesburg • London
Madrid • Melbourne • Mexico City • New York • Paris • Singapore • Tokyo • Toronto • Washington

Sponsoring Editor: *Melissa Henderson*
Marketing Team: *Caroline Croley, Debra Johnston*
Editorial Assistants: *Shelley Gesicki, Joanne Von Zastrow*

Production: *Dorothy Bell*
Cover Design: *Cassandra Chu*
Printing and Binding: *Patterson Printing*

For more information, contact:

BROOKS/COLE PUBLISHING COMPANY
511 Forest Lodge Road
Pacific Grove, CA 93950
USA

International Thomson Publishing Europe
Berkshire House 168-173
High Holborn
London WC1V 7AA
England

Thomas Nelson Australia
102 Dodds Street
South Melbourne, 3205
Victoria, Australia

Nelson Canada
1120 Birchmount Road
Scarborough, Ontario
Canada M1K 5G4

International Thomson Editores
Seneca 53
Col. Polanco
11560 México, D. F., México

International Thomson Publishing GmbH
Königswinterer Strasse 418
53227 Bonn
Germany

International Thomson Publishing Asia
60 Albert Street
#15-01 Albert Complex
Singapore 189969

International Thomson Publishing Japan
Hirakawacho Kyowa Building, 3F
2-2-1 Hirakawacho
Chiyoda-ku, Tokyo 102
Japan

Printed in the United States of America

10   9   8   7   6   5   4   3   2   1

ISBN 0-534-36079-3

PREFACE

This *Student's Solutions Manual* contains selected solutions and strategies for solving typical exercises in the text, *Fundamentals of Trigonometry, Ninth Edition*, by Earl W. Swokowski and Jeffery A. Cole.

In each exercise set, nearly all odd-numbered solutions are included, with an emphasis on the solutions of the applied "word" problems. In the review exercise sections at the end of each chapter, odd- and even-numbered solutions are included. For the discussion exercises at the end of each chapter, all odd-numbered solutions are included. I have tried to illustrate enough solutions so that the student will be able to obtain an understanding of all types of problems in each section.

A significant number of today's students are involved in various outside activities, and find it difficult, if not impossible, to attend all class sessions. This manual should help meet the needs of these students. In addition, it is my hope that this manual's solutions will enhance the understanding of all readers of the material and provide insights to solving other exercises.

I would appreciate any feedback concerning errors, solution correctness, solution style, or manual style—comments from students using previous editions have greatly strengthened the text's supplements as well as the text itself. These and any other comments may be sent directly to me at the address below or in care of the publisher (there is a comment card in the back of this manual).

I would like to thank: Gary Rockswold, of Mankato State University, for supplying solutions for many of the new applied problems and calculator exercises; Joan Cole, my wife, for proofing various features of the manual; George Morris, of Scientific Illustrators, for creating the mathematically precise art package; and Sally Lifland and Gail Magin, of Lifland et al., Bookmakers, for assembling the final manuscript. I dedicate this book to my children, Becky and Brad.

Jeffery A. Cole

Anoka-Ramsey Community College

11200 Mississippi Blvd. NW

Coon Rapids, MN 55433

Table of Contents

Notations

The following notations are used in the manual.

Note: { Notes to the student pertaining to hints on solutions, common mistakes, or

conventions to follow. }

{ }   { comments to the reader are in braces }

LS   { Left Side of an equation }

RS   { Right Side of an equation }

$\approx$ { approximately equal to }

$\Rightarrow$ { implies, next equation, logically follows }

$\Leftrightarrow$ { if and only if, is equivalent to }

$\bullet$ { bullet, used to separate problem statement from solution or explanation }

$\star$ { used to identify the answer to the problem }

$\S$ { *section* references }

$\forall$ { For all, i.e., $\forall x$ means "for all x". }

$\mathbb{R} - \{a\}$ { The set of all real numbers except a. }

$\therefore$ { therefore }

QI–QIV   { quadrants I, II, III, IV }

To the Student

This manual is a text supplement and should be read along *with* the text. Read all exercise solutions in this manual since explanations of concepts are given and then appear in subsequent solutions. All concepts necessary to solve a particular problem are not reviewed for every exercise. If you are having difficulty with a previously covered concept, look back to the section where it was covered for more complete help. The writing style I have used in this manual reflects the way I explain concepts to my own students. It is not as mathematically precise as that of the text, including phrases such as "goes down" or "touches and turns around." My students have told me that these terms help them understand difficult concepts with ease. Lengthier explanations and more steps are given for the more difficult problems.

In the review sections, the solutions are somewhat abbreviated since more detailed solutions were given in previous sections. However, this is not true for the word problems in these sections since they are unique. In easier groups of exercises, representative solutions are shown. Occasionally, alternate solutions are also given.

All figures have been plotted using computer software, offering a high degree of precision. The calculator graphs are from the TI-82 screen, and any specific instructions are for the TI-81/82/83. When possible, each piece of art was made with the same scale to show a realistic and consistent graph.

This manual was done using EXP: *The Scientific Word Processor*. I have used a variety of display formats for the mathematical equations, including centering, vertical alignment, and flushing text to the right. I hope that these make reading and comprehending the material easier for you.

Chapter 1: Topics from Algebra

[1] (a) Since x and y have opposite signs, the product xy is negative.

(b) Since $x^2 > 0$ and $y > 0$, $x^2 y > 0$.

(c) Since $x < 0$ { x is negative } and $y > 0$ { y is positive }, $\frac{x}{y}$ is negative.

$\quad$ Thus, $\frac{x}{y} + x$ is the sum of two negatives, which is *negative*.

(d) Since $y > 0$ and $x < 0$, $y - x > 0$.

[3] (a) Since -7 is to the left of -4 on a coordinate line, $-7 \boxed{<} -4$.

(b) Using a calculator, we see that $\frac{\pi}{2} \approx 1.5708$. Hence, $\frac{\pi}{2} \boxed{>} 1.57$.

(c) $\sqrt{225} \boxed{=} 15$ *Note:* $\sqrt{225} \neq \pm 15$

[5] (a) Since $\frac{1}{11} = 0.\overline{09}$, $\frac{1}{11} \boxed{>} 0.09$. $\qquad$ (b) Since $\frac{2}{3} = 0.\overline{6}$, $\frac{2}{3} \boxed{>} 0.6666$.

(c) Since $\frac{22}{7} = 3.\overline{142857}$ and $\pi \approx 3.141593$, $\frac{22}{7} \boxed{>} \pi$.

Note: An informal definition of absolute value that may be helpful is

$$| \, something \, | = \begin{cases} itself & \text{if } itself \text{ is positive or zero} \\ -(itself) & \text{if } itself \text{ is negative} \end{cases}$$

[9] (a) $|-3 - 2| = |-5| = -(-5)$ { since $-5 < 0$ } $= 5$

(b) $|-5| - |2| = -(-5) - 2 = 5 - 2 = 3$

(c) $|7| + |-4| = 7 + [-(-4)] = 7 + 4 = 11$

[11] (a) $(-5)|3 - 6| = (-5)|-3| = (-5)[-(-3)] = (-5)(3) = -15$

(b) $|-6|/(-2) = -(-6)/(-2) = 6/(-2) = -3$

(c) $|-7| + |4| = -(-7) + 4 = 7 + 4 = 11$

[13] (a) Since $(4 - \pi)$ is positive, $|4 - \pi| = 4 - \pi$.

(b) Since $(\pi - 4)$ is negative, $|\pi - 4| = -(\pi - 4) = 4 - \pi$.

(c) Since $(\sqrt{2} - 1.5)$ is negative, $|\sqrt{2} - 1.5| = -(\sqrt{2} - 1.5) = 1.5 - \sqrt{2}$.

[17] (a) $d(A, B) = |1 - (-9)| = |10| = 10$ (b) $d(B, C) = |10 - 1| = |9| = 9$

(c) $d(C, B) = d(B, C) = 9$ $\qquad\qquad$ (d) $d(A, C) = |10 - (-9)| = |19| = 19$

Note: Exer. 19–24: Since $|a| = |-a|$, the answers could have a different form.

$\quad$ For example, $|-3 - x| \geq 8$ is equivalent to $|x + 3| \geq 8$.

[19] $d(A, B) = |7 - x| \Rightarrow |7 - x| < 5$

[21] $d(A, B) = |-3 - x| \Rightarrow |-3 - x| \geq 8$

[25] Pick an arbitrary value for x that is less than -3, say -5.

$\quad$ Since $3 + (-5) = -2$ is negative, we conclude that if $x < -3$, then $3 + x$ is negative.

$\quad$ Hence, $|3 + x| = -(3 + x) = -x - 3$.

[27] If $x < 2$, then $2 - x > 0$, and $|2 - x| = 2 - x$.

[29] If $a < b$, then $a - b < 0$, and $|a - b| = -(a - b) = b - a$.

[31] Since $x^2 + 4 > 0$ for every x, $|x^2 + 4| = x^2 + 4$.

[33] $\quad 4(2x + 5) = 3(5x - 2)$ $\qquad$ given

$\quad\quad 8x + 20 = 15x - 6$ $\qquad$ multiply terms

$\quad\quad\quad\quad 26 = 7x$ $\qquad$ get constants on one side, variables on the other

$\quad\quad\quad\quad x = \frac{26}{7}$ $\qquad$ solve for x

[35] $3x(x - 2)(4x + 3) = 0 \Rightarrow$

$\quad 3x = 0$ or $(x - 2) = 0$ or $(4x + 3) = 0$ { by the Zero Factor Theorem } $\Rightarrow x = 0, 2, -\frac{3}{4}$

[37] The least common denominator of all terms is x. Multiply each term of the equation

$$8 - \frac{5}{x} = 2 + \frac{3}{x}$$

by the lcd to obtain $8x - 5 = 2x + 3$ and then solve that equation $\{ 6x = 8 \}$ to get $x = \frac{4}{3}$. Remember that if we multiply an equation by a variable expression, we must make sure that any solutions do not make that expression equal to 0, since then the original equation would be undefined. In this particular case, since x equals $\frac{4}{3}$ and x is not equal to 0, $x = \frac{4}{3}$ is a valid solution.

[39] $6x^2 + x - 12 = 0 \Rightarrow (2x + 3)(3x - 4) = 0 \Rightarrow x = -\frac{3}{2}, \frac{4}{3}$

[41] $15x^2 - 12 = -8x$ { get all terms on one side of the equals sign, zero on the other side }

$\quad\quad \Rightarrow 15x^2 + 8x - 12 = 0$ { factor } $\Rightarrow (5x + 6)(3x - 2) = 0 \Rightarrow x = -\frac{6}{5}, \frac{2}{3}$

[43] $25x^2 = 9 \Rightarrow x^2 = \frac{9}{25} \Rightarrow x = \pm\sqrt{\frac{9}{25}} = \pm\frac{3}{5}$

[45] $(x - 3)^2 = 17 \Rightarrow x - 3 = \pm\sqrt{17} \Rightarrow x = 3 \pm \sqrt{17}$

[47] For the equation $x^2 + 4x + 2 = 0$, we use the quadratic formula, $x = \dfrac{-b \pm \sqrt{b^2 - 4ac}}{2a}$,

with $a = 1$, $b = 4$, and $c = 2$. Thus, the solutions are

$$x = \frac{-(4) \pm \sqrt{(4)^2 - 4(1)(2)}}{2(1)} = \frac{-4 \pm \sqrt{16 - 8}}{2} = \frac{-4 \pm 2\sqrt{2}}{2} = -2 \pm \sqrt{2}.$$

Note: A common mistake is to not divide 2 into *both* terms of the numerator.

[49] For the equation $2x^2 - 3x - 4 = 0$, we use the quadratic formula, $x = \dfrac{-b \pm \sqrt{b^2 - 4ac}}{2a}$,

with $a = 2$, $b = -3$, and $c = -4$. Thus, the solutions are

$$x = \frac{-(-3) \pm \sqrt{(-3)^2 - 4(2)(-4)}}{2(2)} = \frac{3 \pm \sqrt{9 + 32}}{4} = \frac{3}{4} \pm \frac{1}{4}\sqrt{41}.$$

Note: A common mistake is to not divide 4 into *both* terms of the numerator.

[51] $x^4 - 25x^2 + 144 = 0 \Rightarrow (x^2 - 9)(x^2 - 16) = 0 \Rightarrow x^2 = 9, 16 \Rightarrow x = \pm 3, \pm 4$

Alternative Solution: Use the quadratic formula and solve for x^2.

Note: The bracket symbols "[" and "]", are used with $\le$ or $\ge$ to denote that the endpoint of the interval is part of the solution. Parentheses, "(" and ")" are used with $<$ or $>$ and denote that the endpoint is *not* part of the solution.

[53] $x < -2 \Leftrightarrow (-\infty, -2)$ [57] $-2 < x \le 4 \Leftrightarrow (-2, 4]$

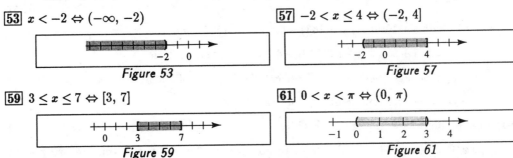

Figure 53   Figure 57

[59] $3 \le x \le 7 \Leftrightarrow [3, 7]$ [61] $0 < x < \pi \Leftrightarrow (0, \pi)$

Figure 59   Figure 61

[63] $(-5, 8] \Leftrightarrow -5 < x \le 8$ [65] $[-4, -1] \Leftrightarrow -4 \le x \le -1$

[67] $[4, \infty) \Leftrightarrow x \ge 4$ [69] $(-\infty, -5) \Leftrightarrow x < -5$

[73] $2x + 5 < 3x - 7 \Rightarrow -x < -12$ { Remember to change the direction of the inequality when multiplying or dividing by a negative value.} $\Rightarrow x > 12 \Leftrightarrow (12, \infty)$

[75] $\left[3 \le \dfrac{2x - 3}{5} < 7\right] \cdot 5$ { given inequality } $\Rightarrow$

$15 \le 2x - 3 < 35$ { multiply by the lcd, 5 } $\Rightarrow$

$18 \le 2x < 38$ { add 3 to all three parts } $\Rightarrow$

$9 \le x < 19$ { divide all three parts by 2 } $\Leftrightarrow [9, 19)$ { equivalent interval notation }

[77] By the law of signs, a quotient is positive if the sign of the numerator and the sign of the denominator are the same. Since the numerator is positive, $\dfrac{4}{3x + 2} > 0 \Rightarrow$

$3x + 2 > 0 \Rightarrow x > -\frac{2}{3} \Leftrightarrow (-\frac{2}{3}, \infty)$. The expression is never equal to 0 since the numerator is never 0. Thus, the solution of $\dfrac{4}{3x + 2} \ge 0$ is $(-\frac{2}{3}, \infty)$.

[79] $\dfrac{-2}{4 - 3x} > 0 \Rightarrow 4 - 3x < 0$ { denominator must also be negative } $\Rightarrow x > \frac{4}{3} \Leftrightarrow (\frac{4}{3}, \infty)$

[81] $(1 - x)^2 > 0 \ \forall x$ except 1. Thus, $\dfrac{2}{(1 - x)^2} > 0$ has solution $\mathbb{R} - \{1\}$.

[83] Construct a right triangle with sides of lengths $\sqrt{2}$ and 1. The hypotenuse will have length $\sqrt{(\sqrt{2})^2 + 1^2} = \sqrt{3}$. Next construct a right triangle with sides of lengths $\sqrt{3}$ and $\sqrt{2}$. The hypotenuse will have length $\sqrt{(\sqrt{3})^2 + (\sqrt{2})^2} = \sqrt{5}$.

[85] $W = 0.1166 h^{1.7}$

Height	64	65	66	67	68	69	70	71
Weight	137	141	145	148	152	156	160	164
Height	72	73	74	75	76	77	78	79
Weight	168	172	176	180	184	188	192	196

87 Let x denote the number of years before A becomes more economical than B.

The costs are the initial costs plus the yearly costs times the number of years.

$\text{Cost}_A < \text{Cost}_B \Rightarrow 50,000 + 4000x < 40,000 + 5500x \Rightarrow 10,000 < 1500x \Rightarrow$

$$x > \tfrac{20}{3}, \text{ or } 6\tfrac{2}{3} \text{ yr.}$$

89 (a) $C_f = 66.5 + 13.8(59) + 5(163) - 6.8(25) = 1525.7$ calories

$C_m = 655 + 9.6(75) + 1.9(178) - 4.7(55) = 1454.7$ calories

(b) As people age they require fewer calories. The coefficients of w and h are positive because large people require more calories.

91 The y-values are increasing rapidly and can best be described by equation (4), $y = x^3 - x^2 + x - 10$. Don't forget to go through Appendix I if you are using a TI-82/83 graphing calculator in your course.

93 (a) Let $Y_1 = D_1 = 6.096L + 685.7$ and

$Y_2 = D_2 = 0.00178L^3 - 0.072L^2 + 4.37L + 719$.

Table each equation and compare them to the actual values.

x (L)	Y_1	Y_2	Summer
0	686	719	720
10	747	757	755
20	808	792	792
30	869	833	836
40	930	893	892
50	991	980	978
60	1051	1106	1107

Comparing Y_1 (D_1) with Y_2 (D_2) we can see that the linear equation D_1 is not as accurate as the cubic equation D_2.

(b) $L = 35 \Rightarrow D_2 = 0.00178(35)^3 - 0.072(35)^2 + 4.37(35) + 719 \approx 860$ min.

1.2 Exercises

1 (a) $x = -2$ is the line parallel to the y-axis that intersects the x-axis at $(-2, 0)$.

(b) $y = 3$ is the line parallel to the x-axis that intersects the y-axis at $(0, 3)$.

(c) $x \geq 0$ { x is zero or positive }

is the set of all points to the right of and on the y-axis.

(d) $xy > 0$ { x and y have the same sign, that is, either both are positive or both are negative } is the set of all points in quadrants I and III.

(e) $y < 0$ { y is negative } is the set of all points below the x-axis.

(f) $x = 0$ is the set of all points on the y-axis.

3 (a) $A(4, -3)$, $B(6, 2) \Rightarrow d(A, B) = \sqrt{(6-4)^2 + [2-(-3)]^2} = \sqrt{4+25} = \sqrt{29}$

(b) $M_{AB} = \left(\dfrac{4+6}{2}, \dfrac{-3+2}{2}\right) = (5, -\tfrac{1}{2})$

5 We need to show that the sides satisfy the Pythagorean theorem. Finding the distances, we have $d(A, B) = \sqrt{98}$, $d(B, C) = \sqrt{32}$, and $d(A, C) = \sqrt{130}$. Since $d(A, C)$ is the largest of the three values, it must be the hypotenuse, hence, we need to check if $d(A, C)^2 = d(A, B)^2 + d(B, C)^2$. Since $(\sqrt{130})^2 = (\sqrt{98})^2 + (\sqrt{32})^2$, we know that $\triangle ABC$ is a right triangle. The area of a triangle is given by $A = \tfrac{1}{2}(\text{base})(\text{height})$. We can use $d(B, C)$ for the base and $d(A, B)$ for the height. Hence, area $= \tfrac{1}{2}bh = \tfrac{1}{2}(\sqrt{32})(\sqrt{98}) = \tfrac{1}{2}(4\sqrt{2})(7\sqrt{2}) = \tfrac{1}{2}(28)(2) = 28$.

7 We need to show that all 4 sides are the same length. Checking, we find that $d(A, B) = d(B, C) = d(C, D) = d(D, A) = \sqrt{29}$. This guarantees that we have a rhombus {a parallelogram with 4 equal sides}. Thus, we also need to show that adjacent sides meet at right angles. This can be done by showing that two adjacent sides and a diagonal form a right triangle. Using $\triangle ABC$, we see that $d(A, C) = \sqrt{58}$ and hence $d(A, C)^2 = d(A, B)^2 + d(B, C)^2$. We conclude that $ABCD$ is a square.

9 Let $B = (x, y)$. $A(-3, 8) \Rightarrow M_{AB} = \left(\dfrac{-3+x}{2}, \dfrac{8+y}{2}\right)$. $M_{AB} = C(5, -10) \Rightarrow$

$-3 + x = 2(5)$ and $8 + y = 2(-10) \Rightarrow x = 13$ and $y = -28$. $B = (13, -28)$.

11 The perpendicular bisector of AB is the line that passes through the midpoint of segment AB and intersects segment AB at a right angle. The points on the perpendicular bisector are all equidistant from A and B. Thus, we need to show that $d(A, C) = d(B, C)$. Since each of these is $\sqrt{145}$, we conclude that C is on the perpendicular bisector of AB.

13 Let $Q(0, y)$ be an arbitrary point on the y-axis.

Applying the distance formula with Q and $P(5, 3)$, we have $6 = d(P, Q) \Rightarrow$

$6 = \sqrt{(0-5)^2 + (y-3)^2} \Rightarrow 36 = 25 + y^2 - 6y + 9 \Rightarrow$
$y^2 - 6y - 2 = 0 \Rightarrow y = 3 \pm \sqrt{11}$. The points are $(0, 3 + \sqrt{11})$ and $(0, 3 - \sqrt{11})$.

$\boxed{15}$ With $P(a, 3)$ and $Q(5, 2a)$, $d(P, Q) > \sqrt{26} \Rightarrow \sqrt{(5-a)^2 + (2a-3)^2} > \sqrt{26} \Rightarrow$

$25 - 10a + a^2 + 4a^2 - 12a + 9 > 26 \Rightarrow 5a^2 - 22a + 8 > 0 \Rightarrow (5a - 2)(a - 4) > 0.$

Using *Diagram 15*, we see that $a < \frac{2}{5}$ or $a > 4$ will assure us that $d(P, Q) > \sqrt{26}$.

Resulting sign:	$\oplus$	$\ominus$	$\oplus$
Sign of $5a - 2$:	$-$	$+$	$+$
Sign of $a - 4$:	$-$	$-$	$+$
a values:	2/5	4	

$\boxed{\text{Diagram 15}}$

$\boxed{17}$ As in Example 2, we expect the graph of $y = 2x - 3$ to be a line.

Creating a table of values similar to those in the text, we have:

x	-2	-1	0	1	2
y	-7	-5	-3	-1	1

By plotting these points and connecting them, we obtain *Figure 17*.

To find the x-intercept, let $y = 0$ in $y = 2x - 3$, and solve for x. (1.5, 0)

To find the y-intercept, let $x = 0$ in $y = 2x - 3$, and solve for y. (0, −3)

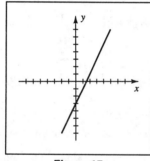

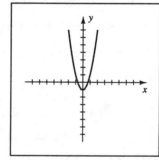

Figure 17   *Figure 23*

$\boxed{23}$ $y = 2x^2 - 1$ • x-intercepts: $(\pm\frac{1}{2}\sqrt{2}, 0)$ y-intercept: $(0, -1)$

Since we can substitute $-x$ for x in the equation and obtain an equivalent equation, we know the graph is symmetric with respect to the y-axis. We will make use of this fact when constructing our table. As in Example 3, we obtain a parabola.

x	± 2	$\pm\frac{3}{2}$	± 1	$\pm\frac{1}{2}$	0
y	7	$\frac{7}{2}$	1	$-\frac{1}{2}$	-1

27 $x = -y^2 + 3$ • x-intercept: $(3, 0)$ y-intercepts: $(0, \pm\sqrt{3})$

Since we can substitute $-y$ for y in the equation and obtain an equivalent equation, we know the graph is symmetric with respect to the x-axis. We will make use of this fact when constructing our table. As in Example 5, we obtain a parabola.

x	-13	-6	-1	2	3
y	± 4	± 3	± 2	± 1	0

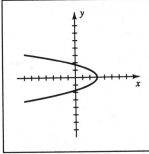

Figure 27

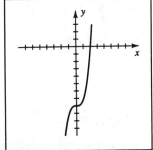

Figure 31

31 $y = x^3 - 8$ • x-intercept: $(2, 0)$ y-intercept: $(0, -8)$

x	-2	-1	0	1	2
y	-16	-9	-8	-7	0

35 $y = \sqrt{x} - 4$ • x-intercept: $(16, 0)$ y-intercept: $(0, -4)$

x	0	1	4	9	16
y	-4	-3	-2	-1	0

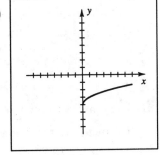

Figure 35

37 You may be able to do this exercise mentally. For example (using Exercise 17 with $y = 2x - 3$), we see that substituting $-x$ for x gives us $y = -2x - 3$; substituting $-y$ for y gives us $-y = 2x - 3$ or, equivalently, $y = -2x + 3$; and substituting $-x$ for x and $-y$ for y gives us $-y = -2x - 3$ or, equivalently, $y = 2x + 3$. None of the resulting equations are equivalent to the original equation, so there is no symmetry with respect to the y-axis, x-axis, or the origin.

(a) The graphs of the equations in Exercises 21 and 23 are symmetric with respect to the y-axis.

37 (b) The graphs of the equations in Exercises 25 and 27 are symmetric with respect to the x-axis.

(c) The graph of the equation in Exercise 29 is symmetric with respect to the origin.

41 $(x+3)^2 + (y-2)^2 = 9$ is a circle of radius $r = \sqrt{9} = 3$ with center $C(-3, 2)$.

45 $4x^2 + 4y^2 = 25 \Rightarrow x^2 + y^2 = \frac{25}{4}$ is a circle of radius $r = \sqrt{\frac{25}{4}} = \frac{5}{2}$ with center $C(0, 0)$.

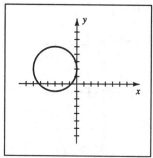
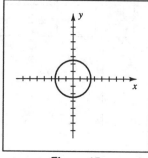
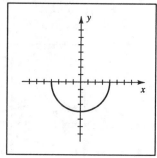

| Figure 41 | Figure 45 | Figure 47 |

47 As in Example 9, $y = -\sqrt{16 - x^2}$ is the lower half of the circle $x^2 + y^2 = 16$.

55 An equation of a circle with center $C(-4, 6)$ is $(x+4)^2 + (y-6)^2 = r^2$.

Since the circle passes through $P(1, 2)$, we know that $x = 1$ and $y = 2$ is one solution of the general equation. Letting $x = 1$ and $y = 2$ yields $5^2 + (-4)^2 = r^2 \Rightarrow r^2 = 41$.

An equation is $(x+4)^2 + (y-6)^2 = 41$.

57 "Tangent to the y-axis" means that the circle will intersect the y-axis at exactly one point. The distance from the center $C(-3, 6)$ to this point of tangency is 3 units—this is the length of the radius of the circle. An equation is $(x+3)^2 + (y-6)^2 = 9$.

59 Since the radius is 4 and $C(h, k)$ is in QII, $h = -4$ and $k = 4$.

An equation is $(x+4)^2 + (y-4)^2 = 16$.

61 The center of the circle is the midpoint M of $A(4, -3)$ and $B(-2, 7)$. $M = (1, 2)$.

The radius of the circle is $\frac{1}{2} \cdot d(A, B) = \frac{1}{2}\sqrt{136} = \sqrt{34}$.

An equation is $(x-1)^2 + (y-2)^2 = 34 \; \{(\sqrt{34})^2\}$.

63 $x^2 + y^2 - 4x + 6y = 36$ {complete the square on x and y} $\Rightarrow$

$x^2 - 4x + \underline{4} + y^2 + 6y + \underline{9} = 36 + \underline{4} + \underline{9} \Rightarrow$

$(x-2)^2 + (y+3)^2 = 49.$ This is a circle with center $C(2, -3)$ and radius $r = 7$.

67 $2x^2 + 2y^2 - 12x + 4y = 15$ {divide both sides by 2} $\Rightarrow$

$x^2 + y^2 - 6x + 2y = \frac{15}{2}$ {complete the square on x and y} $\Rightarrow$

$x^2 - 6x + \underline{9} + y^2 + 2y + \underline{1} = \frac{15}{2} + \underline{9} + \underline{1} \Rightarrow$

$(x-3)^2 + (y+1)^2 = \frac{35}{2}.$ This is a circle with center $C(3, -1)$ and radius $r = \frac{1}{2}\sqrt{70}$.

71 $x^2 + y^2 - 2x - 8y + 19 = 0 \Rightarrow x^2 - 2x + \underline{1} + y^2 - 8y + \underline{16} = -19 + \underline{1} + \underline{16} \Rightarrow$
$(x-1)^2 + (y-4)^2 = -2.$ This is not a circle since r^2 cannot equal -2.

75 To obtain equations for the upper and lower halves, we solve the given equation for y in terms of x. $(x-2)^2 + (y+1)^2 = 49 \Rightarrow (y+1)^2 = 49 - (x-2)^2 \Rightarrow$

$y + 1 = \pm\sqrt{49 - (x-2)^2} \Rightarrow y = -1 \pm \sqrt{49 - (x-2)^2}.$

The upper half is $y = -1 + \sqrt{49 - (x-2)^2}$ and

the lower half is $y = -1 - \sqrt{49 - (x-2)^2}.$

To obtain equations for the right and left halves, we solve for x in terms of y.

$(x-2)^2 + (y+1)^2 = 49 \Rightarrow (x-2)^2 = 49 - (y+1)^2 \Rightarrow$

$x - 2 = \pm\sqrt{49 - (y+1)^2} \Rightarrow x = 2 \pm \sqrt{49 - (y+1)^2}.$ The right half is

$x = 2 + \sqrt{49 - (y+1)^2}$ and the left half is $x = 2 - \sqrt{49 - (y+1)^2}.$

77 We need to determine if the distance from P to C is *less than* r, *greater than* r, or *equal to* r and hence, P will be *inside* the circle, *outside* the circle, or *on* the circle, respectively.

(a) $P(2, 3)$, $C(4, 6) \Rightarrow d(P, C) = \sqrt{4 + 9} = \sqrt{13} < r \; \{r = 4\} \Rightarrow P$ is *inside* C.

(b) $P(4, 2)$, $C(1, -2) \Rightarrow d(P, C) = \sqrt{9 + 16} = 5 = r \; \{r = 5\} \Rightarrow P$ is *on* C.

(c) $P(-3, 5)$, $C(2, 1) \Rightarrow d(P, C) = \sqrt{25 + 16} = \sqrt{41} > r \; \{r = 6\} \Rightarrow P$ is *outside* C.

79 (a) To find the x-intercepts, let $y = 0$ and solve the resulting equation for x.

$$x^2 - 4x + 4 = 0 \Rightarrow (x-2)^2 = 0 \Rightarrow x = 2.$$

(b) To find the y-intercepts, let $x = 0$ and solve the resulting equation for y.

$$y^2 - 6y + 4 = 0 \Rightarrow y = \frac{6 \pm \sqrt{36 - 16}}{2} = 3 \pm \sqrt{5}.$$

81 $x^2 + y^2 + 4x - 6y + 4 = 0 \Leftrightarrow (x+2)^2 + (y-3)^2 = 9.$ This is a circle with center $C(-2, 3)$ and radius 3. The circle we want has the same center, $C(-2, 3)$, and radius that is equal to the distance from C to $P(2, 6)$.

$d(P, C) = \sqrt{16 + 9} = 5$ and an equation is $(x+2)^2 + (y-3)^2 = 25.$

83 The equation of circle C_2 is $(x-h)^2 + (y-2)^2 = 2^2.$ If we draw a line from the origin to the center of C_2, we form a right triangle with hypotenuse $5 - 2 \; \{C_2$ radius $- C_1$ radius$\} = 3$ and sides of length 2 and h. Thus, $h^2 + 2^2 = 3^2 \Rightarrow h = \sqrt{5}.$

85 Assuming that the x- and y-values of each point of intersection are integers, and that the ticks each represent one unit, we see that the intersection points are $(-3, -5)$ and $(2, 0)$. The viewing rectangle (VR) is $[-15, 15]$ by $[-10, 10]$.

$Y_1 < Y_2$ on $[-15, -3) \cup (2, 15].$

87 VR: $[-5, 5]$ by $[-1.3, 5.3]$. The intersection points are $(-1, 1)$, $(0, 0)$, and $(1, 1)$.

$$Y_1 > Y_2 \text{ on } [-5, -1) \cup (1, 5]. \quad Y_1 < Y_2 \text{ on } (-1, 0) \cup (0, 1).$$

89–90   For point plotting instructions on the TI-82/83, see Example 2 in Appendix I.

89 (a) Plot $(1990, 54{,}871{,}330)$, $(1991, 55{,}786{,}390)$, $(1992, 57{,}211{,}600)$,

$(1993, 58{,}834{,}440)$, and $(1994, 59{,}332{,}200)$.

$[1988, 1996]$ by $[54 \times 10^6, 61 \times 10^6]$

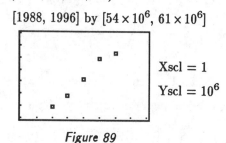

$\text{Xscl} = 1$
$\text{Yscl} = 10^6$

Figure 89

(b) The number of cable subscribers is increasing each year.

91 The viewing rectangles significantly affect the shape of the circle. Viewing rectangle
(2) results in a graph that most looks like a circle.

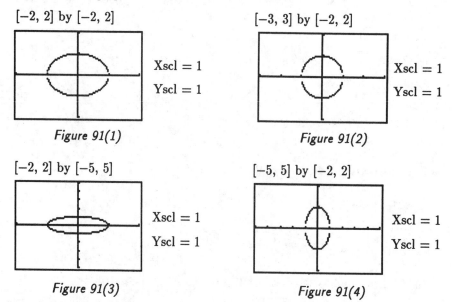

$[-2, 2]$ by $[-2, 2]$

$\text{Xscl} = 1$
$\text{Yscl} = 1$

Figure 91(1)

$[-3, 3]$ by $[-2, 2]$

$\text{Xscl} = 1$
$\text{Yscl} = 1$

Figure 91(2)

$[-2, 2]$ by $[-5, 5]$

$\text{Xscl} = 1$
$\text{Yscl} = 1$

Figure 91(3)

$[-5, 5]$ by $[-2, 2]$

$\text{Xscl} = 1$
$\text{Yscl} = 1$

Figure 91(4)

93 Assign $x^3 - \frac{9}{10}x^2 - \frac{43}{25}x + \frac{24}{25}$ to Y_1. After trying a standard viewing rectangle, we see
that the x-intercepts are near the origin and we choose the viewing rectangle $[-6, 6]$
by $[-4, 4]$. This is simply one choice, not necessarily the best choice. For most C
exercises, we have selected viewing rectangles that are in a $3:2$ proportion
(horizontal : vertical) to maintain a true proportion. From the graph, there are three
x-intercepts. Use a zoom-in feature to determine that they are approximately -1.2,
0.5, and 1.6. See *Figure 93* on the next page.

[−6, 6] by [−4, 4]

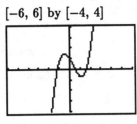

Xscl = 1

Yscl = 1

[−3, 3] by [−2, 2]

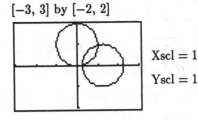

Xscl = 1

Yscl = 1

Figure 93   *Figure 97*

97 Depending on the type of graphing utility used, you may need to solve for y first.

$$x^2 + (y-1)^2 = 1 \Rightarrow y = 1 \pm \sqrt{1-x^2}; \qquad (x - \tfrac{5}{4})^2 + y^2 = 1 \Rightarrow y = \pm \sqrt{1 - (x - \tfrac{5}{4})^2}.$$

Make the assignments $Y_1 = \sqrt{1 - x^2}$, $Y_2 = 1 + Y_1$, $Y_3 = 1 - Y_1$, $Y_4 = \sqrt{1 - (x - \tfrac{5}{4})^2}$, and $Y_5 = -Y_4$. If a Y_5 is not available, you will need to use other function assignments or alternate methods. For example, on the TI-81, you can graph Y_5 by using DrawF $-Y_4$. Be sure to "turn off" Y_1 before graphing. From the graph, there are two points of intersection.

They are approximately (0.999, 0.968) and (0.251, 0.032).

99 The cars are initially 4 miles apart. Their distance decreases to 0 when they meet on the highway after 2 minutes. Then, their distance starts to increase until it is 4 miles after a total of 4 minutes.

[0, 4] by [0, 4]

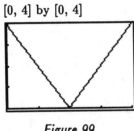

Xscl = 1

Yscl = 1

[−50, 50] by [900, 1200]

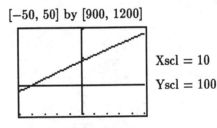

Xscl = 10

Yscl = 100

Figure 99   *Figure 101*

101   (a) $v = 1087\sqrt{(20 + 273)/273} \approx 1126$ ft/sec.

(b) Algebraically: $v = 1087\sqrt{\dfrac{T + 273}{273}} \Rightarrow 1000 = 1087\sqrt{\dfrac{T + 273}{273}} \Rightarrow$

$$T = \frac{1000^2 \times 273}{1087^2} - 273 \approx -42^\circ\text{C}.$$

Graphically: Graph $Y_1 = 1087\sqrt{(T + 273)/273}$ and $Y_2 = 1000$.

At the point of their intersection, $T \approx -42^\circ\text{C}$.

3 $f(x) = \sqrt{x-4} - 3x \Rightarrow f(4) = \sqrt{4-4} - 3(4) = \sqrt{0} - 12 = 0 - 12 = -12.$ Similarly,

$f(8) = -22$ and $f(13) = -36.$ Note that $f(a)$, with $a < 4$, would be undefined.

5 (a) $f(x) = 5x - 2 \Rightarrow f(a) = 5(a) - 2 = 5a - 2$ (b) $f(-a) = 5(-a) - 2 = -5a - 2$

(c) $-f(a) = -1 \cdot (5a - 2) = -5a + 2$ (d) $f(a+h) = 5(a+h) - 2 = 5a + 5h - 2$

(e) $f(a) + f(h) = (5a - 2) + (5h - 2) = 5a + 5h - 4$

(f) $\dfrac{f(a+h) - f(a)}{h} = \dfrac{(5a + 5h - 2) - (5a - 2)}{h} = \dfrac{5h}{h} = 5$

7 (a) $f(x) = x^2 - x + 3 \Rightarrow f(a) = (a)^2 - (a) + 3 = a^2 - a + 3$

(b) $f(-a) = (-a)^2 - (-a) + 3 = a^2 + a + 3$

(c) $-f(a) = -1 \cdot (a^2 - a + 3) = -a^2 + a - 3$

(d) $f(a+h) = (a+h)^2 - (a+h) + 3 = a^2 + 2ah + h^2 - a - h + 3$

(e) $f(a) + f(h) = (a^2 - a + 3) + (h^2 - h + 3) = a^2 + h^2 - a - h + 6$

(f) $\dfrac{f(a+h) - f(a)}{h} = \dfrac{(a^2 + 2ah + h^2 - a - h + 3) - (a^2 - a + 3)}{h} = \dfrac{2ah + h^2 - h}{h} =$

$\dfrac{h(2a + h - 1)}{h} = 2a + h - 1$

9 $\dfrac{f(x+h) - f(x)}{h} = \dfrac{[(x+h)^2 + 5] - [x^2 + 5]}{h} = \dfrac{(x^2 + 2xh + h^2 + 5) - (x^2 + 5)}{h} =$

$\dfrac{2xh + h^2}{h} = \dfrac{h(2x + h)}{h} = 2x + h$

11 $\dfrac{f(x) - f(a)}{x - a} = \dfrac{\sqrt{x-3} - \sqrt{a-3}}{x - a} = \dfrac{\sqrt{x-3} - \sqrt{a-3}}{x - a} \cdot \dfrac{\sqrt{x-3} + \sqrt{a-3}}{\sqrt{x-3} + \sqrt{a-3}} =$

$\dfrac{(x-3) - (a-3)}{(x-a)(\sqrt{x-3} + \sqrt{a-3})} = \dfrac{x - a}{(x-a)(\sqrt{x-3} + \sqrt{a-3})} = \dfrac{1}{\sqrt{x-3} + \sqrt{a-3}}$

13 (a) $g\left(\dfrac{1}{a}\right) = 4\left(\dfrac{1}{a}\right)^2 = \dfrac{4}{a^2}$ (b) $\dfrac{1}{g(a)} = \dfrac{1}{4(a)^2} = \dfrac{1}{4a^2}$

(c) $g(\sqrt{a}) = 4(\sqrt{a})^2 = 4a$ (d) $\sqrt{g(a)} = \sqrt{4a^2} = 2\,|\,a\,| = 2a$ since $a > 0$

15 (a) $g(x) = \dfrac{2x}{x^2 + 1} \Rightarrow g\left(\dfrac{1}{a}\right) = \dfrac{2(1/a)}{(1/a)^2 + 1} = \dfrac{2/a}{1/a^2 + 1} \cdot \dfrac{a^2}{a^2} = \dfrac{2a}{1 + a^2} = \dfrac{2a}{a^2 + 1}$

(b) $\dfrac{1}{g(a)} = \dfrac{1}{\dfrac{2a}{a^2 + 1}} = \dfrac{a^2 + 1}{2a}$ (c) $g(\sqrt{a}) = \dfrac{2\sqrt{a}}{(\sqrt{a})^2 + 1} = \dfrac{2\sqrt{a}}{a + 1}$

(d) $\sqrt{g(a)} = \sqrt{\dfrac{2a}{a^2 + 1}} \cdot \dfrac{\sqrt{a^2 + 1}}{\sqrt{a^2 + 1}} = \dfrac{\sqrt{2a(a^2 + 1)}}{a^2 + 1}$, or, equivalently, $\dfrac{\sqrt{2a^3 + 2a}}{a^2 + 1}$

[17] (a) The domain of a function f is the set of all x-values for which the function is defined. In this case, the graph extends from $x = -3$ to $x = 4$. Hence, the domain is $[-3, 4]$.

(b) The range of a function f is the set of all y-values that the function takes on. In this case, the graph includes all values from $y = -2$ to $y = 2$. Hence, the range is $[-2, 2]$.

(c) $f(1)$ is the y-value of f corresponding to $x = 1$. In this case, $f(1) = 0$.

(d) If we were to draw the horizontal line $y = 1$ on the same coordinate plane, it would intersect the graph at $x = -1, \frac{1}{2}$, and 2. Hence, $f(x) = 1 \Rightarrow x = -1, \frac{1}{2}, 2$.

(e) The function is above 1 between $x = -1$ and $x = \frac{1}{2}$, and also to the right of $x = 2$. Hence, $f(x) > 1 \Rightarrow x \in (-1, \frac{1}{2}) \cup (2, 4]$.

[19–30] We need to make sure that the radicand { the expression under the radical sign } is greater than or equal to zero and that the denominator is not equal to zero.

[21] $f(x) = \sqrt{9 - x^2}$ • $9 - x^2 \geq 0 \Rightarrow 3 \geq |x| \Rightarrow -3 \leq x \leq 3$ ★ $[-3, 3]$

[23] $f(x) = \dfrac{x + 1}{x^3 - 4x}$ • $x^3 - 4x = 0 \Rightarrow x(x + 2)(x - 2) = 0$ ★ $\mathbb{R} - \{ \pm 2, 0 \}$

[25] $f(x) = \dfrac{\sqrt{2x - 3}}{x^2 - 5x + 4}$ • For this function we must have the radicand greater than or equal to 0 *and* the denominator not equal to 0. The radicand is greater than or equal to 0 if $2x - 3 \geq 0$, or, equivalently, $x \geq \frac{3}{2}$. The denominator is $(x - 1)(x - 4)$, so $x \neq 1, 4$. The solution is then all real numbers greater than or equal to $\frac{3}{2}$, excluding 4. In interval notation, we have $[\frac{3}{2}, 4) \cup (4, \infty)$.

[27] $f(x) = \dfrac{x - 4}{\sqrt{x - 2}}$ • $x - 2 > 0 \Rightarrow x > 2$

{ > must be used since the denominator cannot equal 0 } ★ $(2, \infty)$

[29] $f(x) = \sqrt{x + 2} + \sqrt{2 - x}$ • $x + 2 \geq 0 \Rightarrow x \geq -2; 2 - x \geq 0 \Rightarrow x \leq 2$.

The domain is the intersection of $x \geq -2$ and $x \leq 2$, that is, $[-2, 2]$.

[33] (a) To sketch the graph of $f(x) = 4 - x^2$, we can make use of the symmetry with respect to the y-axis. See *Figure 33* on the next page.

x	± 4	± 3	± 2	± 1	0
y	-12	-5	0	3	4

(b) Since we can substitute any number for x, the domain is all real numbers, that is, $D = \mathbb{R}$. By examining *Figure 33*, we see that the values of y are at most 4. Hence, the range of f is all reals less than or equal to 4, that is, $R = (-\infty, 4]$.

(c) A common mistake is to confuse the function values, the y's, with the input values, the x's. We are not interested in the specific y-values for determining if the function is increasing, decreasing, or constant. We are only interested if the y-values are going up, going down, or staying the same. For the function $f(x) = 4 - x^2$, we say f *is increasing on* $(-\infty, 0]$ since the y-values are getting larger as we move from left to right over the x-values from $-\infty$ to 0. Also, f *is decreasing on* $[0, \infty)$ since the y-values are getting smaller as we move from left to right over the x-values from 0 to ∞. Note that this answer would have been the same if the function was $f(x) = 500 - x^2$, $f(x) = -300 - x^2$, or any function of the form $f(x) = a - x^2$, where a is any real number.

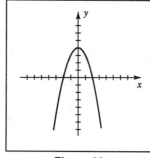

Figure 33

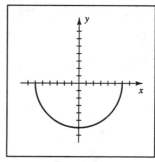

Figure 39

39 (a) We recognize $y = f(x) = -\sqrt{36 - x^2}$ as the lower half of the circle $x^2 + y^2 = 36$.

(b) To find the domain, we solve $36 - x^2 \geq 0$. $36 - x^2 \geq 0 \Rightarrow x^2 \leq 36 \Rightarrow$

$|x| \leq 6 \Rightarrow D = [-6, 6]$. From *Figure 39*, we see that the y-values vary from

$$y = -6 \text{ to } y = 0. \text{ Hence, the range } R \text{ is } [-6, 0].$$

(c) As we move from left to right, $x = -6$ to $x = 0$, the y-values are decreasing.

From $x = 0$ to $x = 6$, the y-values increase.

$$\text{Hence, } f \text{ is decreasing on } [-6, 0] \text{ and increasing on } [0, 6].$$

41 As in Example 7, $a = \dfrac{2 - 1}{3 - (-3)} = \dfrac{1}{6}$ and f has the form $f(x) = \frac{1}{6}x + b$.

$$f(3) = \tfrac{1}{6}(3) + b = \tfrac{1}{2} + b. \text{ But } f(3) = 2, \text{ so } \tfrac{1}{2} + b = 2 \Rightarrow b = \tfrac{3}{2}, \text{ and } f(x) = \tfrac{1}{6}x + \tfrac{3}{2}.$$

43-52 *Note:* A good question to consider is "Given a particular value of x, can a unique value of y be found?" If the answer is yes, the value of y (general formula) is given. If no, two ordered pairs satisfying the relation having x in the first position are given.

43 $2y = x^2 + 5 \Rightarrow y = \dfrac{x^2 + 5}{2}$, a function

45 $x^2 + y^2 = 4 \Rightarrow y^2 = 4 - x^2 \Rightarrow y = \pm\sqrt{4 - x^2}$, not a function, $(0, \pm 2)$

47 $y = 3$ is a function since for any x,

$(x, 3)$ is the only ordered pair in W having x in the first position.

49 Any ordered pair with x-coordinate 0 satisfies $xy = 0$.

Two such ordered pairs are $(0, 0)$ and $(0, 1)$. Not a function

53 $V = lwh = (30 - 2x)(20 - 2x)(x) = 4x(15 - x)(10 - x)$

55 (a) The formula for the area of a rectangle is $A = lw$ { Area = length × width }.

$$A = 500 \Rightarrow xy = 500 \Rightarrow y = \frac{500}{x}$$

(b) We need to determine the number of linear feet (P) first. There are two walls of length y, two walls of length $(x - 3)$, and one wall of length x.

Thus, $P = $ Linear feet of wall $= x + 2(y) + 2(x - 3) = 3x + 2\left(\dfrac{500}{x}\right) - 6.$

The cost C is 100 times P, so $C = 100P = 300x + \dfrac{100{,}000}{x} - 600.$

57 The expression $(h - 25)$ represents the number of feet *above* 25 feet.

$$S(h) = 6(h - 25) + 100 = 6h - 150 + 100 = 6h - 50.$$

59 (a) Using $(6, 48)$ and $(7, 50.5)$, we have

$$y - 48 = \frac{50.5 - 48}{7 - 6}(t - 6), \text{ or } y = 2.5t + 33.$$

(b) The slope represents the yearly increase in height, 2.5 in/yr.

(c) $t = 10 \Rightarrow y = 2.5(10) + 33 = 58$ in.

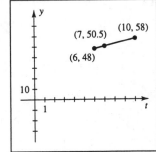

Figure 59

61 The height of the balloon is $2t$. Using the Pythagorean theorem,

$$d^2 = 100^2 + (2t)^2 \Rightarrow d^2 = 2^2(50)^2 + 2^2 t^2 \Rightarrow d = 2\sqrt{t^2 + 2500}.$$

63 (a) CTP forms a right angle, so the Pythagorean theorem may be applied.

$$(CT)^2 + (PT)^2 = (PC)^2 \Rightarrow r^2 + y^2 = (h + r)^2 \Rightarrow r^2 + y^2 = h^2 + 2hr + r^2 \Rightarrow$$
$$y^2 = h^2 + 2hr \ \{y > 0\} \Rightarrow y = \sqrt{h^2 + 2hr}$$

(b) $y = \sqrt{(200)^2 + 2(4000)(200)} = \sqrt{(200)^2(1 + 40)} = 200\sqrt{41} \approx 1280.6$ mi

65 Form a right triangle with the control booth and the beginning of the runway. Let y denote the distance from the control booth to the beginning of the runway and apply the Pythagorean theorem. $y^2 = 300^2 + 20^2 \Rightarrow y^2 = 90{,}400.$ Now form a right triangle, in a different plane, with sides y and x and hypotenuse d.

Then $d^2 = y^2 + x^2 \Rightarrow d^2 = 90{,}400 + x^2 \Rightarrow d = \sqrt{90{,}400 + x^2}.$

[67] Since F is directly proportional to x { x is the number of inches *beyond* the spring's natural length of 10 inches }, $F = kx$. $F = 4$ and $x = 10.3 - 10 = 0.3 \Rightarrow 4 = k(0.3) \Rightarrow$ $k = \frac{4}{0.3} = \frac{40}{3}$. If the spring is stretched to a length of 11.5 inches,

$$\text{then } x = 11.5 - 10 = 1.5 \text{ inches.}   F = \frac{40}{3}(1.5) = 20 \text{ lb.}$$

[69] $R = k\dfrac{l}{d^2} = \dfrac{kl}{d^2}$. $25 = \dfrac{k(100)}{(0.01)^2} \Rightarrow k = \dfrac{25(0.01)^2}{100} = \dfrac{1}{40,000}$.

$$R = \frac{\frac{1}{40,000}(50)}{(0.015)^2} = \frac{50}{(40,000)(0.015)^2} = \frac{50}{9} \approx 5.56 \text{ ohms.}$$

[71] The square of the distance from the origin to the point (x, y) is $x^2 + y^2$. $d = \dfrac{k}{x^2 + y^2}$.

If $(x_1, y_1) = (\frac{1}{3}x, \frac{1}{3}y)$ is the new point that has density d_1, then

$$d_1 = \frac{k}{x_1^2 + y_1^2} = \frac{k}{(\frac{1}{3}x)^2 + (\frac{1}{3}y)^2} = \frac{k}{\frac{1}{9}x^2 + \frac{1}{9}y^2} = \frac{k}{\frac{1}{9}(x^2 + y^2)} = 9 \cdot \frac{k}{x^2 + y^2} = 9d.$$

The density d is multiplied by 9.

[73] (b) The maximum y-value of 0.75 occurs when $x \approx 0.55$ and

the minimum y-value of -0.75 occurs when $x \approx -0.55$.

Therefore, the range of f is approximately $[-0.75, 0.75]$.

(c) f is decreasing on $[-2, -0.55]$ and on $[0.55, 2]$. f is increasing on $[-0.55, 0.55]$.

$[-2, 2]$ by $[-2, 2]$ $[-0.7, 1.4]$ by $[-1.1, 1]$

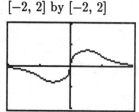
Xscl = 1
Yscl = 1

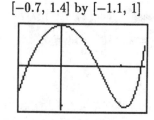
Xscl = 1
Yscl = 1

Figure 73   *Figure 75*

[75] (b) The maximum y-value of 1 occurs when $x = 0$ and

the minimum y-value of -1.03 occurs when $x \approx 1.06$.

Therefore, the range of f is approximately $[-1.03, 1]$.

(c) f is decreasing on $[0, 1.06]$. f is increasing on $[-0.7, 0]$ and on $[1.06, 1.4]$.

[77] For each of (a)–(e), an assignment to Y_1, an appropriate viewing rectangle, and the solution(s) are listed.

(a) $Y_1 = (x\char`^5)\char`^(1/3)$, VR: $[-40, 40]$ by $[-40, 40]$, $x = 8$

(b) $Y_1 = (x\char`^4)\char`^(1/3)$, VR: $[-20, 20]$ by $[-20, 20]$, $x = \pm 8$

(c) $Y_1 = (x\char`^2)\char`^(1/3)$, VR: $[-40, 40]$ by $[-40, 40]$,   no real solutions

(d) $Y_1 = (x\char`^3)\char`^(1/4)$, VR: $[0, 650]$ by $[0, 650]$, $x = 625$

(e) $Y_1 = (x\char`^3)\char`^(1/2)$, VR: $[-30, 30]$ by $[-30, 30]$,   no real solutions

79 (a) First, we must determine the equation of the line that passes through the points (1985, 11,450) and (1994, 20,021).

$$y - 11,450 = \frac{20,021 - 11,450}{1994 - 1985}(x - 1985) = \frac{2857}{3}(x - 1985) \Rightarrow$$

$$y = \frac{2857}{3}x - \frac{5,636,795}{3}. \text{ Thus, let } f(x) = \frac{2857}{3}x - \frac{5,636,795}{3} \text{ and graph } f.$$

(b) The average annual increase in the price paid for a new car is equal to the slope:

$$\frac{2857}{3} \approx \$952.33.$$

(c) Graph $y = \frac{2857}{3}x - \frac{5,636,795}{3}$ and $y = 25,000$ on the same coordinate axes. Their point of intersection is approximately (1999.2, 25,000). Thus, according to this model, in the year 1999 the average price paid for a new car will be \$25,000.

[1984, 2005] by [10,000, 30,000]   [1984, 2005] by [10,000, 30,000]

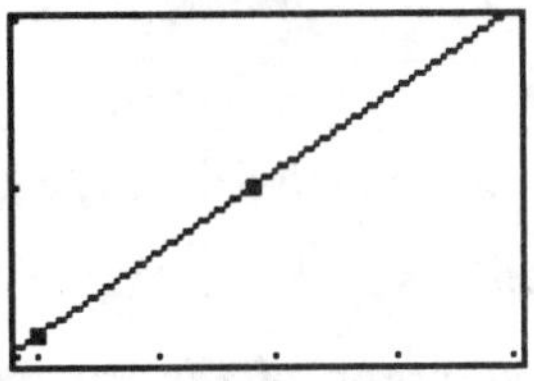

Xscl = 5

Yscl = 10,000

Figure 79(a)

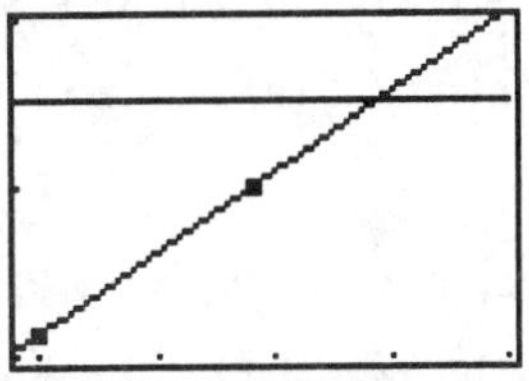

Xscl = 5

Yscl = 10,000

Figure 79(c)

81 (a)   [0, 75] by [0, 600]

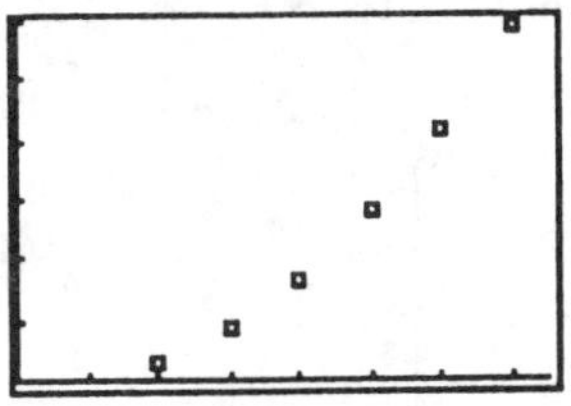

Xscl = 10

Yscl = 100

Figure 81

(b) The data and plot show that stopping distance is not a linear function of the speed. The distance required to stop a car traveling at 30 mi/hr is 86 ft whereas the distance required to stop a car traveling at 60 mi/hr is 414 ft. $\frac{414}{86} \approx 4.81$ rather than double.

(c) If you double the speed of a car, it requires almost *five* times the stopping distance. If stopping distance were a linear function of speed, doubling the speed would require approximately twice the stopping distance.

3 $f(x) = 3x^4 + 2x^2 - 5 \Rightarrow f(-x) = 3(-x)^4 + 2(-x)^2 - 5 = 3x^4 + 2x^2 - 5 = f(x)$

Since $f(-x) = f(x)$, f is even and its graph is symmetric with respect to the y-axis.

Note that this means if (a, b) is a point on the graph of f,

then the point $(-a, b)$ is also on the graph.

5 $f(x) = 8x^3 - 3x^2 \Rightarrow f(-x) = 8(-x)^3 - 3(-x)^2 = -8x^3 - 3x^2$

$-f(x) = -1 \cdot f(x) = -1(8x^3 - 3x^2) = -8x^3 + 3x^2$

Since $f(-x) \neq f(x)$ and $f(-x) \neq -f(x)$, f is neither even nor odd.

9 $f(x) = \sqrt[3]{x^3 - x} \Rightarrow f(-x) = \sqrt[3]{(-x)^3 - (-x)} = \sqrt[3]{-x^3 + x} = \sqrt[3]{-1(x^3 - x)} =$

$\sqrt[3]{-1}\sqrt[3]{x^3 - x} = -\sqrt[3]{x^3 - x}; \; -f(x) = -1 \cdot f(x) = -1 \cdot \sqrt[3]{x^3 - x} = -\sqrt[3]{x^3 - x}$

Since $f(-x) = -f(x)$, f is odd and its graph is symmetric with respect to the origin.

Note that this means if (a, b) is a point on the graph of f,

then the point $(-a, -b)$ is also on the graph.

15 $f(x) = 2\sqrt{x} + c$, $c = -3, 0, 2$ • The graph of $y^2 = x$ is shown in Figure 16 in Section 1.2 in the text. The top half of this graph is the graph of the **square root function**, $h(x) = \sqrt{x}$. The second value of c, 0, gives us the graph of $g(x) = 2\sqrt{x}$, which is a vertical stretching of h by a factor of 2. The effect of *adding* -3 and 2 is to vertically shift g down 3 units and up 2 units, respectively.

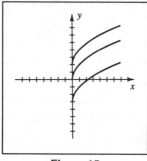   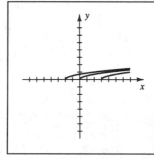

Figure 15   *Figure 17*

17 $f(x) = \frac{1}{2}\sqrt{x - c}$, $c = -2, 0, 3$ • The graph of $g(x) = \frac{1}{2}\sqrt{x}$ is a vertical compression of the square root function by a factor of $1/(1/2) = 2$. The effect of *subtracting* -2 and 3 from x will be to horizontally shift g left 2 units and right 3 units, respectively. If you forget which way to shift the graph, it is helpful to find the domain of the function. For example, if $h(x) = \sqrt{x - 2}$, then $x - 2$ must be nonnegative. $x - 2 \geq 0 \Rightarrow x \geq 2$, which also indicates a shift of 2 units to the right.

19 $f(x) = c\sqrt{4-x^2}$, $c = -2$, 1, 3 • For $c = 1$, the graph of $g(x) = \sqrt{4-x^2}$ is the upper half of the circle $x^2 + y^2 = 4$. For $c = -2$, reflect g through the x-axis and vertically stretch it by a factor of 2. For $c = 3$, vertically stretch g by a factor of 3.

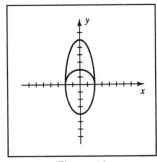

Figure 19

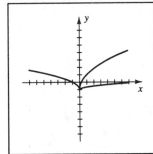

Figure 23

23 $f(x) = \sqrt{cx} - 1$, $c = -1$, $\frac{1}{9}$, 4 • If $c = 1$, then the graph of $g(x) = \sqrt{x} - 1$ is the graph of the square root function vertically shifted down one unit. For $c = -1$, reflect g through the y-axis. For $c = \frac{1}{9}$, horizontally stretch g by a factor $1/(1/9) = 9$ { x-intercept changes from 1 to 9 }. For $c = 4$, horizontally compress g by a factor 4 { x-intercept changes from 1 to $\frac{1}{4}$ }.

25 $f(x) = \frac{2}{x-c}$, $c = -3$, 0, 2 •

The graph of $h(x) = \frac{1}{x}$ was sketched in Example 3.

We may think of $g(x) = \frac{2}{x}$ as $g(x) = 2 \cdot \frac{1}{x}$, which

is just a vertical stretch of h by a factor of 2. To

sketch f with the given values of c, we need only

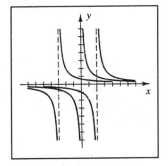

shift $g(x) = \frac{2}{x}$ left 3 units and right 2 units, respectively.

Figure 25

27 To determine what happens to a point P under this transformation, think of how you would evaluate $y = 2f(x-4) + 1$ for a particular value of x. You would first subtract 4 from x and then put that value into the function, obtaining a corresponding y-value. Next, you would multiply that y-value by 2 and finally, add 1. Summarizing these steps using the given point P, we have the following:

$$P(3, -2) \quad \{\, x - 4 \text{ [add 4 to the x-coordinate]}\,\} \qquad \rightarrow (7, -2)$$
$$\{\, \times 2 \text{ [multiply the y-coordinate by 2]}\,\} \quad \rightarrow (7, -4)$$
$$\{\, +1 \text{ [add 1 to the y-coordinate]}\,\} \qquad \rightarrow (7, -3)$$

29 (a) $y = f(x + 3)$ •   shift f left 3 units

(b) $y = f(x - 3)$ •   shift f right 3 units

(c) $y = f(x) + 3$ •   shift f up 3 units

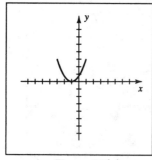 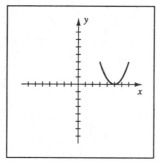 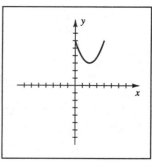

Figure 29(a)   Figure 29(b)   Figure 29(c)

(d) $y = f(x) - 3$ •   shift f down 3 units

(e) $y = -3f(x)$ •

reflect f through the x-axis and vertically stretch it by a factor of 3

(f) $y = -\frac{1}{3}f(x)$ •   reflect f through the x-axis { the effect of the negative sign

in front of $\frac{1}{3}$} and vertically compress it by a factor of $1/(1/3) = 3$

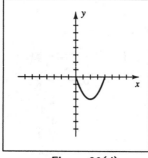 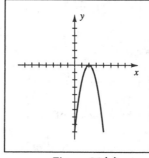 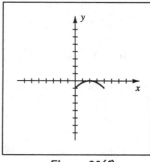

Figure 29(d)   Figure 29(e)   Figure 29(f)

(g) $y = f(-\frac{1}{2}x)$ •   reflect f through the y-axis { the effect of the negative sign

inside the parentheses } and horizontally stretch it by a factor of $1/(1/2) = 2$

(h) $y = f(2x)$ •   horizontally compress f by a factor of 2

(i) $y = -f(x + 2) - 3$ •   shift f left 2 units, reflect it about the x-axis,

and then shift it down 3 units

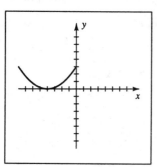 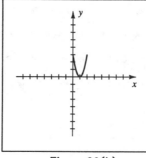 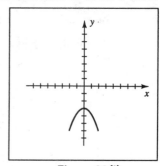

Figure 29(g)   Figure 29(h)   Figure 29(i)

(j) $y = f(x-2)+3$ • shift f right 2 units and up 3

(k) $y = |f(x)|$ • since no portion of the graph lies below the x-axis,

the graph is unchanged

(l) $y = f(|x|)$ • include the reflection of the given graph through the y-axis

since all points have positive x-coordinates

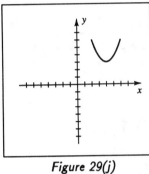

Figure 29(j)

Figure 29(k)

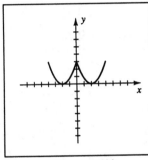

Figure 29(l)

31 (a) The minimum point on $y = f(x)$ is $(2, -1)$.

On the graph labeled (a), the minimum point is $(-7, 0)$.

It has been shifted left 9 units and up 1. Hence, $y = f(x+9)+1$.

(b) f is reflected about the x-axis $\Rightarrow y = -f(x)$

(c) f is reflected about the x-axis and shifted left 7 units and down 1 $\Rightarrow$

$$y = -f(x+7) - 1$$

33 (a) f is shifted left 4 units $\Rightarrow y = f(x+4)$

(b) f is shifted up 1 unit $\Rightarrow y = f(x)+1$

(c) f is reflected about the y-axis $\Rightarrow y = f(-x)$

$\boxed{35}$ $f(x) = \begin{cases} 3 & \text{if } x \leq -1 \\ -2 & \text{if } x > -1 \end{cases}$

We can think of f as 2 functions: If $x \leq -1$, then $y = 3$ {include the point $(-1, 3)$},

and if $x > -1$, then $y = -2$ {exclude the point $(-1, -2)$}.

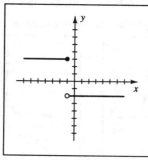

Figure 35

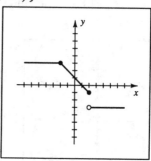

Figure 37

$\boxed{37}$ $f(x) = \begin{cases} 3 & \text{if } x < -2 \\ -x + 1 & \text{if } |x| \leq 2 \\ -3 & \text{if } x > 2 \end{cases}$

For the second part of the function, we have $|x| \leq 2$, or, equivalently, $-2 \leq x \leq 2$.

On this part of the domain, we want to graph $f(x) = -x + 1$, a line with slope -1

and y-intercept 1. Include both endpoints, $(-2, 3)$ and $(2, -1)$.

$\boxed{39}$ $f(x) = \begin{cases} x + 2 & \text{if } x \leq -1 \\ x^3 & \text{if } |x| < 1 \\ -x + 3 & \text{if } x \geq 1 \end{cases}$

If $x \leq -1$, we want the graph of $y = x + 2$. To determine the endpoint of this part of

the graph, merely substitute $x = -1$ in $y = x + 2$, obtaining $y = 1$. If $|x| < 1$, or,

equivalently, $-1 < x < 1$, we want the graph of $y = x^3$. We do not include the

endpoints $(-1, -1)$ and $(1, 1)$. If $x \geq 1$, we want the graph of $y = -x + 3$ and

include its endpoint $(1, 2)$.

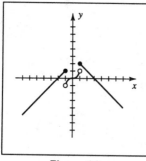

Figure 39

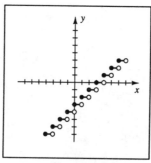

Figure 41(a)

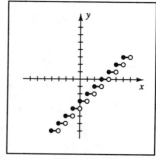

Figure 41(b)

$\boxed{41}$ (a) $f(x) = [\![x - 3]\!]$ • shift $g(x) = [\![x]\!]$ right 3 units

(b) $f(x) = [\![x]\!] - 3$ • shift g down 3 units, which is the same graph as in part (a)

(c) $f(x) = 2[\![x]\!]$ • vertically stretch g by a factor of 2

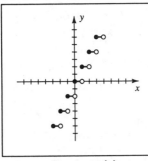 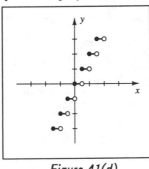 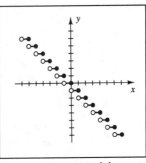

Figure 41(c)   *Figure 41(d)*   *Figure 41(e)*

(d) $f(x) = [\![2x]\!]$ • horizontally compress g by a factor of 2

Alternatively, we could determine the pattern of "steps" for this function by finding the values of x that make $f(x)$ change from 0 to 1, then from 1 to 2, etc. If $2x = 0$, then $x = 0$, and if $2x = 1$, then $x = \frac{1}{2}$.

Thus, the function will equal 0 from $x = 0$ to $x = \frac{1}{2}$ and then jump to 1 at $x = \frac{1}{2}$.

If $2x = 2$, then $x = 1$. The pattern is established: each step will be $\frac{1}{2}$ unit long.

(e) $f(x) = [\![-x]\!]$ • reflect g through the y-axis

43 A question you can ask to help determine if a relationship is a function is "If x is a particular value, can I find a unique y-value?" In this case, if x was 16, then $16 = y^2 \Rightarrow y = \pm 4$. Since we cannot find a unique y-value, this is not a function. Graphically, {see Figure 16 in §1.2 in the text} given any x-value greater than 0, there are two points on the graph and a vertical line intersects the graph in more than one point.

45 Reflect each portion of the graph that is below the x-axis through the x-axis.

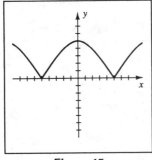 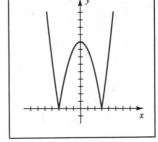

Figure 45   *Figure 47*

47 $y = |\,9 - x^2\,|$ • First sketch $y = 9 - x^2$,

then reflect the portions of the graph below the x-axis through the x-axis.

51 (a) For $y = -2f(x)$, multiply the y-coordinates by -2.

The domain { x-coordinates } remains the same. $D = [-2, 6]$, $R = [-16, 8]$

(b) For $y = f(\frac{1}{2}x)$, multiply the x-coordinates by 2.

The range { y-coordinates } remains the same. $D = [-4, 12]$, $R = [-4, 8]$

(c) For $y = f(x - 3) + 1$, add 3 to the x-coordinates and add 1 to the y-coordinates.

$$D = [1, 9], \quad R = [-3, 9]$$

(d) For $y = f(x + 2) - 3$, subtract 2 from the x-coordinates and subtract 3 from the y-coordinates. $D = [-4, 4]$, $R = [-7, 5]$

(e) For $y = f(-x)$, negate all x-coordinates. $D = [-6, 2]$, $R = [-4, 8]$

(f) For $y = -f(x)$, negate all y-coordinates. $D = [-2, 6]$, $R = [-8, 4]$

(g) $y = f(\,|x|\,)$ •   Graphically, we can reflect all points with positive x-coordinates through the y-axis, so the domain $[-2, 6]$ becomes $[-6, 6]$. Algebraically, we are replacing x with $|x|$, so $-2 \le x \le 6$ becomes $-2 \le |x| \le 6$, which is equivalent $|x| \le 6$, or, equivalently, $-6 \le x \le 6$. The range stays the same because of the given assumptions: $f(2) = 8$ and $f(6) = -4$; that is, the full range is taken on for $x \ge 0$. Note that the range could not be determined if $f(-2)$ was equal to 8. $D = [-6, 6]$, $R = [-8, 4]$

(h) $y = |\,f(x)\,|$ •   The points with y-coordinates having values from -4 to 0 will have values from 0 to 4, so the range will be $[0, 8]$. $D = [-2, 6]$, $R = [0, 8]$

53 If $x \le 20{,}000$, then $T(x) = 0.15x$. If $x > 20{,}000$, then the tax is 15% of the first 20,000, which is 3000, plus 20% of the amount over 20,000, which is $(x - 20{,}000)$. We may summarize and simplify as follows:

$$T(x) = \begin{cases} 0.15x & \text{if } x \le 20{,}000 \\ 3000 + 0.20(x - 20{,}000) & \text{if } x > 20{,}000 \end{cases} = \begin{cases} 0.15x & \text{if } x \le 20{,}000 \\ 0.20x - 1000 & \text{if } x > 20{,}000 \end{cases}$$

55 The author receives \$1.20 on the first 10,000 copies,

\$1.50 on the next 5000, and \$1.80 on each additional copy.

$$R(x) = \begin{cases} 1.20x & \text{if } 0 \le x \le 10{,}000 \\ 12{,}000 + 1.50(x - 10{,}000) & \text{if } 10{,}000 < x \le 15{,}000 \\ 19{,}500 + 1.80(x - 15{,}000) & \text{if } x > 15{,}000 \end{cases}$$

$$= \begin{cases} 1.20x & \text{if } 0 \le x \le 10{,}000 \\ 1.50x - 3000 & \text{if } 10{,}000 < x \le 15{,}000 \\ 1.80x - 7500 & \text{if } x > 15{,}000 \end{cases}$$

59 Assign ABS$(1.2x^2 - 10.8)$ to Y_1 and $1.36x + 4.08$ to Y_2. The standard viewing rectangle $[-15, 15]$ by $[-10, 10]$ shows intersection points at approximately 1.87 and 4.13. The solution is $(-\infty, -3) \cup (-3, 1.87) \cup (4.13, \infty)$.

63 Since $g(x) = f(\frac{1}{2}x)$, the graph of g can be obtained by stretching the graph of f horizontally by a factor of 2.

$[-12, 12]$ by $[-8, 8]$

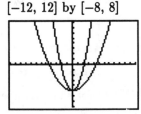

Xscl $= 1$
Yscl $= 1$

Figure 63

$[-12, 12]$ by $[-8, 8]$

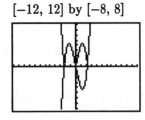

Xscl $= 1$
Yscl $= 1$

Figure 65

65 Since $g(x) = |f(x)|$, the graph of g is the same as the graph of f if f is non-negative. If $f(x) < 0$, then the graph of f will be reflected about the x-axis.

67 (a) Option I gives $C_1 = 4(\$29.95) + \$0.25(500 - 200) = 119.80 + 75.00 = \$194.80$.

Option II gives $C_2 = 4(\$39.95) + \$0.15(500) = 159.80 + 75.00 = \$234.80$.

(b) Let x represent the mileage. The cost function for Option I is the piecewise linear function

$$C_1(x) = \begin{cases} 119.80 & \text{if } 0 \le x \le 200 \\ 119.80 + 0.25(x - 200) & \text{if } x > 200 \end{cases}$$

Option II is the linear function $C_2(x) = 159.8 + 0.15x$ for $x \ge 0$.

(c) Let $C_1 = Y_1$ and $C_2 = Y_2$.

Table $Y_1 = 119.80 + 0.25(x - 200)*(x > 200)$ and $Y_2 = 159.80 + 0.15x$

x	Y_1	Y_2	x	Y_1	Y_2
100	119.8	174.8	700	244.8	264.8
200	119.8	189.8	800	269.8	279.8
300	144.8	204.8	900	294.8	294.8
400	169.8	219.8	1000	319.8	309.8
500	194.8	234.8	1100	344.8	324.8
600	219.8	249.8	1200	369.8	339.8

(d) From the table, we see that the options are equal in cost for $x = 900$ miles. Option I is preferable if $x \in [0, 900)$ and Option II is preferable if $x > 900$.

Note: To help determine if you should try to prove the function is one-to-one or look for a counterexample to show that it is not one-to-one, consider the question: "If y was a particular value, could I find a unique x?" If the answer is yes, try to prove the function is one-to-one. Also, consider the Horizontal Line Test listed on page 68 in the text.

$\boxed{1}$ If y was a particular value, say 5, we would have $5 = 3x - 7$. Trying to solve for x would yield $12 = 3x \Rightarrow x = 4$. Since we could find a unique x, we will try to prove that the function is one-to-one. **Proof** Suppose that $f(a) = f(b)$ for some numbers a and b in the domain. This gives us $3a - 7 = 3b - 7 \Rightarrow 3a = 3b \Rightarrow a = b$. Since $f(a) = f(b)$ implies that $a = b$, we conclude that f is one-to-one.

$\boxed{3}$ If y was a particular value, say 7, we would have $7 = x^2 - 9$. Trying to solve for x would yield $16 = x^2 \Rightarrow x = \pm 4$. Since we could not find a *unique* x, we will show that two different numbers have the same function value. Using the information already obtained, $f(4) = 7$ and $f(-4) = 7$, but $4 \neq -4$ and hence, f is *not* one-to-one.

$\boxed{5}$ Suppose $f(a) = f(b)$ with $a, b \geq 0$. $\sqrt{a} = \sqrt{b} \Rightarrow (\sqrt{a})^2 = (\sqrt{b})^2 \Rightarrow a = b$.

f is one-to-one.

$\boxed{9}$ For $f(x) = \sqrt{4 - x^2}$, $f(-1) = \sqrt{3} = f(1)$. f is *not* one-to-one.

Note: For Exercises 13–16, we need to show that $f(g(x)) = x = g(f(x))$.

$\boxed{15}$ $f(g(x)) = -(\sqrt{3-x})^2 + 3 = -(3 - x) + 3 = x$.

$g(f(x)) = \sqrt{3 - (-x^2 + 3)} = \sqrt{x^2} = |x| = x$ { since $x \geq 0$ }.

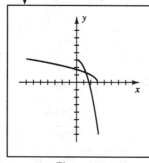

Figure 15

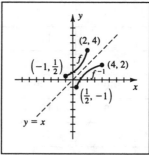

Figure 35

$\boxed{19}$ $f(x) = \dfrac{1}{3x-2} \Rightarrow 3xy - 2y = 1 \Rightarrow 3xy = 2y + 1 \Rightarrow x = \dfrac{2y+1}{3y} \Rightarrow f^{-1}(x) = \dfrac{2x+1}{3x}$

$\boxed{21}$ $f(x) = \dfrac{3x+2}{2x-5} \Rightarrow 2xy - 5y = 3x + 2 \Rightarrow 2xy - 3x = 5y + 2 \Rightarrow$

$$x(2y - 3) = 5y + 2 \Rightarrow x = \dfrac{5y+2}{2y-3} \Rightarrow f^{-1}(x) = \dfrac{5x+2}{2x-3}$$

$\boxed{23}$ $f(x) = 2 - 3x^2$, $x \le 0 \Rightarrow y + 3x^2 = 2 \Rightarrow$

$$x^2 = \frac{2-y}{3} \Rightarrow x = \pm\sqrt{\frac{2-y}{3}} \text{ \{ choose minus since } x \le 0 \} \Rightarrow f^{-1}(x) = -\sqrt{\frac{2-x}{3}}$$

$\boxed{27}$ $f(x) = \sqrt{3-x} \Rightarrow y^2 = 3 - x \Rightarrow$

$$x = 3 - y^2 \text{ \{ Since } y \ge 0 \text{ for } f, x \ge 0 \text{ for } f^{-1}. \} \Rightarrow f^{-1}(x) = 3 - x^2, \ x \ge 0$$

$\boxed{29}$ $f(x) = \sqrt[3]{x} + 1 \Rightarrow y - 1 = \sqrt[3]{x} \Rightarrow x = (y-1)^3 \Rightarrow f^{-1}(x) = (x-1)^3$

$\boxed{33}$ $f(x) = x^2 - 4 \Rightarrow y = x^2 - 4 \Rightarrow y + 4 = x^2 \Rightarrow x = \pm\sqrt{y+4} \Rightarrow$

$$f^{-1}(x) = \sqrt{x+4} \quad \text{or} \quad f^{-1}(x) = -\sqrt{x+4}.$$

Hence, $f^{-1}(5) = 3$ or $f^{-1}(5) = -3$. Since we are given $f^{-1}(5) = -3$ {and *not* 3 },

$$\text{we choose } f^{-1}(x) = -\sqrt{x+4}.$$

$\boxed{35}$ Remember that the domain of f is the range of f^{-1} and

that the range of f is the domain of f^{-1}. See *Figure 35* on the preceding page.

(b) $D = [-1, 2]$; $R = [\frac{1}{2}, 4]$ **(c)** $D_1 = R = [\frac{1}{2}, 4]$; $R_1 = D = [-1, 2]$

$\boxed{41}$ **(a)** $f(x) = -x + b \Rightarrow y = -x + b \Rightarrow x = -y + b$, or $f^{-1}(x) = -x + b$.

(b) $f(x) = \frac{ax+b}{cx-a}$ for $c \ne 0 \Rightarrow y = \frac{ax+b}{cx-a} \Rightarrow cyx - ya = ax + b \Rightarrow$

$$cyx - ax = ay + b \Rightarrow x(cy - a) = ay + b \Rightarrow x = \frac{ay+b}{cy-a}, \text{ or } f^{-1}(x) = \frac{ax+b}{cx-a}.$$

(c) The graph of f is symmetric about the line $y = x$. Thus, $f(x) = f^{-1}(x)$.

$\boxed{43}$ From *Figure 43*, we see that f is always increasing. Thus, f is one-to-one.

[−6, 6] by [−4, 4]   [−12, 12] by [−8, 8]

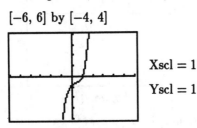

Xscl = 1

Yscl = 1

Figure 43

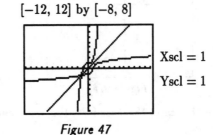

Xscl = 1

Yscl = 1

Figure 47

$\boxed{47}$ The graph of f will be reflected about the line $y = x$.

$$y = \sqrt[3]{x-1} \Rightarrow y^3 = x - 1 \Rightarrow x = y^3 + 1 \Rightarrow f^{-1}(x) = x^3 + 1.$$

$$\text{Graph } Y_1 = \sqrt[3]{x-1}, \ Y_2 = x^3 + 1, \text{ and } Y_3 = x.$$

$\boxed{49}$ **(a)** $V(23) = 35(23) = 805 \text{ ft}^3/\text{min}$

(b) $V^{-1}(x) = \frac{1}{35}x$. Given an air circulation of x cubic feet per minute,

$V^{-1}(x)$ computes the maximum number of people that should be in the

restaurant at one time.

(c) $V^{-1}(2350) = \frac{1}{35}(2350) \approx 67.1 \Rightarrow$ the maximum number of people is 67.

1.6 Exercises

$\boxed{1}$ See the solution to Exercise 9 in Section 5.1 of this manual.

$\boxed{3}$ $f(x) = \left(\frac{2}{5}\right)^{-x} = \left[\left(\frac{2}{5}\right)^{-1}\right]^{x} = \left(\frac{5}{2}\right)^{x}$ • goes through $\left(-1, \frac{2}{5}\right)$, $(0, 1)$, and $\left(1, \frac{5}{2}\right)$

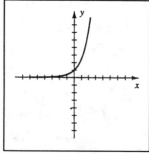

Figure 3

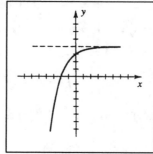

Figure 5

$\boxed{5}$ $f(x) = -\left(\frac{1}{2}\right)^{x} + 4$ • reflect $y = \left(\frac{1}{2}\right)^{x}$ through the x-axis and shift up 4 units

Note: Examine Figure 60 in this section to reinforce the idea that $y = e^{x}$ is just a special

   case of $y = a^{x}$ with $a > 1$.

$\boxed{7}$ (a) $f(x) = e^{x+4}$ • shift $y = e^{x}$ left 4 units

   (b) $f(x) = e^{x} + 4$ • shift $y = e^{x}$ up 4 units

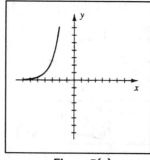

Figure 7(a)

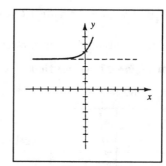

Figure 7(b)

$\boxed{9}$ (b) In this case, the *base* is 4, the *exponent* is -3, and the *argument* is $\frac{1}{64}$. Thus,

$$4^{-3} = \tfrac{1}{64} \quad \text{is equivalent to} \quad \log_4 \tfrac{1}{64} = -3.$$

$\boxed{11}$ (a) In this case, the *base* is 2, the *argument* is 32, and the *exponent* is 5. Thus,

$$\log_2 32 = 5 \quad \text{is equivalent to} \quad 2^5 = 32.$$

$\boxed{13}$ (a) Changing $10^5 = 100{,}000$ to logarithmic form gives us $\log_{10} 100{,}000 = 5$.

   Since this is a common logarithm, we denote it as $\log 100{,}000 = 5$.

$\boxed{15}$ (b) Remember that $\log x = 20t$ is the same as $\log_{10} x = 20t$.

   Changing to exponential form, we have $10^{20t} = x$.

15 (c) Remember that $\ln x = 0.1$ is the same as $\log_e x = 0.1$.

Changing to exponential form, we have $e^{0.1} = x$.

17 (c) Remember that you cannot take the logarithm, any base, of a negative number.

Hence, $\log_4(-2)$ is undefined.

(g) We will change the form of $\frac{1}{16}$ so that it can be written as an exponential expression with the same base as the logarithm—in this case, that base is 4.

$$\log_4 \tfrac{1}{16} = \log_4 4^{-2} = -2$$

19 Parts (a)–(f) are direct applications of the properties in the chart on page 81 using $a = 10$ and $a = e$. For part (g), we use a property of exponents that will enable us to use the property $e^{\ln x} = x$. (g) $e^{2 + \ln 3} = e^2 e^{\ln 3} = e^2(3) = 3e^2$

21 See the solution to Exercise 31 in Section 5.3 of this manual.

23 The graph of $f(x) = \log x - 1$ is the graph of $y = \log x$ shifted down 1 unit.

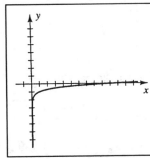

Figure 23

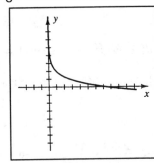

Figure 25

25 The graph of $f(x) = 2 - \ln x = -\ln x + 2$ is the graph of

$y = \ln x$ reflected through the x-axis and shifted up 2 units.

29 $\log_a \dfrac{x^3 w}{y^2 z^4} = \log_a x^3 w - \log_a y^2 z^4 = \log_a x^3 + \log_a w - (\log_a y^2 + \log_a z^4) =$

$$3 \log_a x + \log_a w - 2 \log_a y - 4 \log_a z$$

The most common mistake is to not have the minus sign in front of $4 \log_a z$.

This error results from not having the parentheses in the correct place.

31 $\log_5 6 = \dfrac{\log 6}{\log 5} \left\{ \text{or, equivalently, } \dfrac{\ln 6}{\ln 5} \right\} \approx 1.1133$.

Note that either log or ln can be used here. You should get comfortable using both.

33 $\log_9 0.2 = \dfrac{\log 0.2}{\log 9} \left\{ \text{or, equivalently, } \dfrac{\ln 0.2}{\ln 9} \right\} \approx -0.7325$.

35 $\dfrac{\log_5 16}{\log_5 4} = \log_4 16 = \log_4 4^2 = 2$

$\boxed{37}$ $f(x) = \log_2(x+3)$ • $x = 0 \Rightarrow$ y-intercept $= \log_2 3 = \dfrac{\log 3}{\log 2} \approx 1.5850$

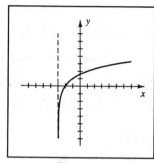

Figure 37

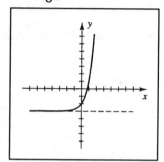

Figure 39

$\boxed{39}$ $f(x) = 4^x - 3$ • $y = 0 \Rightarrow 4^x = 3 \Rightarrow$ x-intercept $= \log_4 3 = \dfrac{\log 3}{\log 4} \approx 0.7925$

Chapter 1 Review Exercises

$\boxed{3}$ (c) $\left| 3^{-1} - 2^{-1} \right| = \left| \frac{1}{3} - \frac{1}{2} \right| = \left| \frac{2}{6} - \frac{3}{6} \right| = \left| -\frac{1}{6} \right| = -(-\frac{1}{6}) = \frac{1}{6}$

$\boxed{4}$ (a) $d(A, C) = \left| -3 - (-8) \right| = \left| 5 \right| = 5$

$\boxed{6}$ $-2(4x - 7) = 5(8 - x) \Rightarrow -8x + 14 = 40 - 5x \Rightarrow -3x = 26 \Rightarrow x = -\frac{26}{3}$

$\boxed{8}$ $(x - 2)(x + 1) = 3 \Rightarrow x^2 - x - 2 = 3 \Rightarrow x^2 - x - 5 = 0 \Rightarrow x = \dfrac{1 \pm \sqrt{1 + 20}}{2} = \frac{1}{2} \pm \frac{1}{2}\sqrt{21}$

$\boxed{10}$ $\left[-\frac{1}{2} < \dfrac{2x + 3}{5} < \frac{3}{2} \right] \cdot 10 \Rightarrow -5 < 4x + 6 < 15 \Rightarrow -11 < 4x < 9 \Rightarrow -\frac{11}{4} < x < \frac{9}{4} \Leftrightarrow$

$$(-\tfrac{11}{4}, \tfrac{9}{4})$$

$\boxed{13}$ If $y/x < 0$, then y and x must have opposite signs, and hence,

the set consists of all points in quadrants II and IV.

$\boxed{14}$ (a) $P(-5, 9)$, $Q(-8, -7) \Rightarrow$

$$d(P, Q) = \sqrt{[-8 - (-5)]^2 + (-7 - 9)^2} = \sqrt{9 + 256} = \sqrt{265}.$$

(b) $P(-5, 9)$, $Q(-8, -7) \Rightarrow M_{PQ} = \left(\dfrac{-5 + (-8)}{2}, \dfrac{9 + (-7)}{2} \right) = (-\tfrac{13}{2}, 1).$

17 $2y + 5x - 8 = 0 \Leftrightarrow y = -\frac{5}{2}x + 4$, a line with y-intercept 4 and x-intercept $(1.6, 0)$

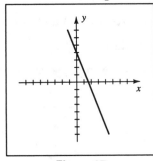

Figure 17

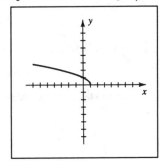

Figure 21

21 $y = \sqrt{1-x}$ ●   The radicand must be nonnegative for the radical to be defined.

$1 - x \geq 0 \Rightarrow 1 \geq x$ or $x \leq 1$. The domain is $(-\infty, 1]$ and the range is $[0, \infty)$.

x-intercept: $(1, 0)$, y-intercept: $(0, 1)$

22 $y = (x-1)^3$, shift $y = x^3$ right one unit; x-intercept: $(1, 0)$, y-intercept: $(0, -1)$

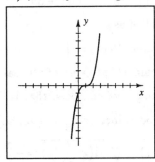

Figure 22

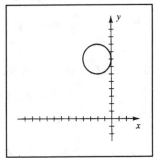

Figure 24

24 $x^2 + y^2 + 4x - 16y + 64 = 0 \Rightarrow x^2 + 4x + \underline{4} + y^2 - 16y + \underline{64} = -64 + \underline{4} + \underline{64} \Rightarrow$

$(x+2)^2 + (y-8)^2 = 4$; $C(-2, 8)$, $r = \sqrt{4} = 2$; x-intercept: None, y-intercept: $(0, 8)$

26 The center of the circle is the midpoint of $A(8, 10)$ and $B(-2, -14)$.

$M_{AB} = \left(\dfrac{8 + (-2)}{2}, \dfrac{10 + (-14)}{2}\right) = (3, -2)$. The radius of the circle is

$\frac{1}{2} \cdot d(A, B) = \frac{1}{2}\sqrt{(-2-8)^2 + (-14-10)^2} = \frac{1}{2}\sqrt{100 + 576} = \frac{1}{2} \cdot 26 = 13$.

An equation is $(x-3)^2 + (y+2)^2 = 13^2 = 169$.

27 $x^2 + y^2 - 12y + 31 = 0 \Rightarrow x^2 + y^2 - 12y + \underline{36} = -31 + \underline{36} \Rightarrow x^2 + (y-6)^2 = 5$.

$C(0, 6); r = \sqrt{5}$

31 $\dfrac{f(a+h) - f(a)}{h} = \dfrac{\left[-(a+h)^2 + (a+h) + 5\right] - \left[-a^2 + a + 5\right]}{h} =$

$\dfrac{-a^2 - 2ah - h^2 + a + h + 5 + a^2 - a - 5}{h} = \dfrac{-2ah - h^2 + h}{h} = \dfrac{h(-2a - h + 1)}{h} =$

$-2a - h + 1$

$\boxed{33}$ (a) $f(x) = \dfrac{1 - 3x}{2} = \dfrac{1}{2} - \dfrac{3}{2}x.$

The graph of f is a line with y-intercept $(0, \frac{1}{2})$ and x-intercept $(\frac{1}{3}, 0)$.

(b) We may substitute any value for x, so the domain is $D = \mathbf{R}.$

As shown in *Figure 33*, all y values are taken on, so the range is $R = \mathbf{R}.$

(c) f is decreasing on its entire domain, $(-\infty, \infty).$

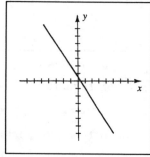

Figure 33

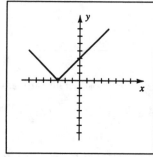

Figure 35

$\boxed{35}$ (a) The graph of $f(x) = |x + 3|$ can be thought of as

the graph of $g(x) = |x|$ shifted left 3 units.

(b) The function is defined for all x, so the domain D is the set of all real numbers.

The range is the set of all nonnegative numbers, that is, $R = [0, \infty).$

(c) The function f is decreasing on $(-\infty, -3]$ and is increasing on $[-3, \infty).$

$\boxed{37}$ (a) $f(x) = \sqrt[3]{x^3 + 4x} \Rightarrow f(-x) = \sqrt[3]{(-x)^3 + 4(-x)} = \sqrt[3]{-1(x^3 + 4x)} = -\sqrt[3]{x^3 + 4x} =$

$-f(x)$, f is odd

(b) $f(x) = \sqrt[3]{3x^2 - x^3} \Rightarrow f(-x) = \sqrt[3]{3(-x)^2 - (-x)^3} = \sqrt[3]{3x^2 + x^3} \ne \pm f(x),$

f is neither even nor odd

(c) $f(x) = \sqrt[3]{x^4 + 3x^2 + 5} \Rightarrow f(-x) = \sqrt[3]{(-x)^4 + 3(-x)^2 + 5} = \sqrt[3]{x^4 + 3x^2 + 5} = f(x),$

f is even

$\boxed{39}$ (a) $y = f(x - 2)$ •   shift f right 2 units

(b) $y = f(x) - 2$ •   shift f down 2 units

(c) $y = f(-x)$ •   reflect f through the y-axis

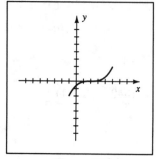

Figure 39(a)

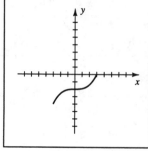

Figure 39(b)

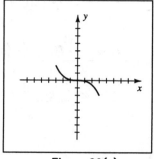

Figure 39(c)

(d) $y = f(2x)$ • horizontally compress f by a factor of 2

(e) $y = f(\frac{1}{2}x)$ • horizontally stretch f by a factor of $1/(1/2) = 2$

(f) $y = f^{-1}(x)$ • reflect f through the line $y = x$

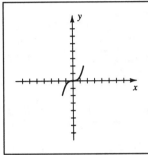

Figure 39(d)

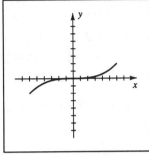

Figure 39(e)

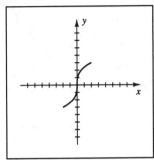

Figure 39(f)

(g) $y = |f(x)|$ •

reflect the portion of the graph below the x-axis through the x-axis.

(h) $y = f(|x|)$ • include the reflection of all points with positive x-coordinates

through the y-axis—results in the same graph as in part (g).

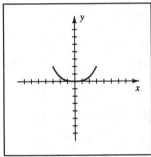

Figure 39(g)

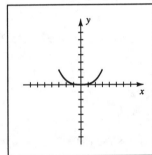

Figure 39(h)

42 $f(x) = 9 - 2x^2$, $x \le 0 \Rightarrow$

$y + 2x^2 = 9 \Rightarrow x^2 = \dfrac{9 - y}{2} \Rightarrow$

$x = \pm\sqrt{\dfrac{9 - y}{2}}$ { choose minus since $x \le 0$ } $\Rightarrow$

$f^{-1}(x) = -\sqrt{\dfrac{9 - x}{2}}$

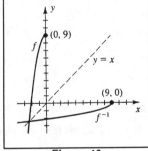

Figure 42

43 (a) $f(1) = 2$ since the point $(1, 2)$ in on the graph of f

(b) $(f \circ f)(1) = f(f(1)) = f(2) = 4$ since the point $(2, 4)$ in on the graph of f

(c) $f(2) = 4$ and f is one-to-one $\Rightarrow f^{-1}(4) = 2$

(d) $f(x) = 4 \Rightarrow x = 2$

(e) $f(x) > 4 \Rightarrow x > 2$

44 Since f and g are one-to-one functions, we know that $f(2) = 7$, $f(4) = 2$, and

$g(2) = 5$ imply that $f^{-1}(7) = 2$, $f^{-1}(2) = 4$, and $g^{-1}(5) = 2$, respectively.

(a) $(g \circ f^{-1})(7) = g(f^{-1}(7)) = g(2) = 5$

(b) $(f \circ g^{-1})(5) = f(g^{-1}(5)) = f(2) = 7$

(c) $(f^{-1} \circ g^{-1})(5) = f^{-1}(g^{-1}(5)) = f^{-1}(2) = 4$

(d) $(g^{-1} \circ f^{-1})(2) = g^{-1}(f^{-1}(2)) = g^{-1}(4)$, which is not known.

47 $f(x) = 1 - 3^{-x} = 1 - (\frac{1}{3})^x = -(\frac{1}{3})^x + 1$ •

reflect $y = (\frac{1}{3})^x$ through the x-axis and shift it up 1 unit

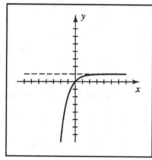
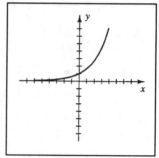

Figure 47   *Figure 48*

48 $f(x) = e^{x/2} = (e^{1/2})^x \approx (1.65)^x$ •   goes through $(-1, 1/\sqrt{e})$, $(0, 1)$, and $(1, \sqrt{e})$;

or approximately $(-1, 0.61)$, $(0, 1)$, and $(1, 1.65)$

50 (a) $\log_2 \frac{1}{16} = \log_2 2^{-4} = -4$ (b) $\log_\pi 1 = 0$ (c) $\ln e = 1$

(d) $6^{\log_6 4} = 4$ (e) $\log 1{,}000{,}000 = \log 10^6 = 6$

(f) $10^{3\log 2} = 10^{\log 2^3} = 2^3 = 8$ (g) $\log_4 2 = \log_4 4^{1/2} = \frac{1}{2}$

52 $\log x^4 \sqrt[3]{y^2/z} = \log \left(x^4 y^{2/3} z^{-1/3} \right) = \log x^4 + \log y^{2/3} + \log z^{-1/3} =$

$4 \log x + \frac{2}{3} \log y - \frac{1}{3} \log z$

53 (a) For $y = \log_2(x + 1)$, $D = (-1, \infty)$ and $R = \mathbf{R}$.

(b) $y = \log_2(x + 1) \Rightarrow x = \log_2(y + 1) \Rightarrow 2^x = y + 1 \Rightarrow y = 2^x - 1$,

$D = \mathbf{R}$, $R = (-1, \infty)$

55 The slope of the ramp should be between $\frac{1}{12}$ and $\frac{1}{20}$. If the rise of the ramp is 3 feet, then the run should be between $3 \times 12 = 36$ ft and $3 \times 20 = 60$ ft. The range of the ramp lengths should be from $L = \sqrt{3^2 + 36^2} \approx 36.1$ ft to $L = \sqrt{3^2 + 60^2} \approx 60.1$ ft.

57 (a) $V = at + b$ is the desired form. $V = 89{,}000$ when $t = 0 \Rightarrow V = at + 89{,}000$.

$V = 125{,}000$ when $t = 6 \Rightarrow 125{,}000 = 6a + 89{,}000 \Rightarrow a = \frac{36{,}000}{6} = 6000$ and hence,

$V = 6000t + 89{,}000$.

(b) $V = 103{,}000 \Rightarrow 103{,}000 = 6000t + 89{,}000 \Rightarrow t = \frac{7}{3}$, or $2\frac{1}{3}$.

$\boxed{58}$ (a) $F = aC + b$ is the desired form. $F = 32$ when $C = 0 \Rightarrow F = aC + 32$. $F = 212$

when $C = 100 \Rightarrow 212 = 100a + 32 \Rightarrow a = \frac{180}{100} = \frac{9}{5}$ and hence, $F = \frac{9}{5}C + 32$.

(b) If C increases $1°$, F increases $\left(\frac{9}{5}\right)°$, or $1.8°$.

$\boxed{59}$ (a) $h = 5280$ and $T_0 = 70 \Rightarrow T = 70 - \left(\frac{5.5}{1000}\right)5280 = 40.96°\,\text{F}$.

(b) $T = 32 \Rightarrow 32 = 70 - \left(\frac{5.5}{1000}\right)h \Rightarrow h = (70 - 32)\left(\frac{1000}{5.5}\right) \approx 6909$ ft.

$\boxed{61}$ $B = 55$ and $h = 10{,}000 - 4000 = 6000 \Rightarrow T = 55 - \left(\frac{3}{1000}\right)(6000) = 37°\,\text{F}$.

$\boxed{62}$ (a) $\frac{y}{b} = \frac{y + h}{a} \Rightarrow ay = by + bh \Rightarrow y(a - b) = bh \Rightarrow y = \frac{bh}{a - b}$

(b) $V = \frac{1}{3}\pi a^2(y + h) - \frac{1}{3}\pi b^2 y = \frac{\pi}{3}\left[(a^2 - b^2)y + a^2 h\right] =$

$\frac{\pi}{3}\left[(a^2 - b^2)\frac{bh}{a - b} + a^2 h\right] = \frac{\pi}{3}h\left[(a + b)b + a^2\right] = \frac{\pi}{3}h(a^2 + ab + b^2)$

(c) $a = 6$, $b = 3$, $V = 600 \Rightarrow \frac{\pi}{3}h(6^2 + 6 \cdot 3 + 3^2) = 600 \Rightarrow h = \frac{1800}{63\pi} = \frac{200}{7\pi} \approx 9.1$ ft

$\boxed{63}$ $C = \frac{kDE}{Vt} \Rightarrow D = \left(\frac{Ct}{k}\right)\frac{V}{E}$, where $\frac{Ct}{k}$ is constant. If V is twice its original value and

E is 0.8 of its original value (reduced by 20%), then D becomes $D_1 = \left(\frac{Ct}{k}\right)\frac{2V}{0.8E}$.

Comparing D_1 to D, we have $\dfrac{D_1}{D} = \dfrac{\left(\frac{Ct}{k}\right)\frac{2V}{0.8E}}{\left(\frac{Ct}{k}\right)\frac{V}{E}} = \dfrac{2}{0.8} = 2.5 = 250\%$ of its original

value. Thus, D increases by 250%.

$\boxed{64}$ Let h denote the height of the triangle. Using the Pythagorean theorem and one of

the right triangles with sides h and $\frac{1}{2}s$, we have

$h^2 + \left(\frac{1}{2}s\right)^2 = s^2 \Rightarrow h^2 + \frac{1}{4}s^2 = s^2 \Rightarrow h^2 = \frac{3}{4}s^2 \Rightarrow h = \frac{\sqrt{3}}{2}s$.

The area A of a triangle is $\frac{1}{2}(\text{base})(\text{height}) = \frac{1}{2}(s)\left(\frac{\sqrt{3}}{2}s\right) = \frac{\sqrt{3}}{4}s^2$.

$\boxed{65}$ The y-values are increasing slowly and can best be described by equation (3),

$y = 3\sqrt{x - 0.5}$.

$\boxed{1}$ 1 gallon ≈ 0.13368 ft^3 is a conversion factor that would help. The volume of the tank is 10,000 gallons ≈ 1336.8 ft^3. Use $V = \frac{4}{3}\pi r^3$ to determine the radius. $1336.8 = \frac{4}{3}\pi r^3 \Rightarrow r^3 = \frac{1002.6}{\pi} \Rightarrow r \approx 6.83375$ ft. Then use $S = 4\pi r^2$ to find the surface area. $S = 4\pi(6.83375)^2 \approx 586.85$ ft^2.

$\boxed{3}$ (a) $\dfrac{1}{\frac{a+bi}{c+di}} = \dfrac{c+di}{a+bi} \cdot \dfrac{a-bi}{a-bi} = \dfrac{ac+bd+(ad-bc)i}{a^2+b^2} = \dfrac{ac+bd}{a^2+b^2} + \dfrac{ad-bc}{a^2+b^2}i = p+qi$

     (b) Yes, try an example such as 3/4. Let $a = 3$, $b = 0$, $c = 4$, and $d = 0$. Then, from part (a), $p + qi = \frac{12}{9} + \frac{0}{9}i = \frac{12}{9} = \frac{4}{3}$, which is the multiplicative inverse of 3/4.

     (c) a and b cannot both be 0 because then the denominator would be 0.

$\boxed{5}$ To determine the x-coordinate of R,

we want to start at x_1 and go $\frac{m}{n}$ of the way to x_2. We could write this as

$$x_3 = x_1 + \tfrac{m}{n}\Delta x = x_1 + \tfrac{m}{n}(x_2 - x_1) = x_1 + \tfrac{m}{n}x_2 - \tfrac{m}{n}x_1 = \left(1 - \tfrac{m}{n}\right)x_1 + \tfrac{m}{n}x_2.$$

$$\text{Similarly, } y_3 = \left(1 - \tfrac{m}{n}\right)y_1 + \tfrac{m}{n}y_2.$$

$\boxed{7}$ $f(x) = ax^2 + bx + c \Rightarrow$

$$\frac{f(x+h) - f(x)}{h} = \frac{\left[a(x+h)^2 + b(x+h) + c\right] - \left[ax^2 + bx + c\right]}{h}$$

$$= \frac{ax^2 + 2ahx + ah^2 + bx + bh + c - ax^2 - bx - c}{h}$$

$$= \frac{2ahx + ah^2 + bh}{h} = \frac{h(2ax + ah + b)}{h} = 2ax + ah + b$$

$\boxed{9}$ The graph from Exercise 42(e) of Section 1.4 ($y = -[\![-x]\!]$) illustrates the concept of one of the most common billing methods with the open and closed endpoints reversed from those of the greatest integer function. Starting with $y = -[\![-x]\!]$ and adjusting for jumps every 15 minutes gives us $y = -[\![-x/15]\!]$. Since each quarter-hour charge is \$20, we multiply by 20 to obtain $y = -20[\![-x/15]\!]$. Because of the initial \$40 charge, we must add 40 to obtain the function $f(x) = 40 - 20[\![-x/15]\!]$.

$\boxed{11}$ (a) Let January correspond to 1, February to 2, ... , and December to 12.

[0.5, 12.5] by [0, 5]

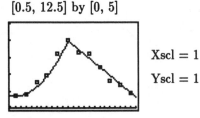

Xscl = 1

Yscl = 1

Figure 11

(b) The data points are (approximately) parabolic on the interval [1, 6] and linear on
[6, 12]. Let $f_1(x) = a(x - h)^2 + k$ on [1, 6] and $f_2(x) = mx + b$ on [6, 12]. On
[1, 6], let the vertex $(h,\ k) = (1,\ 0.7)$. Since (6, 4) is on the graph of f_1,
$f_1(6) = a(6 - 1)^2 + 0.7 = 4 \Rightarrow a = 0.132$. Thus, $f_1(x) = 0.132(x - 1)^2 + 0.7$ on
[1, 6]. Now, let $f_2(x) = mx + b$ pass through the points (6, 4) and (12, 0.9). An
equation of this line is approximately $(y - 4) = -0.517(x - 6)$. Thus,
let $f_2(x) = -0.517x + 7.102$ on [6, 12].

$$f(x) = \begin{cases} 0.132(x - 1)^2 + 0.7 & \text{if } 1 \le x \le 6 \\ -0.517x + 7.102 & \text{if } 6 < x \le 12 \end{cases}$$

(c) To plot the piecewise function, let $Y_1 = (0.132(x - 1)^2 + 0.7)/(x \le 6)$ and
$Y_2 = (-0.517x + 7.102)/(x > 6)$. These assignments use the concept of Boolean
division. For example, when $(x \le 6)$ is false, the expression Y_1 will be undefined
(division by 0) and the calculator will not plot any values.

Chapter 2: The Trigonometric Functions

Note: Exercises 1 and 3: The answers listed are the smallest (in magnitude) two positive coterminal angles and two negative coterminal angles.

1. (a) $120° + 1(360°) = 480°,$ $120° + 2(360°) = 840°;$

   $120° - 1(360°) = -240°,$ $120° - 2(360°) = -600°$

   (b) $135° + 1(360°) = 495°,$ $135° + 2(360°) = 855°;$

   $135° - 1(360°) = -225°,$ $135° - 2(360°) = -585°$

   (c) $-30° + 1(360°) = 330°,$ $-30° + 2(360°) = 690°;$

   $-30° - 1(360°) = -390°,$ $-30° - 2(360°) = -750°$

3. (a) $620° - 1(360°) = 260°,$ $620° + 1(360°) = 980°;$

   $620° - 2(360°) = -100°,$ $620° - 3(360°) = -460°$

   (b) $\frac{5\pi}{6} + 1(2\pi) = \frac{5\pi}{6} + \frac{12\pi}{6} = \frac{17\pi}{6},$ $\frac{5\pi}{6} + 2(2\pi) = \frac{5\pi}{6} + \frac{24\pi}{6} = \frac{29\pi}{6};$

   $\frac{5\pi}{6} - 1(2\pi) = \frac{5\pi}{6} - \frac{12\pi}{6} = -\frac{7\pi}{6},$ $\frac{5\pi}{6} - 2(2\pi) = \frac{5\pi}{6} - \frac{24\pi}{6} = -\frac{19\pi}{6}$

   (c) $-\frac{\pi}{4} + 1(2\pi) = -\frac{\pi}{4} + \frac{8\pi}{4} = \frac{7\pi}{4},$ $-\frac{\pi}{4} + 2(2\pi) = -\frac{\pi}{4} + \frac{16\pi}{4} = \frac{15\pi}{4};$

   $-\frac{\pi}{4} - 1(2\pi) = -\frac{\pi}{4} - \frac{8\pi}{4} = -\frac{9\pi}{4},$ $-\frac{\pi}{4} - 2(2\pi) = -\frac{\pi}{4} - \frac{16\pi}{4} = -\frac{17\pi}{4}$

5. (a) $90° - 5°17'34'' = 84°42'26''$ (b) $90° - 32.5° = 57.5°$

7. (a) $180° - 48°51'37'' = 131°8'23''$ (b) $180° - 136.42° = 43.58°$

Note: Multiply each degree measure by $\frac{\pi}{180}$ to obtain the listed radian measure.

9. (a) $150° \cdot \frac{\pi}{180} = \frac{5 \cdot 30\pi}{6 \cdot 30} = \frac{5\pi}{6}$ (b) $-60° \cdot \frac{\pi}{180} = -\frac{60\pi}{3 \cdot 60} = -\frac{\pi}{3}$

   (c) $225° \cdot \frac{\pi}{180} = \frac{5 \cdot 45\pi}{4 \cdot 45} = \frac{5\pi}{4}$

Note: Multiply each radian measure by $\frac{180}{\pi}$ to obtain the listed degree measure.

15. (a) $-\frac{7\pi}{2} \cdot \left(\frac{180}{\pi}\right)° = -\left(\frac{7 \cdot 90 \cdot 2\pi}{2\pi}\right)° = -630°$ (b) $7\pi \cdot \left(\frac{180}{\pi}\right)° = (7 \cdot 180)° = 1260°$

    (c) $\frac{\pi}{9} \cdot \left(\frac{180}{\pi}\right)° = \left(\frac{20 \cdot 9\pi}{9\pi}\right)° = 20°$

17. *Note:* Some calculators can easily change radians to degrees, minutes, and seconds by pressing a few keys. For **TI-82/83** users, enter

    2 | $\times$ | 180 | $\div$ | 2nd | π | 2nd | ANGLE | 4 | ENTER .

    Check your calculator manual for this feature.

    We first convert 2 radians to degrees. $2 \cdot \left(\frac{180}{\pi}\right)° \approx 114.59156° = 114° + 0.59156°.$

    We now use the decimal portion, $0.59156°$, and convert it to minutes.

    Since $60' = 1°$, we have $0.59156° = 0.59156(60') = 35.4936'.$

    Using the decimal portion, $0.4936'$, we convert it to seconds.

    Since $60'' = 1'$, we have $0.4936' = 0.4936(60'') \approx 30''.$ $\therefore$ 2 radians $\approx 114°35'30''$

21 Since $1' = \left(\frac{1}{60}\right)^{\circ}$, $41' = \left(\frac{41}{60}\right)^{\circ}$. Thus, $37°41' = \left(37 + \frac{41}{60}\right)^{\circ} \approx 37.6833°$.

23 Since $1'' = \left(\frac{1}{3600}\right)^{\circ}$, $27'' = \left(\frac{27}{3600}\right)^{\circ}$. Thus, $115°26'27'' = \left(115 + \frac{26}{60} + \frac{27}{3600}\right)^{\circ} \approx 115.4408°$.

25 We have $63°$ and a portion of one more degree. Since $1° = 60'$,

$0.169° = 0.169\,(60') = 10.14'$. We now have $10'$ and a portion of one more minute.

Since $1' = 60''$, $0.14' = 0.14\,(60'') = 8.4''$. $\therefore 63.169° \approx 63°10'8''$

29 We will use the formula for the length of a circular arc.

$$s = r\theta \Rightarrow r = \frac{s}{\theta} = \frac{10}{4} = 2.5 \text{ cm}.$$

31 (a) $s = r\theta = 8 \cdot \left(45 \cdot \frac{\pi}{180}\right) = 8 \cdot \frac{\pi}{4} = 2\pi \approx 6.28$ cm

(b) $A = \frac{1}{2}r^2\theta = \frac{1}{2}(8)^2\left(\frac{\pi}{4}\right) = 8\pi \approx 25.13$ cm^2

33 (a) Remember that θ is measured in radians. $s = r\theta \Rightarrow \theta = \frac{s}{r} = \frac{7}{4} = 1.75$ radians.

Converting to degrees, we have $\frac{7}{4} \cdot \left(\frac{180}{\pi}\right)^{\circ} = \left(\frac{315}{\pi}\right)^{\circ} \approx 100.27°$.

(b) $A = \frac{1}{2}r^2\theta = \frac{1}{2}(4)^2\left(\frac{7}{4}\right) = 14$ cm^2

35 (a) A measure of $50°$ is equivalent to $\left(50 \cdot \frac{\pi}{180}\right)$ radians. The radius is one-half of the

diameter. Thus, $s = r\theta = \left(\frac{1}{2} \cdot 16\right)\left(50 \cdot \frac{\pi}{180}\right) = 8 \cdot \frac{5\pi}{18} = \frac{20\pi}{9} \approx 6.98$ m.

(b) $A = \frac{1}{2}r^2\theta = \frac{1}{2}(8)^2\left(\frac{5\pi}{18}\right) = \frac{80\pi}{9} \approx 27.93$ m^2

37 radius $= \frac{1}{2} \cdot 8000$ miles $= 4000$ miles

(a) $s = r\theta = 4000\left(60 \cdot \frac{\pi}{180}\right) = \frac{4000\pi}{3} \approx 4189$ miles

(b) $s = r\theta = 4000\left(45 \cdot \frac{\pi}{180}\right) = 1000\pi \approx 3142$ miles

(c) $s = r\theta = 4000\left(30 \cdot \frac{\pi}{180}\right) = \frac{2000\pi}{3} \approx 2094$ miles

(d) $s = r\theta = 4000\left(10 \cdot \frac{\pi}{180}\right) = \frac{2000\pi}{9} \approx 698$ miles

(e) $s = r\theta = 4000\left(1 \cdot \frac{\pi}{180}\right) = \frac{200\pi}{9} \approx 70$ miles

39 $\theta = \frac{s}{r} = \frac{500}{4000} = \frac{1}{8}$ radian; $\left(\frac{1}{8}\right)\left(\frac{180}{\pi}\right)^{\circ} = \left(\frac{45}{2\pi}\right)^{\circ} \approx 7°10'$

41 23 hours, 56 minutes, and 4 seconds $= 23(60)^2 + 56(60) + 4 = 86{,}164$ sec.

Since the earth turns through 2π radians in 86,164 seconds,

it rotates through $\frac{2\pi}{86{,}164} \approx 7.29 \times 10^{-5}$ radians in one second.

43 (a) $\left(40 \dfrac{\text{revolutions}}{\text{minute}}\right)\left(2\pi \dfrac{\text{radians}}{\text{revolution}}\right) = 80\pi \dfrac{\text{radians}}{\text{minute}}$.

Note: Remember to write out and "cancel" the units if you are unsure about

what units your answer is measured in.

(b) The distance that a point on the circumference travels is

$$s = r\theta = (5 \text{ in}) \cdot 80\pi = 400\pi \text{ in}.$$

Hence, its linear speed is

400π in/min $= \frac{100\pi}{3}$ ft/min $\left\{400\pi \cdot \frac{1}{12}\right\} \approx 104.72$ ft/min.

45 (a) As in Exercise 43, we take the number of

revolutions per minute times the number of radians per revolution and

obtain $(33\frac{1}{3})(2\pi) = \frac{200\pi}{3}$ and $45(2\pi) = 90\pi$.

(b) $s = r\theta = (\frac{1}{2} \cdot 12)(\frac{200\pi}{3}) = 400\pi$ in. Linear speed $= 400\pi$ in/min $= \frac{100\pi}{3}$ ft/min.

$s = r\theta = (\frac{1}{2} \cdot 7)(90\pi) = 315\pi$ in. Linear speed $= 315\pi$ in/min $= \frac{105\pi}{4}$ ft/min.

47 (a) The distance that the cargo is lifted is equal to the arc length that the cable is

moved through. $s = r\theta = (\frac{1}{2} \cdot 3)(\frac{7\pi}{4}) = \frac{21\pi}{8} \approx 8.25$ ft.

(b) $s = r\theta \Rightarrow d = (\frac{1}{2} \cdot 3)\theta \Rightarrow \theta = (\frac{2}{3}d)$ radians. For example, to lift the cargo 6 feet,

the winch must rotate $\frac{2}{3} \cdot 6 = 4$ radians, or about 229°.

49 $\text{Area}_{\text{small}} = \frac{1}{2}r^2\theta = \frac{1}{2}(\frac{1}{2} \cdot 18)^2 \cdot (\frac{2\pi}{6}) = \frac{27\pi}{2}$. $\text{Area}_{\text{large}} = \frac{1}{2}(\frac{1}{2} \cdot 26)^2 \cdot (\frac{2\pi}{8}) = \frac{169\pi}{8}$.

$\text{Cost}_{\text{small}} = \text{Area}_{\text{small}} \div \text{Cost} = \frac{27\pi}{2} \div 2 \approx 21.21$ in^2/dollar.

$\text{Cost}_{\text{large}} = \text{Area}_{\text{large}} \div \text{Cost} = \frac{169\pi}{8} \div 3 \approx 22.12$ in^2/dollar.

The large slice provides slightly more pizza per dollar.

51 $\frac{40 \text{ miles}}{\text{hour}} = \frac{40 \text{ miles}}{\text{hour}} \cdot \frac{1 \text{ hour}}{3600 \text{ seconds}} \cdot \frac{5280 \text{ feet}}{\text{mile}} \cdot \frac{12 \text{ inches}}{\text{foot}} = \frac{704 \text{ inches}}{\text{second}}$.

The circumference of the wheel is $2\pi(14)$ inches. The back sprocket then rotates

$\frac{704 \text{ inches}}{\text{second}} \cdot \frac{1 \text{ revolution}}{28\pi \text{ inches}} = \frac{704 \text{ revolutions}}{28\pi \text{ second}}$ or $\frac{704}{28\pi} \cdot 2\pi = \frac{352}{7}$ radians per second.

The front sprocket's angular speed is given by { from Exercise 50 }

$\theta_1 = \frac{r_2\theta_2}{r_1} = \frac{2 \cdot \frac{352}{7}}{5} = \frac{704}{35} \approx 20.114$ radians per second

or 3.2 revolutions per second or 192.08 revolutions per minute.

2.2 Exercises

Note: Answers are in the order *sin, cos, tan, cot, sec, csc* for any exercises that require
the values of the six trigonometric functions.

1 Using the definition of the trigonometric functions in terms of a unit circle with
$P(-\frac{15}{17}, \frac{8}{17})$, we have

$\sin t = y = \frac{8}{17},$ $\qquad \cos t = x = -\frac{15}{17},$ $\qquad \tan t = \frac{y}{x} = \frac{8/17}{-15/17} = -\frac{8}{15},$

$\csc t = \frac{1}{y} = \frac{1}{8/17} = \frac{17}{8},$ $\quad \sec t = \frac{1}{x} = \frac{1}{-15/17} = -\frac{17}{15},$ $\quad \cot t = \frac{x}{y} = \frac{-15/17}{8/17} = -\frac{15}{8}.$

5   See *Figure 5.* $P(t) = (\frac{3}{5}, \frac{4}{5})$.

(a) $(t + \pi)$ will be $\frac{1}{2}$ revolution from (t) in the counterclockwise direction.

Hence, $P(t + \pi) = (-\frac{3}{5}, -\frac{4}{5})$.

(b) $(t - \pi)$ will be $\frac{1}{2}$ revolution from (t) in the clockwise direction.

Hence, $P(t - \pi) = (-\frac{3}{5}, -\frac{4}{5})$. Observe that $P(t + \pi) = P(t - \pi)$ for any t.

(c) $(-t)$ is the angle measure in the opposite direction of (t).

Hence, $P(-t) = (\frac{3}{5}, -\frac{4}{5})$.

(d) $(-t - \pi)$ will be $\frac{1}{2}$ revolution from $(-t)$ in the clockwise direction.

Hence, $P(-t - \pi) = (-\frac{3}{5}, \frac{4}{5})$.

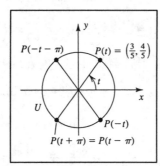

Figure 5

9   (a) The point P on the unit circle U that corresponds to $t = 2\pi$ has coordinates (1, 0). Thus, we choose $x = 1$ and $y = 0$ and use the definition of the trigonometric functions in terms of a unit circle.

$\sin t = y \Rightarrow \sin 2\pi = 0$. $\cos t = x \Rightarrow \cos 2\pi = 1$.

$\tan t = \frac{y}{x} \Rightarrow \tan 2\pi = \frac{0}{1} = 0$. $\cot t = \frac{x}{y} \Rightarrow \cot 2\pi = \frac{1}{0}$ and is undefined.

$\sec t = \frac{1}{x} \Rightarrow \sec 2\pi = \frac{1}{1} = 1$. $\csc t = \frac{1}{y} \Rightarrow \csc 2\pi = \frac{1}{0}$ and is undefined.

(b) $t = -3\pi$ is coterminal with $t = \pi$. The point P on the unit circle U that corresponds to $t = -3\pi$ has coordinates (−1, 0). Thus, we choose $x = -1$ and $y = 0$.

$\sin t = y \Rightarrow \sin(-3\pi) = 0$. $\cos t = x \Rightarrow \cos(-3\pi) = -1$.

$\tan t = \frac{y}{x} \Rightarrow \tan(-3\pi) = \frac{0}{-1} = 0$. $\cot t = \frac{x}{y} \Rightarrow \cot(-3\pi) = \frac{-1}{0}$ and is und.

$\sec t = \frac{1}{x} \Rightarrow \sec(-3\pi) = \frac{1}{-1} = -1$. $\csc t = \frac{1}{y} \Rightarrow \csc(-3\pi) = \frac{1}{0}$ and is und.

$\boxed{11}$ (a) The point P on the unit circle U that corresponds to $t = \frac{3\pi}{2}$ has coordinates $(0, -1)$. Thus, we choose $x = 0$ and $y = -1$.

$$\sin t = y \Rightarrow \sin\frac{3\pi}{2} = -1. \qquad\qquad \cos t = x \Rightarrow \cos\frac{3\pi}{2} = 0.$$

$$\tan t = \frac{y}{x} \Rightarrow \tan\frac{3\pi}{2} = \frac{-1}{0} \text{ and is und.} \qquad \cot t = \frac{x}{y} \Rightarrow \cot\frac{3\pi}{2} = \frac{0}{-1} = 0.$$

$$\sec t = \frac{1}{x} \Rightarrow \sec\frac{3\pi}{2} = \frac{1}{0} \text{ and is und.} \qquad \csc t = \frac{1}{y} \Rightarrow \csc\frac{3\pi}{2} = \frac{1}{-1} = -1.$$

(b) $t = -\frac{7\pi}{2}$ is coterminal with $t = \frac{\pi}{2}$. The point P on the unit circle U that corresponds to $t = -\frac{7\pi}{2}$ has coordinates $(0, 1)$. Thus, we choose $x = 0$ and $y = 1$.

$$\sin t = y \Rightarrow \sin\left(-\frac{7\pi}{2}\right) = 1. \qquad\qquad \cos t = x \Rightarrow \cos\left(-\frac{7\pi}{2}\right) = 0.$$

$$\tan t = \frac{y}{x} \Rightarrow \tan\left(-\frac{7\pi}{2}\right) = \frac{1}{0} \text{ and is und.} \qquad \cot t = \frac{x}{y} \Rightarrow \cot\left(-\frac{7\pi}{2}\right) = \frac{0}{1} = 0.$$

$$\sec t = \frac{1}{x} \Rightarrow \sec\left(-\frac{7\pi}{2}\right) = \frac{1}{0} \text{ and is und.} \qquad \csc t = \frac{1}{y} \Rightarrow \csc\left(-\frac{7\pi}{2}\right) = \frac{1}{1} = 1.$$

$\boxed{13}$ (a) $t = \frac{9\pi}{4}$ is coterminal with $t = \frac{\pi}{4}$. The point P on the unit circle U that corresponds to $t = \frac{9\pi}{4}$ has coordinates $\left(\frac{\sqrt{2}}{2}, \frac{\sqrt{2}}{2}\right)$. Thus, we choose $x = \frac{\sqrt{2}}{2}$ and $y = \frac{\sqrt{2}}{2}$.

$$\sin t = y \Rightarrow \sin\frac{9\pi}{4} = \frac{\sqrt{2}}{2}. \qquad\qquad \cos t = x \Rightarrow \cos\frac{9\pi}{4} = \frac{\sqrt{2}}{2}.$$

$$\tan t = \frac{y}{x} \Rightarrow \tan\frac{9\pi}{4} = \frac{\sqrt{2}/2}{\sqrt{2}/2} = 1.$$

We now use the reciprocal relationships to find the values of the 3 remaining trigonometric functions.

$$\cot t = \frac{1}{\tan t} = \frac{1}{1} = 1. \qquad\qquad \sec t = \frac{1}{\cos t} = \frac{1}{\sqrt{2}/2} = \frac{2}{\sqrt{2}} = \frac{2^1}{2^{1/2}} = 2^{1/2} = \sqrt{2}.$$

$$\csc t = \frac{1}{\sin t} = \frac{1}{\sqrt{2}/2} = \sqrt{2}.$$

(b) $t = -\frac{5\pi}{4}$ is coterminal with $t = \frac{3\pi}{4}$. The point P on the unit circle U that corresponds to $t = -\frac{5\pi}{4}$ has coordinates $\left(-\frac{\sqrt{2}}{2}, \frac{\sqrt{2}}{2}\right)$. Thus, we choose $x = -\frac{\sqrt{2}}{2}$ and $y = \frac{\sqrt{2}}{2}$.

$$\sin t = y \Rightarrow \sin\left(-\frac{5\pi}{4}\right) = \frac{\sqrt{2}}{2}. \qquad\qquad \cos t = x \Rightarrow \cos\left(-\frac{5\pi}{4}\right) = -\frac{\sqrt{2}}{2}.$$

$$\tan t = \frac{y}{x} \Rightarrow \tan\left(-\frac{5\pi}{4}\right) = \frac{\sqrt{2}/2}{-\sqrt{2}/2} = -1. \qquad \cot t = \frac{1}{\tan t} = \frac{1}{-1} = -1.$$

$$\sec t = \frac{1}{\cos t} = \frac{1}{-\sqrt{2}/2} = -\sqrt{2}. \qquad\qquad \csc t = \frac{1}{\sin t} = \frac{1}{\sqrt{2}/2} = \sqrt{2}.$$

15 (a) The point P on the unit circle U that corresponds to $t = \frac{5\pi}{4}$ has coordinates $\left(-\frac{\sqrt{2}}{2}, -\frac{\sqrt{2}}{2} \right)$. Thus, we choose $x = -\frac{\sqrt{2}}{2}$ and $y = -\frac{\sqrt{2}}{2}$.

$$\sin t = y \Rightarrow \sin \frac{5\pi}{4} = -\frac{\sqrt{2}}{2}. \qquad \cos t = x \Rightarrow \cos \frac{5\pi}{4} = -\frac{\sqrt{2}}{2}.$$

$$\tan t = \frac{y}{x} \Rightarrow \tan \frac{5\pi}{4} = \frac{-\sqrt{2}/2}{-\sqrt{2}/2} = 1. \qquad \cot t = \frac{1}{\tan t} = \frac{1}{1} = 1.$$

$$\sec t = \frac{1}{\cos t} = \frac{1}{-\sqrt{2}/2} = -\sqrt{2}. \qquad \csc t = \frac{1}{\sin t} = \frac{1}{-\sqrt{2}/2} = -\sqrt{2}.$$

(b) $t = -\frac{\pi}{4}$ is coterminal with $t = \frac{7\pi}{4}$. The point P on the unit circle U that corresponds to $t = -\frac{\pi}{4}$ has coordinates $\left(\frac{\sqrt{2}}{2}, -\frac{\sqrt{2}}{2} \right)$. Thus, we choose $x = \frac{\sqrt{2}}{2}$ and $y = -\frac{\sqrt{2}}{2}$.

$$\sin t = y \Rightarrow \sin \left(-\frac{\pi}{4} \right) = -\frac{\sqrt{2}}{2}. \qquad \cos t = x \Rightarrow \cos \left(-\frac{\pi}{4} \right) = \frac{\sqrt{2}}{2}.$$

$$\tan t = \frac{y}{x} \Rightarrow \tan \left(-\frac{\pi}{4} \right) = \frac{-\sqrt{2}/2}{\sqrt{2}/2} = -1. \qquad \cot t = \frac{1}{\tan t} = \frac{1}{-1} = -1.$$

$$\sec t = \frac{1}{\cos t} = \frac{1}{\sqrt{2}/2} = \sqrt{2}. \qquad \csc t = \frac{1}{\sin t} = \frac{1}{-\sqrt{2}/2} = -\sqrt{2}.$$

17 (a) $\cos t > 0 \ \{x > 0\}$ implies that the point P corresponding to t is in quadrant I or quadrant IV. $\sin t < 0 \ \{y < 0\}$ implies that the point P corresponding to t is in quadrant III or quadrant IV. P must be in quadrant IV to satisfy both conditions.

(b) $\sin t < 0 \Rightarrow P$ is in QIII or QIV. $\cot t > 0 \Rightarrow P$ is in QI or QIII.

Hence, P is in QIII.

(c) $\csc t > 0 \Rightarrow P$ is in QI or QII. $\sec t < 0 \Rightarrow P$ is in QII or QIII.

Hence, P is in QII.

(d) $\sec t < 0 \Rightarrow P$ is in QII or QIII. $\tan t > 0 \Rightarrow P$ is in QI or QIII.

Hence, P is in QIII.

19 $\cot t = \frac{\cos t}{\sin t}$ {cotangent identity} $= \frac{\sqrt{1 - \sin^2 t}}{\sin t}$ $\{\sin^2 t + \cos^2 t = 1\}$ Note that the "$\pm$" notation is not needed in front of the square root since all of the trigonometric functions of an acute angle are positive.

21 $\sec t = \frac{1}{\cos t}$ {reciprocal identity} $= \frac{1}{\sqrt{1 - \sin^2 t}}$

23 One solution is $\sin t = \sqrt{1 - \cos^2 t} = \sqrt{1 - \frac{1}{\sec^2 t}} = \frac{\sqrt{\sec^2 t - 1}}{\sec t}$.

Alternatively, $\sin t = \frac{\sin t / \cos t}{1 / \cos t} = \frac{\tan t}{\sec t} = \frac{\sqrt{\sec^2 t - 1}}{\sec t}$ $\{1 + \tan^2 t = \sec^2 t\}$.

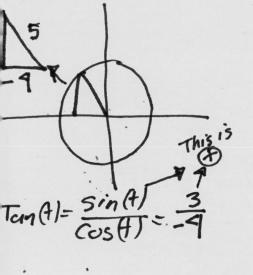

$\text{Tan}(t) = \dfrac{\sin(t)}{\cos(t)} = \dfrac{3}{-4}$

This is ⊕

...e 2 function values using only the fundamental identities are

... 3 function values are just the reciprocals of those given and are

...er.

$> 0 \Rightarrow t$ is in QII.

$\sec t = \pm\sqrt{1 + \tan^2 t}$ { Use the "−" since the secant is negative in

$\} = -\sqrt{1 + \tan^2 t} = -\sqrt{1 + \frac{9}{16}} = -\frac{5}{4}.$

$\dfrac{\sin t}{-4/5} \Rightarrow \sin t = \left(-\frac{3}{4}\right)\left(-\frac{4}{5}\right) = \frac{3}{5}.$ ★ $\frac{3}{5}, -\frac{4}{5}, -\frac{3}{4}, -\frac{4}{3}, -\frac{5}{4}, \frac{5}{3}$

$\boxed{27}$ $\sin t = -\frac{5}{13}$ and $\sec t > 0 \Rightarrow t$ is in QIV.

$\sin^2 t + \cos^2 t = 1 \Rightarrow \cos^2 t = 1 - \sin^2 t \Rightarrow \cos t = \pm\sqrt{1 - \sin^2 t} =$

{ Use the "+" since the cosine is positive in the fourth quadrant. } $= \sqrt{1 - \frac{25}{169}} = \frac{12}{13}.$

$\tan t = \dfrac{\sin t}{\cos t} = \dfrac{-5/13}{12/13} = -\frac{5}{12}.$ ★ $-\frac{5}{13}, \frac{12}{13}, -\frac{5}{12}, -\frac{12}{5}, \frac{13}{12}, -\frac{13}{5}$

$\boxed{29}$ $\cos t = -\frac{1}{3}$ and $\sin t < 0 \Rightarrow t$ is in QIII. $\sin t = -\sqrt{1 - \cos^2 t} = -\sqrt{1 - \frac{1}{9}} = -\dfrac{\sqrt{8}}{3}.$

$\tan t = \dfrac{\sin t}{\cos t} = \dfrac{-\sqrt{8}/3}{-1/3} = \sqrt{8},$ or $2\sqrt{2}.$ ★ $-\dfrac{\sqrt{8}}{3}, -\frac{1}{3}, \sqrt{8}, \dfrac{1}{\sqrt{8}}, -3, -\dfrac{3}{\sqrt{8}}$

$\boxed{31}$ $\sec t = -4$ and $\csc t > 0 \Rightarrow t$ is in QII. $\tan t = -\sqrt{\sec^2 t - 1} = -\sqrt{16 - 1} = -\sqrt{15}.$

$\tan t = \dfrac{\sin t}{\cos t} \Rightarrow -\sqrt{15} = \dfrac{\sin t}{-1/4} \Rightarrow \sin t = \dfrac{\sqrt{15}}{4}.$

★ $\dfrac{\sqrt{15}}{4}, -\frac{1}{4}, -\sqrt{15}, -\dfrac{1}{\sqrt{15}}, -4, \dfrac{4}{\sqrt{15}}$

$\boxed{33}$ (a) Since $1 + \tan^2 4\beta = \sec^2 4\beta$, $\tan^2 4\beta - \sec^2 4\beta = -1.$

(b) $4\tan^2\beta - 4\sec^2\beta = 4(\tan^2\beta - \sec^2\beta) = 4(-1) = -4$

$\boxed{35}$ (a) $5\sin^2\theta + 5\cos^2\theta = 5(\sin^2\theta + \cos^2\theta) = 5(1) = 5$

(b) $5\sin^2(\theta/4) + 5\cos^2(\theta/4) = 5\left[\sin^2(\theta/4) + \cos^2(\theta/4)\right] = 5(1) = 5$

$\boxed{37}$ $\dfrac{\sin^3 t + \cos^3 t}{\sin t + \cos t}$ $= \dfrac{(\sin t + \cos t)(\sin^2 t - \sin t \cos t + \cos^2 t)}{\sin t + \cos t}$ { factor, sum of cubes }

$= \sin^2 t - \sin t \cos t + \cos^2 t$ { cancel $(\sin t + \cos t)$ }

$= (\sin^2 t + \cos^2 t) - \sin t \cos t$ { group terms }

$= 1 - \sin t \cos t$ { Pythagorean identity }

$\boxed{39}$ $\dfrac{2 - \tan t}{2\csc t - \sec t}$ $= \dfrac{2 - \dfrac{\sin t}{\cos t}}{2\cdot\dfrac{1}{\sin t} - \dfrac{1}{\cos t}}$ { tangent and reciprocal identities }

$= \dfrac{\dfrac{2\cos t - \sin t}{\cos t}}{\dfrac{2\cos t - \sin t}{\sin t \cos t}}$ { combine terms } $= \dfrac{\dfrac{1}{1}}{\dfrac{1}{\sin t}}$ { cancel like terms } $= \sin t$

$\boxed{41}$ $\cos t \sec t = \cos t \, (1/\cos t) \, \{ \text{reciprocal identity} \} = 1$

$\boxed{43}$ $\sin t \sec t = \sin t \, (1/\cos t) = \sin t / \cos t = \tan t \, \{ \text{tangent identity} \}$

$\boxed{45}$ $\dfrac{\csc t}{\sec t} = \dfrac{1/\sin t}{1/\cos t} \, \{ \text{reciprocal identities} \} = \dfrac{\cos t}{\sin t} = \cot t \, \{ \text{cotangent identity} \}$

$\boxed{47}$ $(1 + \cos 2t)(1 - \cos 2t) = 1 - \cos^2 2t = \sin^2 2t \, \{ \text{Pythagorean identity} \}$

$\boxed{49}$ $\cos^2 t \, (\sec^2 t - 1) = \cos^2 t \, (\tan^2 t) \, \{ \text{Pythagorean identity} \}$

$$= \cos^2 t \cdot \frac{\sin^2 t}{\cos^2 t} \, \{ \text{tangent identity} \} = \sin^2 t$$

$\boxed{51}$ $\dfrac{\sin (t/2)}{\csc (t/2)} + \dfrac{\cos (t/2)}{\sec (t/2)} = \dfrac{\sin (t/2)}{1/\sin (t/2)} + \dfrac{\cos (t/2)}{1/\cos (t/2)} = \sin^2 (t/2) + \cos^2 (t/2) = 1$

$\boxed{53}$ $(1 + \sin t)(1 - \sin t) = 1 - \sin^2 t \, \{ \text{multiply as a difference of squares} \} = \cos^2 t = \dfrac{1}{\sec^2 t}$

$\boxed{55}$ $\sec t - \cos t = \dfrac{1}{\cos t} - \cos t = \dfrac{1 - \cos^2 t}{\cos t} = \dfrac{\sin^2 t}{\cos t} = \dfrac{\sin t}{\cos t} \cdot \sin t = \tan t \sin t$

$\boxed{57}$ $(\cot t + \csc t)(\tan t - \sin t)$

$$= \cot t \tan t - \cot t \sin t + \csc t \tan t - \csc t \sin t \, \{ \text{multiply binomials} \}$$

$$= \frac{1}{\tan t} \tan t - \frac{\cos t}{\sin t} \sin t + \frac{1}{\sin t} \frac{\sin t}{\cos t} - \frac{1}{\sin t} \sin t$$

$$\{ \text{reciprocal, cotangent, and tangent identities} \}$$

$$= 1 - \cos t + \frac{1}{\cos t} - 1 \, \{ \text{cancel terms} \}$$

$$= -\cos t + \sec t = \sec t - \cos t$$

$\boxed{59}$ $\sec^2 3t \, \csc^2 3t = (1 + \tan^2 3t)(1 + \cot^2 3t)$ $\qquad \{ \text{Pythagorean identities} \}$

$$= 1 + \tan^2 3t + \cot^2 3t + 1 \qquad \{ \text{multiply binomials} \}$$

$$= (1 + \tan^2 3t) + (\cot^2 3t + 1) \qquad \{ \text{group terms} \}$$

$$= \sec^2 3t + \csc^2 3t \qquad \{ \text{Pythagorean identities} \}$$

$\boxed{61}$ $\log \csc t = \log\!\left(\dfrac{1}{\sin t}\right)$ $\qquad \{ \text{reciprocal identity} \}$

$$= \log 1 - \log \sin t \quad \{ \text{property of logarithms} \}$$

$$= 0 - \log \sin t \qquad \{ \log 1 = 0 \}$$

$$= -\log \sin t$$

$\boxed{63}$ $\sqrt{\sec^2 t - 1} = \sqrt{\tan^2 t} \, \{ \text{Pythagorean identity} \} = |\tan t| \, \left\{ \sqrt{x^2} = |x| \right\} =$

$$-\tan t \text{ since } \tan t < 0 \text{ if } \pi/2 < t < \pi.$$

$\boxed{65}$ $\sqrt{1 + \tan^2 t} = \sqrt{\sec^2 t} \, \{ \text{Pythagorean identity} \} = |\sec t| \, \left\{ \sqrt{x^2} = |x| \right\} =$

$$\sec t \text{ since } \sec t > 0 \text{ if } 3\pi/2 < t < 2\pi.$$

$\boxed{67}$ $\sqrt{\sin^2 (t/2)} = |\sin (t/2)| \, \left\{ \sqrt{x^2} = |x| \right\} =$

$$-\sin (t/2) \text{ since } \sin (t/2) < 0 \text{ if } 2\pi < t < 4\pi \, \{ \pi < t/2 < 2\pi \}.$$

69 (a) From $(1, 0)$, move counterclockwise on the unit circle to the point at the tick marked 4. The projection of this point on the y-axis,

approximately -0.7 or -0.8, is the value of $\sin 4$.

(b) From $(1, 0)$, move clockwise 1.2 units to about 5.1.

As in part (a), $\sin(-1.2)$ is about -0.9.

(c) Draw the horizontal line $y = 0.5$.

This line intersects the circle at about 0.5 and 2.6.

71 (a) From $(1, 0)$, move counterclockwise on the unit circle to the point at the tick marked 4. The projection of this point on the x-axis,

approximately -0.6 or -0.7, is the value of $\cos 4$.

(b) Proceeding as in part (a), $\cos(-1.2)$ is about 0.4.

(c) Draw the vertical line $x = -0.6$.

This line intersects the circle at about 2.2 and 4.1.

73 (a) Note that midnight occurs when $t = -6$.

Time	Temp.	Humidity	Time	Temp.	Humidity
12 A.M.	60	60	12 P.M.	60	60
3 A.M.	52	74	3 P.M.	68	46
6 A.M.	48	80	6 P.M.	72	40
9 A.M.	52	74	9 P.M.	68	46

(b) Since $T(t) = -12\cos\left(\frac{\pi}{12}t\right) + 60$, its maximum is $60 + 12 = 72\,°\text{F}$ at $t = 12$ or 6:00 P.M., and its minimum is $60 - 12 = 48\,°\text{F}$ at $t = 0$ or 6:00 A.M. Since $H(t) = 20\cos\left(\frac{\pi}{12}t\right) + 60$, its maximum is $60 + 20 = 80\%$ at $t = 0$ or 6:00 A.M., and its minimum is $60 - 20 = 40\%$ at $t = 12$ or 6:00 P.M.

(c) When the temperature increases, the relative humidity decreases and vice versa. As the temperature cools, the air can hold less moisture and the relative humidity increases. Because of this phenomenon, fog often occurs during the evening hours rather than in the middle of the day.

2.3 Exercises

1 (a) $\sin(-90°) = -\sin 90°$ $\{$ since $\sin(-t) = -\sin t\,\} = -1$

(b) $\cos\left(-\frac{3\pi}{4}\right) = \cos\frac{3\pi}{4}$ $\{$ since $\cos(-t) = \cos t\,\} = -\frac{\sqrt{2}}{2}$

(c) $\tan(-45°) = -\tan 45°$ $\{$ since $\tan(-t) = -\tan t\,\} = -1$

$\boxed{3}$ (a) $\cot\left(-\frac{3\pi}{4}\right) = -\cot\frac{3\pi}{4}$ { since $\cot(-t) = -\cot t$ } $= -(-1) = 1$

(b) $\sec(-180°) = \sec 180°$ { since $\sec(-t) = \sec t$ } $= -1$

(c) $\csc\left(-\frac{3\pi}{2}\right) = -\csc\frac{3\pi}{2}$ { since $\csc(-t) = -\csc t$ } $= -(-1) = 1$

$\boxed{5}$ $\sin(-t)\sec(-t) = (-\sin t)\sec t$ { formulas for negatives }

$= (-\sin t)(1/\cos t)$ { reciprocal identity }

$= -\tan t$ { tangent identity }

$\boxed{7}$ $\dfrac{\cot(-t)}{\csc(-t)} = \dfrac{-\cot t}{-\csc t}$ { formulas for negatives }

$= \dfrac{\cos t/\sin t}{1/\sin t}$ { cotangent identity and reciprocal identity }

$= \cos t$ { simplify }

$\boxed{9}$ $\dfrac{1}{\cos(-t)} - \tan(-t)\sin(-t) = \dfrac{1}{\cos t} - (-\tan t)(-\sin t)$ { formulas for negatives }

$= \dfrac{1}{\cos t} - \dfrac{\sin t}{\cos t}\sin t$ { tangent identity }

$= \dfrac{1 - \sin^2 t}{\cos t}$ { combine terms }

$= \dfrac{\cos^2 t}{\cos t}$ { Pythagorean identity }

$= \cos t$ { cancel $\cos t$ }

$\boxed{11}$ (a) Using Figure 22 in the text, we see that as t gets close to 0 through numbers greater than 0 (from the *right* of 0), $\sin t$ approaches 0.

(b) As t approaches $-\frac{\pi}{2}$ through numbers less than $-\frac{\pi}{2}$ (from the *left* of $-\frac{\pi}{2}$), $\sin t$ approaches -1.

$\boxed{13}$ (a) Using Figure 24 in the text, we see that as t gets close to $\frac{\pi}{4}$ through numbers greater than $\frac{\pi}{4}$ (from the *right* of $\frac{\pi}{4}$), $\cos t$ approaches $\frac{\sqrt{2}}{2}$. Note that the value $\frac{\sqrt{2}}{2}$ is in the table containing specific values of the cosine function.

(b) As t approaches π through numbers less than π (from the *left* of π), $\cos t$ approaches -1.

$\boxed{15}$ (a) Using Figure 26 in the text, we see that as t gets close to $\frac{\pi}{4}$ through numbers greater than $\frac{\pi}{4}$ (from the *right* of $\frac{\pi}{4}$), $\tan t$ approaches 1.

(b) As t approaches $\frac{\pi}{2}$ through numbers greater than $\frac{\pi}{2}$ (from the *right* of $\frac{\pi}{2}$), $\tan t$ is approaching the vertical asymptote at $t = \frac{\pi}{2}$. $\tan t$ is *decreasing* without bound, and we use the notation $\underline{\tan t \to -\infty}$ to denote this.

$\boxed{17}$ (a) Using Figure 29 in the text, we see that as t gets close to $-\frac{\pi}{4}$ through numbers less than $-\frac{\pi}{4}$ (from the *left* of $-\frac{\pi}{4}$), $\cot t$ approaches -1.

(b) As t approaches 0 through numbers greater than 0 (from the *right* of 0), $\cot t$ is approaching the vertical asymptote at $t = 0$ {the y-axis}. $\cot t$ is *increasing* without bound, and we use the notation $\underline{\cot t \to \infty}$ to denote this.

$\boxed{19}$ (a) Using Figure 28 in the text, we see that as t gets close to $\frac{\pi}{2}$ through numbers less than $\frac{\pi}{2}$ (from the *left* of $\frac{\pi}{2}$), $\sec t \to \infty$.

(b) As t approaches $\frac{\pi}{4}$ through numbers greater than $\frac{\pi}{4}$ (from the *right* of $\frac{\pi}{4}$), $\sec t$ approaches $\sqrt{2}$. Recall that $\cos\frac{\pi}{4} = \frac{1}{\sqrt{2}}$ and that $\sec t = \frac{1}{\cos t}$.

$\boxed{21}$ (a) Using Figure 27 in the text, we see that as t gets close to 0 through numbers less than 0 (from the *left* of 0), $\csc t \to -\infty$.

(b) As t approaches $\frac{\pi}{2}$ through numbers greater than $\frac{\pi}{2}$ (from the *right* of $\frac{\pi}{2}$), $\csc t$ approaches 1.

$\boxed{23}$ Refer to Figure 21 and the accompanying table. We see that $\sin\frac{3\pi}{2} = -1$.

Since the period of the sine is 2π, the second value in $[0, 4\pi]$ is $\frac{3\pi}{2} + 2\pi = \frac{7\pi}{2}$.

$\boxed{29}$ Refer to Figure 23 and the accompanying table.

We see that $\cos\frac{3\pi}{4} = \cos\frac{5\pi}{4} = -\frac{\sqrt{2}}{2}$. Since the period of the cosine is 2π, other values in $[0, 4\pi]$ are $\frac{3\pi}{4} + 2\pi = \frac{11\pi}{4}$ and $\frac{5\pi}{4} + 2\pi = \frac{13\pi}{4}$.

$\boxed{31}$ Refer to Figure 26. In the interval $\left(-\frac{\pi}{2}, \frac{\pi}{2}\right)$, $\tan t = 1$ only if $t = \frac{\pi}{4}$. Since the period of the tangent is π, the desired value in the interval $\left(\frac{\pi}{2}, \frac{3\pi}{2}\right)$ is $\frac{\pi}{4} + \pi = \frac{5\pi}{4}$.

$\boxed{33}$ $y = \sin t$; $[-2\pi, 2\pi]$; $a = \sqrt{2}/2$ •   Refer to Figure 21.

$\sin t = \sqrt{2}/2 \Rightarrow t = \frac{\pi}{4}$ and $\frac{3\pi}{4}$. Also, $\frac{\pi}{4} - 2\pi = -\frac{7\pi}{4}$ and $\frac{3\pi}{4} - 2\pi = -\frac{5\pi}{4}$.

$\sin t > \sqrt{2}/2$ when the graph is *above* the horizontal line $y = \sqrt{2}/2$.

$\sin t < \sqrt{2}/2$ when the graph is *below* the horizontal line $y = \sqrt{2}/2$.

  ★ (a) $-\frac{7\pi}{4}, -\frac{5\pi}{4}, \frac{\pi}{4}, \frac{3\pi}{4}$ (b) $-\frac{7\pi}{4} < t < -\frac{5\pi}{4}, \frac{\pi}{4} < t < \frac{3\pi}{4}$

                                    (c) $-2\pi \leq t < -\frac{7\pi}{4}, -\frac{5\pi}{4} < t < \frac{\pi}{4}$, and $\frac{3\pi}{4} < t \leq 2\pi$

$\boxed{35}$ $y = \cos t$; $[-2\pi, 2\pi]$; $a = -\sqrt{2}/2$ •   Refer to Figure 23.

$\cos t = -\sqrt{2}/2 \Rightarrow t = \frac{3\pi}{4}$ and $\frac{5\pi}{4}$. Also, $\frac{3\pi}{4} - 2\pi = -\frac{5\pi}{4}$ and $\frac{5\pi}{4} - 2\pi = -\frac{3\pi}{4}$.

$\cos t > -\sqrt{2}/2$ when the graph is *above* the horizontal line $y = -\sqrt{2}/2$.

$\cos t < -\sqrt{2}/2$ when the graph is *below* the horizontal line $y = -\sqrt{2}/2$.

  ★ (a) $-\frac{5\pi}{4}, -\frac{3\pi}{4}, \frac{3\pi}{4}, \frac{5\pi}{4}$ (b) $-2\pi \leq t < -\frac{5\pi}{4}, -\frac{3\pi}{4} < t < \frac{3\pi}{4}$, and $\frac{5\pi}{4} < t \leq 2\pi$

                                    (c) $-\frac{5\pi}{4} < t < -\frac{3\pi}{4}$ and $\frac{3\pi}{4} < t < \frac{5\pi}{4}$

$\boxed{37}$ $y = 2 + \sin t$ • Shift $y = \sin t$ up 2 units.

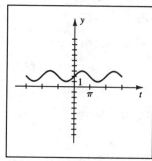

Figure 37

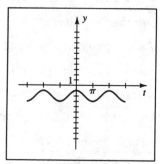

Figure 39

$\boxed{39}$ $y = \cos t - 2$ • Shift $y = \cos t$ down 2 units.

$\boxed{41}$ $y = 1 + \tan t$ • Shift $y = \tan t$ up 1 unit.

Since $1 + \tan\left(-\frac{\pi}{4}\right) = 1 + (-1) = 0$, and the period of the tangent is π,

there are t-intercepts at $t = -\frac{\pi}{4} + \pi n$.

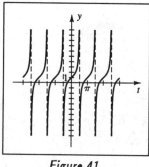

Figure 41

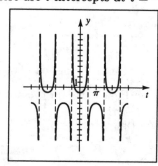

Figure 43

$\boxed{43}$ $y = \sec t - 2$ •

Shift $y = \sec t$ down 2 units, t-intercepts are at $t = \frac{\pi}{3} + 2\pi n, \frac{5\pi}{3} + 2\pi n$.

$\boxed{45}$ (a) As we move from left to right, the function increases { goes up } on the

intervals $[-2\pi, -\frac{3\pi}{2})$, $(-\frac{3\pi}{2}, -\pi]$, $[0, \frac{\pi}{2})$, $(\frac{\pi}{2}, \pi]$.

(b) As we move from left to right, the function decreases { goes down } on the

intervals $[-\pi, -\frac{\pi}{2})$, $(-\frac{\pi}{2}, 0]$, $[\pi, \frac{3\pi}{2})$, $(\frac{3\pi}{2}, 2\pi]$.

$\boxed{47}$ (a) The tangent function increases on *all* intervals on which it is defined.

Between -2π and 2π,

these intervals are $[-2\pi, -\frac{3\pi}{2})$, $(-\frac{3\pi}{2}, -\frac{\pi}{2})$, $(-\frac{\pi}{2}, \frac{\pi}{2})$, $(\frac{\pi}{2}, \frac{3\pi}{2})$, and $(\frac{3\pi}{2}, 2\pi]$.

(b) The tangent function is *never* decreasing on any interval for which it is defined.

$\boxed{49}$ This is good advice.

51 Graph $y = \sin(t^2)$ and $y = 0.5$ on the same coordinate plane. From the graph, we see that $\sin(t^2)$ assumes the value of 0.5 at $x \approx \pm 0.72$, ± 1.62, ± 2.61, ± 2.98.

$[-\pi, \pi]$ by $[-2.09, 2.09]$

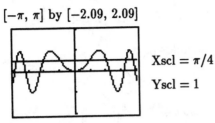

Xscl $= \pi/4$
Yscl $= 1$

Figure 51

$[-2\pi, 2\pi]$ by $[-5.19, 3.19]$

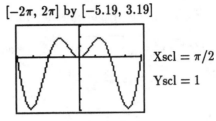

Xscl $= \pi/2$
Yscl $= 1$

Figure 53

53 We see that the graph of $y = t \sin t$ assumes a maximum value of approximately 1.82 at $t \approx \pm 2.03$, and a minimum value of -4.81 at $t \approx \pm 4.91$.

55 As $t \to 0^+$, $f(t) = \dfrac{1 - \cos t}{t} \to 0$.

$[-1, 1]$ by $[-0.67, 0.67]$

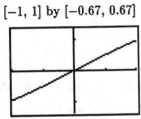

Xscl $= 0.5$
Yscl $= 0.5$

Figure 55

$[-2, 2]$ by $[-1.33, 1.33]$

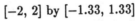
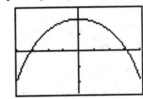
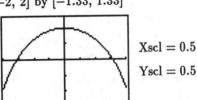

Xscl $= 0.5$
Yscl $= 0.5$

Figure 57

57 As $t \to 0^+$, $f(t) = t \cot t \to 1$.

59 As $t \to 0^+$, $f(t) = \dfrac{\tan t}{t} \to 1$.

$[-1.5, 1.5]$ by $[0, 3]$

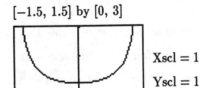

Xscl $= 1$
Yscl $= 1$

Figure 59

2.4 Exercises

$\boxed{1}$ For the point $P(x, y)$, we let r denote the distance from the origin to P. Applying the theorem on trigonometric functions as ratios with $x = 4$, $y = -3$, and $r = \sqrt{x^2 + y^2} = \sqrt{4^2 + (-3)^2} = 5$, we obtain the following:

$$\sin\theta = \frac{y}{r} = \frac{-3}{5} = -\frac{3}{5} \qquad\qquad \cos\theta = \frac{x}{r} = \frac{4}{5}$$

$$\tan\theta = \frac{y}{x} = \frac{-3}{4} = -\frac{3}{4} \qquad\qquad \cot\theta = \frac{x}{y} = \frac{4}{-3} = -\frac{4}{3}$$

$$\sec\theta = \frac{r}{x} = \frac{5}{4} \qquad\qquad \csc\theta = \frac{r}{y} = \frac{5}{-3} = -\frac{5}{3}$$

$\boxed{3}$ $x = -2$ and $y = -5 \Rightarrow r = \sqrt{x^2 + y^2} = \sqrt{(-2)^2 + (-5)^2} = \sqrt{29}$.

$$\sin\theta = \frac{y}{r} = \frac{-5}{\sqrt{29}}, \text{ or } -\frac{5}{29}\sqrt{29} \qquad\qquad \cos\theta = \frac{x}{r} = \frac{-2}{\sqrt{29}}, \text{ or } -\frac{2}{29}\sqrt{29}$$

$$\tan\theta = \frac{y}{x} = \frac{-5}{-2} = \frac{5}{2} \qquad\qquad \cot\theta = \frac{x}{y} = \frac{-2}{-5} = \frac{2}{5}$$

$$\sec\theta = \frac{r}{x} = \frac{\sqrt{29}}{-2}, \text{ or } -\frac{1}{2}\sqrt{29} \qquad\qquad \csc\theta = \frac{r}{y} = \frac{\sqrt{29}}{-5} = -\frac{1}{5}\sqrt{29}$$

Note: In the following exercises, we will only find the values of x, y, and r. The above definitions can then be used to find the values of the trigonometric functions of θ. These values are listed in the usual order in the answer.

$\boxed{5}$ Since the terminal side of θ is in QII, choose x to be negative.

If $x = -1$, then $y = 4$ and $(-1, 4)$ is a point on the terminal side of θ.

$x = -1$ and $y = 4 \Rightarrow r = \sqrt{(-1)^2 + 4^2} = \sqrt{17}$. $\bigstar \dfrac{4}{\sqrt{17}}, -\dfrac{1}{\sqrt{17}}, -4, -\dfrac{1}{4}, -\sqrt{17}, \dfrac{\sqrt{17}}{4}$

$\boxed{7}$ Remember that $y = mx$ is an equation of a line that passes through the origin and has slope m. Thus, an equation of the line is $y = \frac{4}{3}x$.

If $x = 3$, then $y = 4$ and $(3, 4)$ is a point on the terminal side of θ.

$x = 3$ and $y = 4 \Rightarrow r = \sqrt{3^2 + 4^2} = 5$. $\bigstar \dfrac{4}{5}, \dfrac{3}{5}, \dfrac{4}{3}, \dfrac{3}{4}, \dfrac{5}{3}, \dfrac{5}{4}$

$\boxed{9}$ $2y - 7x + 2 = 0 \Leftrightarrow y = \frac{7}{2}x - 1$. Thus, the slope of the given line is $\frac{7}{2}$.

An equation of the line through the origin with that slope is $y = \frac{7}{2}x$.

If $x = -2$, then $y = -7$ and $(-2, -7)$ is a point on the terminal side of θ.

$x = -2$ and $y = -7 \Rightarrow r = \sqrt{(-2)^2 + (-7)^2} = \sqrt{53}$.

$\bigstar -\dfrac{7}{\sqrt{53}}, -\dfrac{2}{\sqrt{53}}, \dfrac{7}{2}, \dfrac{2}{7}, -\dfrac{\sqrt{53}}{2}, -\dfrac{\sqrt{53}}{7}$

⸤11⸥ *Note:* U denotes *undefined.*

(a) For $\theta = 90°$, choose $x = 0$ and $y = 1$. $r = 1$.   ★ 1, 0, U, 0, U, 1

(b) For $\theta = 0°$, choose $x = 1$ and $y = 0$. $r = 1$.   ★ 0, 1, 0, U, 1, U

(c) For $\theta = \frac{7\pi}{2}$, choose $x = 0$ and $y = -1$. $r = 1$.   ★ −1, 0, U, 0, U, −1

(d) For $\theta = 3\pi$, choose $x = -1$ and $y = 0$. $r = 1$.   ★ 0, −1, 0, U, −1, U

Note: It may help to sketch a triangle as shown for Exercises 13 and 17.

⸤13⸥ Use the Pythagorean theorem to find the remaining side.

$$(adj)^2 + (opp)^2 = (hyp)^2 \Rightarrow (adj)^2 + 3^2 = 5^2 \Rightarrow adj = \sqrt{25 - 9} = 4.$$

We now apply the "trigonometric functions of an acute angle of a right triangle."

$$\sin\theta = \frac{opp}{hyp} = \frac{3}{5} \qquad \cos\theta = \frac{adj}{hyp} = \frac{4}{5} \qquad \tan\theta = \frac{opp}{adj} = \frac{3}{4}$$

$$\csc\theta = \frac{hyp}{opp} = \frac{5}{3} \qquad \sec\theta = \frac{hyp}{adj} = \frac{5}{4} \qquad \cot\theta = \frac{adj}{opp} = \frac{4}{3}$$

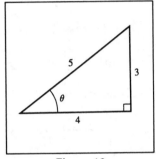

Figure 13

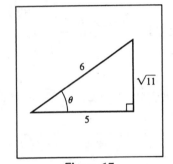

Figure 17

⸤15⸥ $12^2 + 5^2 = (hyp)^2 \Rightarrow hyp = \sqrt{144 + 25} = 13.$ ★ $\frac{5}{13}, \frac{12}{13}, \frac{5}{12}, \frac{12}{5}, \frac{13}{12}, \frac{13}{5}$

⸤17⸥ $5^2 + (opp)^2 = 6^2 \Rightarrow opp = \sqrt{36 - 25} = \sqrt{11}.$ ★ $\frac{\sqrt{11}}{6}, \frac{5}{6}, \frac{\sqrt{11}}{5}, \frac{5}{\sqrt{11}}, \frac{6}{5}, \frac{6}{\sqrt{11}}$

⸤19⸥ Using the theorem on trigonometric functions,

$$\sin\theta = \frac{opp}{hyp} = \frac{4}{5}, \quad \cos\theta = \frac{adj}{hyp} = \frac{3}{5}, \quad and \quad \tan\theta = \frac{opp}{adj} = \frac{4}{3}.$$

We now use the reciprocal identities to find the values of the other trigonometric functions:

$$\cot\theta = \frac{1}{\tan\theta} = \frac{3}{4}, \quad \sec\theta = \frac{1}{\cos\theta} = \frac{5}{3}, \quad and \quad \csc\theta = \frac{1}{\sin\theta} = \frac{5}{4}.$$

⸤21⸥ Using the Pythagorean theorem, $(adj)^2 + (opp)^2 = (hyp)^2 \Rightarrow$

$$(adj)^2 = (hyp)^2 - (opp)^2 \Rightarrow adj = \sqrt{(hyp)^2 - (opp)^2} = \sqrt{5^2 - 2^2} = \sqrt{21}.$$

★ $\frac{2}{5}, \frac{\sqrt{21}}{5}, \frac{2}{\sqrt{21}}, \frac{\sqrt{21}}{2}, \frac{5}{\sqrt{21}}, \frac{5}{2}$

23 Using the Pythagorean theorem, $\text{hyp} = \sqrt{(\text{adj})^2 + (\text{opp})^2} = \sqrt{a^2 + b^2}$.

$$\star \quad \frac{a}{\sqrt{a^2 + b^2}}, \; \frac{b}{\sqrt{a^2 + b^2}}, \; \frac{a}{b}, \; \frac{b}{a}, \; \frac{\sqrt{a^2 + b^2}}{b}, \; \frac{\sqrt{a^2 + b^2}}{a}$$

27 Since we want to find the value of the hypotenuse x, we need to use a trigonometric function which relates x to two given parts of the triangle—in this case, the angle $30°$ and the opposite side of length 4. The sine function relates the opposite side and the hypotenuse. Hence, $\sin 30° = \frac{4}{x} \Rightarrow \frac{1}{2} = \frac{4}{x} \Rightarrow x = 8$. The tangent function relates the opposite and the adjacent side. Hence, $\tan 30° = \frac{4}{y} \Rightarrow \frac{\sqrt{3}}{3} = \frac{4}{y} \Rightarrow y = 4\sqrt{3}$.

31 $\sin 60° = \frac{x}{8} \Rightarrow \frac{\sqrt{3}}{2} = \frac{x}{8} \Rightarrow x = 4\sqrt{3}$ and $\cos 60° = \frac{y}{8} \Rightarrow \frac{1}{2} = \frac{y}{8} \Rightarrow y = 4$.

33 Let h denote the height of the tree.

$$\tan \theta = \frac{\text{opp}}{\text{adj}} \Rightarrow \tan 60° = \frac{h}{200} \Rightarrow h = 200 \tan 60° = 200\sqrt{3} \approx 346.4 \text{ ft.}$$

35 Let d be the distance that the stone was moved. The side opposite $9°$ is 30 feet and the hypotenuse is d. Thus, $\sin 9° = \frac{30}{d} \Rightarrow d = \frac{30}{\sin 9°} \approx 192 \text{ ft.}$

37 $\sin \theta = \frac{1.22\lambda}{D} \Rightarrow D = \frac{1.22\lambda}{\sin \theta} = \frac{1.22 \times 550 \times 10^{-9}}{\sin 0.00003769°} \approx 1.02 \text{ meters}$

2.5 Exercises

Note: Let θ_C denote the coterminal angle of θ such that $0° \le \theta_C < 360° \; \{ \text{or } 0 \le \theta_C < 2\pi \}$. The following formulas are then used in the solutions { on text page 137 }.

(1) If θ_C is in QI, then $\theta_R = \theta_C$.

(2) If θ_C is in QII, then $\theta_R = 180° - \theta_C \; \{ \text{or } \pi - \theta_C \}$.

(3) If θ_C is in QIII, then $\theta_R = \theta_C - 180° \; \{ \text{or } \theta_C - \pi \}$.

(4) If θ_C is in QIV, then $\theta_R = 360° - \theta_C \; \{ \text{or } 2\pi - \theta_C \}$.

1 (a) Since $240°$ is in QIII, $\theta_R = 240° - 180° = 60°$.

   (b) Since $340°$ is in QIV, $\theta_R = 360° - 340° = 20°$.

   (c) $\theta_C = -202° + 1(360°) = 158° \in \text{QII}$. $\theta_R = 180° - 158° = 22°$.

   (d) $\theta_C = -660° + 2(360°) = 60° \in \text{QI}$. $\theta_R = 60°$.

3 (a) Since $\frac{3\pi}{4}$ is in QII, $\theta_R = \pi - \frac{3\pi}{4} = \frac{\pi}{4}$.

   (b) Since $\frac{4\pi}{3}$ is in QIII, $\theta_R = \frac{4\pi}{3} - \pi = \frac{\pi}{3}$.

   (c) $\theta_C = -\frac{\pi}{6} + 1(2\pi) = \frac{11\pi}{6} \in \text{QIV}$. $\theta_R = 2\pi - \frac{11\pi}{6} = \frac{\pi}{6}$.

   (d) $\theta_C = \frac{9\pi}{4} - 1(2\pi) = \frac{\pi}{4} \in \text{QI}$. $\theta_R = \frac{\pi}{4}$.

⑤ (a) Since $\frac{\pi}{2} < 3 < \pi$, θ is in QII and $\theta_R = \pi - 3 \approx 0.14$, or 8.1°.

  (b) $\theta_C = -2 + 1(2\pi) = 2\pi - 2 \approx 4.28$.

   Since $\pi < 4.28 < \frac{3\pi}{2}$, θ_C is in QIII and $\theta_R = (2\pi - 2) - \pi = \pi - 2 \approx 1.14$, or 65.4°.

  (c) Since $\frac{3\pi}{2} < 5.5 < 2\pi$, θ is in QIV and $\theta_R = 2\pi - 5.5 \approx 0.78$, or 44.9°.

  (d) The number of revolutions formed by θ is $\frac{100}{2\pi} \approx 15.92$, so

   $\theta_C = 100 - 15(2\pi) = 100 - 30\pi \approx 5.75$. Since $\frac{3\pi}{2} < 5.75 < 2\pi$,

   θ_C is in QIV and $\theta_R = 2\pi - (100 - 30\pi) = 32\pi - 100 \approx 0.53$, or 30.4°.

   Alternatively, if your calculator is capable of computing trigonometric functions

   of large values, then computing $\sin^{-1}(\sin 100) \approx -0.53 \Rightarrow \theta_R = 0.53$, or 30.4°.

Note: For the following problems, we use the theorem on reference angles before

  evaluating.

⑦ (a) $\sin \frac{2\pi}{3} = \sin \frac{\pi}{3}$ { since the sine is positive in QII

   and $\frac{\pi}{3}$ is the reference angle for $\frac{2\pi}{3}$ } $= \frac{\sqrt{3}}{2}$

  (b) $\sin\left(-\frac{5\pi}{4}\right) = \sin \frac{3\pi}{4}$ { since $\frac{3\pi}{4}$ is coterminal with $-\frac{5\pi}{4}$ } $=$
   $\sin \frac{\pi}{4}$ { since the sine is positive in QII and $\frac{\pi}{4}$ is the reference angle for $\frac{3\pi}{4}$ } $= \frac{\sqrt{2}}{2}$

⑨ (a) $\cos 150° = -\cos 30°$ { since the cosine is negative in QII } $= -\frac{\sqrt{3}}{2}$

  (b) $\cos(-60°) = \cos 300° = \cos 60°$ { since the cosine is positive in QIV } $= \frac{1}{2}$

⑪ (a) $\tan \frac{5\pi}{6} = -\tan \frac{\pi}{6}$ { since the tangent is negative in QII } $= -\frac{\sqrt{3}}{3}$

  (b) $\tan\left(-\frac{\pi}{3}\right) = \tan \frac{5\pi}{3} = -\tan \frac{\pi}{3}$ { since the tangent is negative in QIV } $= -\sqrt{3}$

⑬ (a) $\cot 120° = -\cot 60°$ { since the cotangent is negative in QII } $= -\frac{\sqrt{3}}{3}$

  (b) $\cot(-150°) = \cot 210° = \cot 30°$ { since the cotangent is positive in QIII } $= \sqrt{3}$

⑮ (a) $\sec \frac{2\pi}{3} = -\sec \frac{\pi}{3}$ { since the secant is negative in QII } $= -2$

  (b) $\sec\left(-\frac{\pi}{6}\right) = \sec \frac{11\pi}{6} = \sec \frac{\pi}{6}$ { since the secant is positive in QIV } $= \frac{2}{\sqrt{3}}$

⑰ (a) $\csc 240° = -\csc 60°$ { since the cosecant is negative in QIII } $= -\frac{2}{\sqrt{3}}$

  (b) $\csc(-330°) = \csc 30° = 2$

⑲ (a) Using the *degree* mode on a calculator, $\sin 73°20' \approx 0.958$.

  (b) Using the *radian* mode on a calculator, $\cos 0.68 \approx 0.778$.

㉓ (a) First compute $\cos 67°50'$, obtaining 0.3773. Now use the reciprocal key,

   usually labeled as either $\boxed{1/x}$ or $\boxed{x^{-1}}$, to obtain $\sec 67°50' \approx 2.650$.

  (b) As in part (a), since $\sin 0.32 \approx 0.3146$, we have $\csc 0.32 \approx 3.179$.

㉕ (a) Using the degree mode, calculate $\cos^{-1}(0.8620)$ to obtain 30.46° to

   the nearest one-hundredth of a degree.

25 (b) Using the answer from part (a), subtract 30 and multiply that result by 60

to obtain 30°27′ to the nearest minute.

29 (a) $\sin\theta = 0.4217 \Rightarrow \theta = \sin^{-1}(0.4217) \approx 24.94°$ (b) $24.94° \approx 24°57′$

31 (a) After entering 4.246, use $\boxed{1/x}$ and then $\boxed{\text{INV}}\,\boxed{\text{COS}}$, or, equivalently, $\boxed{\text{COS}^{-1}}$. Similarly, for the cosecant function use $\boxed{1/x}$ and then $\boxed{\text{INV}}\,\boxed{\text{SIN}}$ and for the cotangent function use $\boxed{1/x}$ and then $\boxed{\text{INV}}\,\boxed{\text{TAN}}$. $\sec\theta = 4.246 \Rightarrow \cos\theta = \frac{1}{4.246} \Rightarrow \theta = \cos^{-1}(\frac{1}{4.246}) \approx 76.38°$.

  (b) $76.38° \approx 76°23′$

35 (a) Use the degree mode. $\sin\theta = -0.5640 \Rightarrow \theta = \sin^{-1}(-0.5640) \approx -34.3° \Rightarrow \theta_R \approx 34.3°$. Since the sine is negative in QIII and QIV, we use θ_R in those quadrants. $180° + 34.3° = \underline{214.3°}$ and $360° - 34.3° = \underline{325.7°}$

  (b) $\cos\theta = 0.7490 \Rightarrow \theta = \cos^{-1}(0.7490) \approx 41.5°$. $\theta_R \approx 41.5°$, QI: 41.5°, QIV: 318.5°

  (c) $\tan\theta = 2.798 \Rightarrow \theta = \tan^{-1}(2.798) \approx 70.3°$. $\theta_R \approx 70.3°$, QI: 70.3°, QIII: 250.3°

  (d) $\cot\theta = -0.9601 \Rightarrow \tan\theta = -\frac{1}{0.9601} \Rightarrow \theta = \tan^{-1}(-\frac{1}{0.9601}) \approx -46.2°$.

$\theta_R \approx 46.2°$, QII: 133.8°, QIV: 313.8°

  (e) $\sec\theta = -1.116 \Rightarrow \cos\theta = -\frac{1}{1.116} \Rightarrow \theta = \cos^{-1}(-\frac{1}{1.116}) \approx 153.6°$.

$\theta_R \approx 180° - 153.6° = 26.4°$, QII: 153.6°, QIII: 206.4°

  (f) $\csc\theta = 1.485 \Rightarrow \sin\theta = \frac{1}{1.485} \Rightarrow \theta = \sin^{-1}(\frac{1}{1.485}) \approx 42.3°$.

$\theta_R \approx 42.3°$, QI: 42.3°, QII: 137.7°

37 (a) Use the radian mode. $\sin\theta = 0.4195 \Rightarrow \theta = \sin^{-1}(0.4195) \approx 0.43$.

$\theta_R \approx 0.43$ is one answer. Since the sine is positive in QI and QII, we also use the reference angle for θ in quadrant II. QII: $\pi - 0.43 \approx 2.71$

  (b) $\cos\theta = -0.1207 \Rightarrow \theta = \cos^{-1}(-0.1207) \approx 1.69$ is one answer. Since 1.69 is in QII, $\theta_R \approx \pi - 1.69 \approx 1.45$. The cosine is also negative in QIII. QIII: $\pi + 1.45 \approx 4.59$

  (c) $\tan\theta = -3.2504 \Rightarrow \theta = \tan^{-1}(-3.2504) \approx -1.27 \Rightarrow \theta_R \approx 1.27$.

QII: $\pi - 1.27 \approx 1.87$, QIV: $2\pi - 1.27 \approx 5.01$

  (d) $\cot\theta = 2.6815 \Rightarrow \tan\theta = \frac{1}{2.6815} \Rightarrow \theta = \tan^{-1}(\frac{1}{2.6815}) \approx 0.36 \Rightarrow$

$\theta_R \approx 0.36$ is one answer. QIII: $\pi + 0.36 \approx 3.50$

  (e) $\sec\theta = 1.7452 \Rightarrow \cos\theta = \frac{1}{1.7452} \Rightarrow \theta = \cos^{-1}(\frac{1}{1.7452}) \approx 0.96 \Rightarrow$

$\theta_R \approx 0.96$ is one answer. QIV: $2\pi - 0.96 \approx 5.32$

  (f) $\csc\theta = -4.8521 \Rightarrow \sin\theta = -\frac{1}{4.8521} \Rightarrow \theta = \sin^{-1}(-\frac{1}{4.8521}) \approx -0.21 \Rightarrow \theta_R \approx 0.21$.

QIII: $\pi + 0.21 \approx 3.35$, QIV: $2\pi - 0.21 \approx 6.07$

39 $\ln I_0 - \ln I = kx \sec\theta \Rightarrow \ln\dfrac{I_0}{I} = kx\sec\theta$ {property of logarithms} $\Rightarrow$

$x = \dfrac{1}{k\sec\theta}\ln\dfrac{I_0}{I}$ {solve for x}

$= \dfrac{1}{1.88\sec 12°}\ln 1.72$ {substitute and approximate} ≈ 0.28 cm.

41 (a) Since $\cos\theta$ and $\sin\phi$ are both less than or equal to 1, $R = R_0\cos\theta\sin\phi$, the solar radiation R will equal its maximum R_0 when $\cos\theta = \sin\phi = 1$. This occurs when $\theta = 0°$ and $\phi = 90°$, and corresponds to when the sun is just rising in the east.

(b) The sun located in the southeast corresponds to $\phi = 45°$.

$$\text{percentage of } R_0 = \frac{\text{amount of } R_0}{R_0} = \frac{R_0\cos 60°\sin 45°}{R_0} = \frac{1}{2}\cdot\frac{\sqrt{2}}{2} = \frac{\sqrt{2}}{4} \approx 35\%.$$

43 $\sin\theta = \dfrac{b}{c} \Rightarrow \sin 60° = \dfrac{b}{18} \Rightarrow b = 18\sin 60° = 18\cdot\dfrac{\sqrt{3}}{2} = 9\sqrt{3} \approx 15.6.$

$\cos\theta = \dfrac{a}{c} \Rightarrow \cos 60° = \dfrac{a}{18} \Rightarrow a = 18\cos 60° = 18\cdot\dfrac{1}{2} = 9.$

The hand is located at $(9, 9\sqrt{3})$.

2.6 Exercises

Note: Exercises 1 & 3: We will refer to $y = \sin x$ as just $\sin x$ ($y = \cos x$ as $\cos x$, etc.). For the form $y = a\sin bx$, the amplitude is $|a|$ and the period is $\dfrac{2\pi}{|b|}$.

These are merely listed in the answer along with the values of the x-intercepts.

Let n denote any integer.

1 (a) $y = 4\sin x$ • Vertically stretch $\sin x$ by a factor of 4. The x-intercepts are not affected by a vertical stretch or compression. ★ $4, 2\pi$, x-int. @ πn

(b) $y = \sin 4x$ • Horizontally compress $\sin x$ by a factor of 4. The x-intercepts are affected by a horizontal stretch or compression by the same factor—that is, a horizontal compression by a factor of k will move the x-intercepts of $\sin x$ from πn to $\frac{\pi}{k}n$, and a horizontal stretch by a factor of k will move the x-intercepts of $\sin x$ from πn to $k\pi n$. ★ $1, \frac{\pi}{2}$, x-int. @ $\frac{\pi}{4}n$

(c) $y = \frac{1}{4}\sin x$ • Vertically compress $\sin x$ by a factor of 4. ★ $\frac{1}{4}, 2\pi$, x-int. @ πn

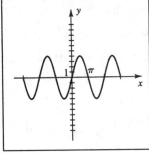

Figure 1(a)

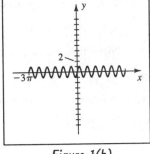

Figure 1(b)

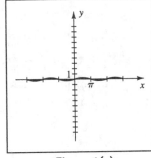

Figure 1(c)

(d) $y = \sin\frac{1}{4}x$ • Horizontally stretch $\sin x$ by a factor of 4.   ★ $1,\ 8\pi,\ x$-int. @ $4\pi n$

(e) $y = 2\sin\frac{1}{4}x$ • Vertically stretch the graph in part (d) by a factor of 2.

★ $2,\ 8\pi,\ x$-int. @ $4\pi n$

(f) $y = \frac{1}{2}\sin 4x$ • Vertically compress the graph in part (b) by a factor of 2.

★ $\frac{1}{2},\ \frac{\pi}{2},\ x$-int. @ $\frac{\pi}{4}n$

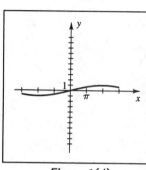   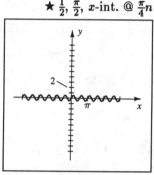

Figure 1(d)   Figure 1(e)   Figure 1(f)

(g) $y = -4\sin x$ • Reflect the graph in part (a) through the x-axis.

★ $4,\ 2\pi,\ x$-int. @ πn

(h) $y = \sin(-4x) = -\sin 4x$ using a formula for negatives. Reflect the graph in part

(b) through the x-axis.   ★ $1,\ \frac{\pi}{2},\ x$-int. @ $\frac{\pi}{4}n$

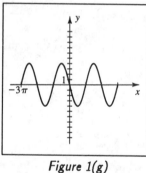   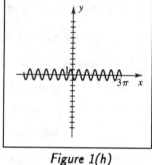

Figure 1(g)   Figure 1(h)

$\boxed{3}$ (a) $y = 3\cos x$ • Vertically stretch $\cos x$ by a factor of 3. Figures for Exercise 3 are on the next page.

★ $3,\ 2\pi,\ x$-int. @ $\frac{\pi}{2}+\pi n$

(b) $y = \cos 3x$ • Horizontally compress $\cos x$ by a factor of 3. The x-intercepts are affected by a horizontal stretch or compression by the same factor—that is, horizontal compression by a factor of k will move the x-intercepts of $\cos x$ from $\frac{\pi}{2}+\pi n$ to $\frac{\pi}{2k}+\frac{\pi}{k}n$, and a horizontal stretch by a factor of k will move the x-intercepts of $\cos x$ from $\frac{\pi}{2}+\pi n$ to $\frac{k\pi}{2}+k\pi n$.   ★ $1,\ \frac{2\pi}{3},\ x$-int. @ $\frac{\pi}{6}+\frac{\pi}{3}n$

(c) $y = \frac{1}{3}\cos x$ • Vertically compress $\cos x$ by a factor of 3.

★ $\frac{1}{3},\ 2\pi,\ x$-int. @ $\frac{\pi}{2}+\pi n$

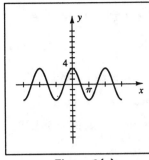

Figure 3(a)

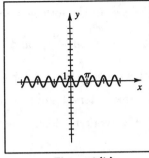

Figure 3(b)

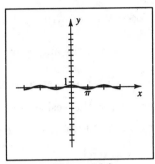

Figure 3(c)

(d) $y = \cos \frac{1}{3}x$ • Horizontally stretch $\cos x$ by a factor of 3.

★ 1, 6π, x-int. @ $\frac{3\pi}{2} + 3\pi n$

(e) $y = 2\cos \frac{1}{3}x$ • Vertically stretch the graph in part (d) by a factor of 2.

★ 2, 6π, x-int. @ $\frac{3\pi}{2} + 3\pi n$

(f) $y = \frac{1}{2}\cos 3x$ • Vertically compress the graph in part (b) by a factor of 2.

★ $\frac{1}{2}$, $\frac{2\pi}{3}$, x-int. @ $\frac{\pi}{6} + \frac{\pi}{3}n$

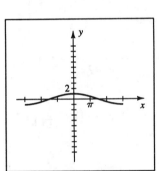

Figure 3(d)

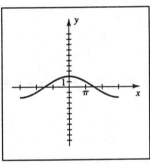

Figure 3(e)

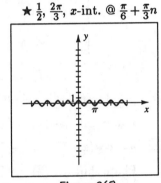

Figure 3(f)

(g) $y = -3\cos x$ • Reflect the graph in part (a) through the x-axis.

★ 3, 2π, x-int. @ $\frac{\pi}{2} + \pi n$

(h) $y = \cos(-3x) = \cos 3x$ using a formula for negatives. This is the same as the graph in part (b).

★ 1, $\frac{2\pi}{3}$, x-int. @ $\frac{\pi}{6} + \frac{\pi}{3}n$

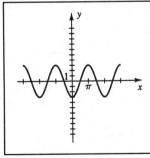

Figure 3(g)

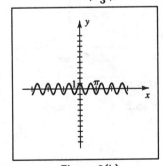

Figure 3(h)

Note: We will write $y = a\sin(bx + c)$ in the form $y = a\sin\left[b(x + \frac{c}{b})\right]$. From this form we have the amplitude, $|a|$, the period, $\frac{2\pi}{|b|}$, and the phase shift, $-\frac{c}{b}$. We will also list the interval that corresponds to $[0, 2\pi]$ for the sine functions and to $[-\frac{\pi}{2}, \frac{3\pi}{2}]$ for the cosine functions—this interval gives us one wave (between zeros) of the graph of the function. The work to determine those intervals is shown in each exercise.

$\boxed{5}$ $y = \sin\left(x - \frac{\pi}{2}\right)$ • $0 \le x - \frac{\pi}{2} \le 2\pi \Rightarrow \frac{\pi}{2} \le x \le \frac{5\pi}{2}$

Phase shift $= -\left(-\frac{\pi}{2}\right) = \frac{\pi}{2}$.

$\bigstar$ $1, 2\pi, \frac{\pi}{2}, [\frac{\pi}{2}, \frac{5\pi}{2}]$

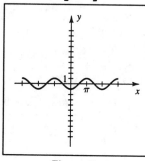

Figure 5

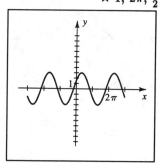

Figure 7

$\boxed{7}$ $y = 3\sin\left(x + \frac{\pi}{6}\right)$ • $0 \le x + \frac{\pi}{6} \le 2\pi \Rightarrow -\frac{\pi}{6} \le x \le \frac{11\pi}{6}$

Phase shift $= -\left(\frac{\pi}{6}\right) = -\frac{\pi}{6}$.

$\bigstar$ $3, 2\pi, -\frac{\pi}{6}, [-\frac{\pi}{6}, \frac{11\pi}{6}]$

$\boxed{9}$ $y = \cos\left(x + \frac{\pi}{2}\right)$ • $-\frac{\pi}{2} \le x + \frac{\pi}{2} \le \frac{3\pi}{2} \Rightarrow -\pi \le x \le \pi$

Phase shift $= -\left(\frac{\pi}{2}\right) = -\frac{\pi}{2}$.

$\bigstar$ $1, 2\pi, -\frac{\pi}{2}, [-\pi, \pi]$

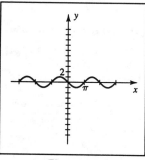

Figure 9

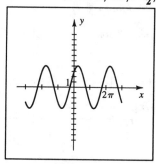

Figure 11

$\boxed{11}$ $y = 4\cos\left(x - \frac{\pi}{4}\right)$ • $-\frac{\pi}{2} \le x - \frac{\pi}{4} \le \frac{3\pi}{2} \Rightarrow -\frac{\pi}{4} \le x \le \frac{7\pi}{4}$

Phase shift $= -\left(-\frac{\pi}{4}\right) = \frac{\pi}{4}$.

$\bigstar$ $4, 2\pi, \frac{\pi}{4}, [-\frac{\pi}{4}, \frac{7\pi}{4}]$

⟦13⟧ $y = \sin(2x - \pi) + 1 = \sin\left[2(x - \frac{\pi}{2})\right] + 1.$ ● Period $= \frac{2\pi}{|2|} = \pi.$ The normal range

of the sine, -1 to 1, is affected by the "$+1$" at the end of the equation. It shifts

the graph up 1 unit and the resulting range is 0 to 2. It may be easiest to graph

$y = \sin\left[2(x - \frac{\pi}{2})\right]$ {1 period of a sine wave with endpoints at $\frac{\pi}{2}$ and $\frac{3\pi}{2}$} and then

make a vertical shift of 1 unit up to complete the graph of $y = \sin\left[2(x - \frac{\pi}{2})\right] + 1.$

$0 \leq 2x - \pi \leq 2\pi \Rightarrow \pi \leq 2x \leq 3\pi \Rightarrow \frac{\pi}{2} \leq x \leq \frac{3\pi}{2}$ ★ $1, \pi, \frac{\pi}{2}, [\frac{\pi}{2}, \frac{3\pi}{2}]$

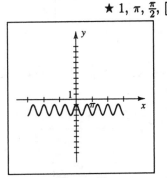

Figure 13 *Figure 15*

⟦15⟧ $y = -\cos(3x + \pi) - 2 = -\cos\left[3(x + \frac{\pi}{3})\right] - 2$ ● The "$-$" in front of cos has the

effect of reflecting the graph of $y = \cos(3x + \pi)$ through the x-axis. The "-2" at the

end of the equation lowers the range 2 units to -3 to -1. Period $= \frac{2\pi}{|3|} = \frac{2\pi}{3}$, phase

shift $= -(\frac{\pi}{3}) = -\frac{\pi}{3}.$ $-\frac{\pi}{2} \leq 3x + \pi \leq \frac{3\pi}{2} \Rightarrow -\frac{3\pi}{2} \leq 3x \leq \frac{\pi}{2} \Rightarrow -\frac{\pi}{2} \leq x \leq \frac{\pi}{6}$

★ $1, \frac{2\pi}{3}, -\frac{\pi}{3}, [-\frac{\pi}{2}, \frac{\pi}{6}]$

⟦17⟧ $y = -2\sin(3x - \pi) = -2\sin\left[3(x - \frac{\pi}{3})\right].$ ● Amplitude $= |-2| = 2.$ The negative

before the "2" has the effect of reflecting the graph of $y = 2\sin(3x - \pi)$ through the

x-axis. Period $= \frac{2\pi}{|3|} = \frac{2\pi}{3}$, phase shift $= -(-\frac{\pi}{3}) = \frac{\pi}{3}, 0 \leq 3x - \pi \leq 2\pi \Rightarrow$

$\pi \leq 3x \leq 3\pi \Rightarrow \frac{\pi}{3} \leq x \leq \pi.$ ★ $2, \frac{2\pi}{3}, \frac{\pi}{3}, [\frac{\pi}{3}, \pi]$

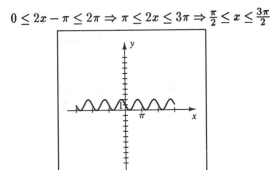

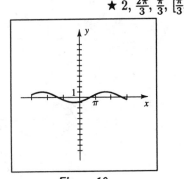

Figure 17 *Figure 19*

⟦19⟧ $y = \sin\left(\frac{1}{2}x - \frac{\pi}{3}\right) = \sin\left[\frac{1}{2}(x - \frac{2\pi}{3})\right].$ ● Period $= \frac{2\pi}{|1/2|} = 4\pi.$

$0 \leq \frac{1}{2}x - \frac{\pi}{3} \leq 2\pi \Rightarrow \frac{\pi}{3} \leq \frac{1}{2}x \leq \frac{7\pi}{3} \Rightarrow \frac{2\pi}{3} \leq x \leq \frac{14\pi}{3}$ ★ $1, 4\pi, \frac{2\pi}{3}, [\frac{2\pi}{3}, \frac{14\pi}{3}]$

$\boxed{21}$ $y = 6\sin \pi x$ • Period $= \dfrac{2\pi}{|\pi|} = 2.$ $0 \le \pi x \le 2\pi \Rightarrow 0 \le x \le 2$ ★ 6, 2, 0, [0, 2]

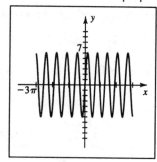

Figure 21

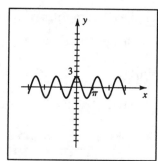

Figure 23

$\boxed{23}$ $y = 2\cos \dfrac{\pi}{2}x$ • Period $= \dfrac{2\pi}{|\pi/2|} = 4.$ $-\dfrac{\pi}{2} \le \dfrac{\pi}{2}x \le \dfrac{3\pi}{2} \Rightarrow -1 \le x \le 3$

                       ★ 2, 4, 0, [−1, 3]

$\boxed{25}$ $y = \dfrac{1}{2}\sin 2\pi x$ • $0 \le 2\pi x \le 2\pi \Rightarrow 0 \le x \le 1$ ★ $\dfrac{1}{2}$, 1, 0, [0, 1]

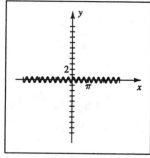

Figure 25

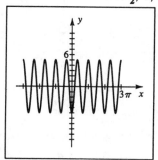

Figure 27

$\boxed{27}$ $y = 5\sin\left(3x - \dfrac{\pi}{2}\right) = 5\sin\left[3\left(x - \dfrac{\pi}{6}\right)\right].$ •   ★ 5, $\dfrac{2\pi}{3}$, $\dfrac{\pi}{6}$, $\left[\dfrac{\pi}{6}, \dfrac{5\pi}{6}\right]$

$0 \le 3x - \dfrac{\pi}{2} \le 2\pi \Rightarrow \dfrac{\pi}{2} \le 3x \le \dfrac{5\pi}{2} \Rightarrow \dfrac{\pi}{6} \le x \le \dfrac{5\pi}{6}.$ To graph this function, draw one period of a sine wave with endpoints at $\dfrac{\pi}{6}$ and $\dfrac{5\pi}{6}$ and an amplitude of 5.

$\boxed{29}$ $y = 3\cos\left(\dfrac{1}{2}x - \dfrac{\pi}{4}\right) = 3\cos\left[\dfrac{1}{2}\left(x - \dfrac{\pi}{2}\right)\right].$ •

$-\dfrac{\pi}{2} \le \dfrac{1}{2}x - \dfrac{\pi}{4} \le \dfrac{3\pi}{2} \Rightarrow -\dfrac{\pi}{4} \le \dfrac{1}{2}x \le \dfrac{7\pi}{4} \Rightarrow -\dfrac{\pi}{2} \le x \le \dfrac{7\pi}{2}$ ★ 3, 4π, $\dfrac{\pi}{2}$, $\left[-\dfrac{\pi}{2}, \dfrac{7\pi}{2}\right]$

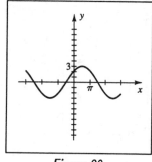

Figure 29

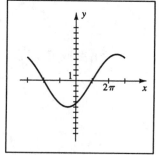

Figure 31

$\boxed{31}$ $y = -5\cos\left(\dfrac{1}{3}x + \dfrac{\pi}{6}\right) = -5\cos\left[\dfrac{1}{3}\left(x + \dfrac{\pi}{2}\right)\right].$ •

$-\dfrac{\pi}{2} \le \dfrac{1}{3}x + \dfrac{\pi}{6} \le \dfrac{3\pi}{2} \Rightarrow -\dfrac{2\pi}{3} \le \dfrac{1}{3}x \le \dfrac{4\pi}{3} \Rightarrow -2\pi \le x \le 4\pi$ ★ 5, 6π, $-\dfrac{\pi}{2}$, [−2π, 4π]

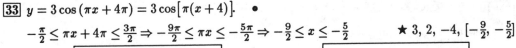

$\boxed{33}$ $y = 3\cos(\pi x + 4\pi) = 3\cos\left[\pi(x+4)\right]$.   •

$-\frac{\pi}{2} \le \pi x + 4\pi \le \frac{3\pi}{2} \Rightarrow -\frac{9\pi}{2} \le \pi x \le -\frac{5\pi}{2} \Rightarrow -\frac{9}{2} \le x \le -\frac{5}{2}$ ★ $3,\ 2,\ -4,\ \left[-\frac{9}{2}, -\frac{5}{2}\right]$

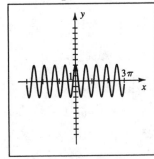

Figure 33

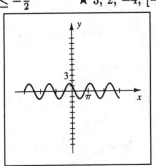

Figure 35

$\boxed{35}$ $y = -\sqrt{2}\sin\left(\frac{\pi}{2}x - \frac{\pi}{4}\right) = -\sqrt{2}\sin\left[\frac{\pi}{2}\left(x - \frac{1}{2}\right)\right]$.   •

$0 \le \frac{\pi}{2}x - \frac{\pi}{4} \le 2\pi \Rightarrow \frac{\pi}{4} \le \frac{\pi}{2}x \le \frac{9\pi}{4} \Rightarrow \frac{1}{2} \le x \le \frac{9}{2}$ ★ $\sqrt{2},\ 4,\ \frac{1}{2},\ \left[\frac{1}{2}, \frac{9}{2}\right]$

$\boxed{37}$ $y = -2\sin(2x - \pi) + 3 = -2\sin\left[2\left(x - \frac{\pi}{2}\right)\right] + 3$.   •

$0 \le 2x - \pi \le 2\pi \Rightarrow \pi \le 2x \le 3\pi \Rightarrow \frac{\pi}{2} \le x \le \frac{3\pi}{2}$. The amplitude of 2 makes the normal sine range of -1 to 1 change to -2 to 2. The "$+3$" at the end of the equation raises the graph up 3 units and the range is 1 to 5.   ★ $2,\ \pi,\ \frac{\pi}{2},\ \left[\frac{\pi}{2}, \frac{3\pi}{2}\right]$

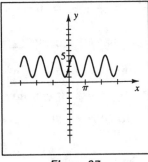

Figure 37

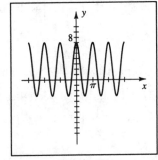

Figure 39

$\boxed{39}$ $y = 5\cos(2x + 2\pi) + 2 = 5\cos\left[2(x + \pi)\right] + 2$.   •

$-\frac{\pi}{2} \le 2x + 2\pi \le \frac{3\pi}{2} \Rightarrow -\frac{5\pi}{2} \le 2x \le -\frac{\pi}{2} \Rightarrow -\frac{5\pi}{4} \le x \le -\frac{\pi}{4}$ ★ $5,\ \pi,\ -\pi,\ \left[-\frac{5\pi}{4}, -\frac{\pi}{4}\right]$

$\boxed{41}$ (a) The amplitude a is 4 and the period { from $-\pi$ to π } is 2π.

The phase shift is the first negative zero that occurs before a maximum, $-\pi$.

(b) Period $= \frac{2\pi}{b} \Rightarrow 2\pi = \frac{2\pi}{b} \Rightarrow b = 1$. Phase shift $= -\frac{c}{b} \Rightarrow -\pi = -\frac{c}{1} \Rightarrow c = \pi$.

Hence, $y = a\sin(bx + c) = 4\sin(x + \pi)$.

$\boxed{43}$ (a) The amplitude a is 2 and the period { from -3 to 1 } is 4.

The phase shift is the first negative zero that occurs before a maximum, -3.

(b) Period $= \frac{2\pi}{b} \Rightarrow 4 = \frac{2\pi}{b} \Rightarrow b = \frac{\pi}{2}$. Phase shift $= -\frac{c}{b} \Rightarrow -3 = -\frac{c}{\pi/2} \Rightarrow c = \frac{3\pi}{2}$.

Hence, $y = a\sin(bx + c) = 2\sin\left(\frac{\pi}{2}x + \frac{3\pi}{2}\right)$.

45 In the first second, there are 2 complete cycles.

Hence 1 cycle is completed in $\frac{1}{2}$ second and thus, the period is $\frac{1}{2}$.

Also, the period is $\frac{2\pi}{b}$. Equating these expressions yields $\frac{2\pi}{b} = \frac{1}{2} \Rightarrow b = 4\pi$.

47 We first note that $\frac{1}{2}$ period takes place in $\frac{1}{4}$ second and thus 1 period in $\frac{1}{2}$ second.

As in Exercise 45, $\frac{2\pi}{b} = \frac{1}{2} \Rightarrow b = 4\pi$. Since the maximum flow rate is 8 liters/minute,

the amplitude is 8. $a = 8$ and $b = 4\pi \Rightarrow y = 8\sin 4\pi t$.

49 $f(t) = \frac{1}{2}\cos\left[\frac{\pi}{6}\left(t - \frac{11}{2}\right)\right]$, amplitude $= \frac{1}{2}$, period $= \frac{2\pi}{\pi/6} = 12$, phase shift $= \frac{11}{2}$

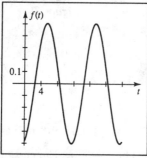

Figure 49

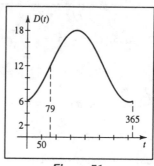

Figure 51

51 $D(t) = 6\sin\left[\frac{2\pi}{365}(t - 79)\right] + 12$, amplitude $= 6$, period $= \frac{2\pi}{2\pi/365} = 365$,

phase shift $= 79$, range $= \underline{12 - 6}$ to $\underline{12 + 6}$ or 6 to 18

53 The temperature is 20°F at 9:00 A.M. $(t = 0)$. It increases to a high of 35°F at 3:00 P.M. $(t = 6)$ and then decreases to 20°F at 9:00 P.M. $(t = 12)$. It continues to decrease to a low of 5°F at 3:00 A.M. $(t = 18)$. It then rises to 20°F at 9:00 A.M. $(t = 24)$.

[0, 24] by [0, 40]

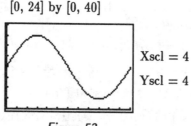

Xscl $= 4$
Yscl $= 4$

Figure 53

Note: Exer. 55–58: The period is 24 hours. Thus, $24 = \frac{2\pi}{b} \Rightarrow b = \frac{\pi}{12}$.

55 A high of 10°C and a low of −10°C imply that $d = \frac{\text{high} + \text{low}}{2} = \frac{10 + (-10)}{2} = 0$ and $a = \text{high} - \text{average} = 10 - 0 = 10$. The average temperature of 0°C will occur 6 hours {one-half of 12} after the low at 4 A.M., which corresponds to $t = 10$. Letting this correspond to the first zero of the sine function, we have

$f(t) = 10 \sin\left[\frac{\pi}{12}(t-10)\right] + 0 = 10 \sin\left(\frac{\pi}{12}t - \frac{5\pi}{6}\right)$ with $a = 10$, $b = \frac{\pi}{12}$, $c = -\frac{5\pi}{6}$, $d = 0$.

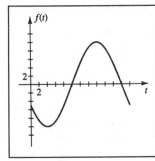

Figure 55

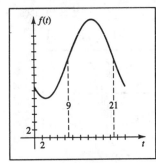

Figure 57

57 A high of 30 °C and a low of 10 °C imply that $d = \dfrac{30+10}{2} = 20$ and $a = 30 - 20 = 10$.

The average temperature of 20 °C at 9 A.M. corresponds to $t = 9$.

Letting this correspond to the first zero of the sine function, we have

$f(t) = 10 \sin\left[\frac{\pi}{12}(t-9)\right] + 20 = 10 \sin\left(\frac{\pi}{12}t - \frac{3\pi}{4}\right) + 20$ with

$$a = 10,\ b = \frac{\pi}{12},\ c = -\frac{3\pi}{4},\ d = 20.$$

59 (b) Since the period is 12 months, $12 = \frac{2\pi}{b} \Rightarrow b = \frac{\pi}{6}$. The maximum precipitation is 6.1 and the minimum is 0.2, so the sine wave is centered vertically at $d = \dfrac{6.1 + 0.2}{2} = 3.15$ and its amplitude is $a = \dfrac{6.1 - 0.2}{2} = 2.95$. Since the maximum precipitation occurs at $t = 1$ (January), we must have $bt + c = \frac{\pi}{2} \Rightarrow \frac{\pi}{6}(1) + c = \frac{\pi}{2} \Rightarrow c = \frac{\pi}{3}$. Thus, $P(t) = a \sin(bt + c) + d = 2.95 \sin\left(\frac{\pi}{6}t + \frac{\pi}{3}\right) + 3.15$.

[0.5, 24.5] by [−1, 8]

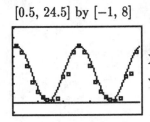

Xscl = 4
Yscl = 1

Figure 59

[0.5, 24.5] by [0, 20]

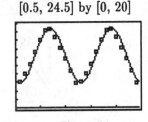

Xscl = 4
Yscl = 4

Figure 61

61 (b) Since the period is 12 months, $b = \frac{2\pi}{12} = \frac{\pi}{6}$. From the table, the maximum number of daylight hours is 18.72 and the minimum is 5.88. Thus, the sine wave is centered vertically at $d = \dfrac{18.72 + 5.88}{2} = 12.3$ and its amplitude is $a = \dfrac{18.72 - 5.88}{2} = 6.42$. Since the maximum daylight occurs at $t = 7$ (July), we must have $bt + c = \frac{\pi}{2} \Rightarrow \frac{\pi}{6}(7) + c = \frac{\pi}{2} \Rightarrow c = -\frac{2\pi}{3}$.

Thus, $D(t) = a \sin(bt + c) + d = 6.42 \sin\left(\frac{\pi}{6}t - \frac{2\pi}{3}\right) + 12.3$.

63 As $x \to 0^-$ or as $x \to 0^+$,

y oscillates between -1 and 1 and does not approach a unique value.

$[-2, 2]$ by $[-1.33, 1.33]$ $[-2, 2]$ by $[-0.33, 2.33]$

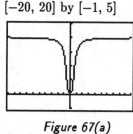   Xscl $= 0.5$
                        Yscl $= 0.5$

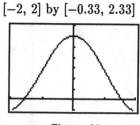   Xscl $= 0.5$
                        Yscl $= 0.5$

Figure 63   *Figure 65*

65 As $x \to 0^-$ or as $x \to 0^+$, y appears to approach 2.

67 From the graph, we see that there is a horizontal asymptote of $y = 4$.

$[-20, 20]$ by $[-1, 5]$ $[-1, 1]$ by $[-0.67, 0.67]$

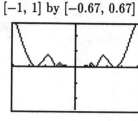   Xscl $= 2$
                        Yscl $= 1$

                        Xscl $= 0.25$
                        Yscl $= 0.25$

Figure 67(a)   *Figure 67(b)*

69 Graph $Y_1 = \cos 3x$ and $Y_2 = \frac{1}{2}x - \sin x$.

From the graph, Y_1 intersects Y_2 at $x \approx -1.63, -0.45, 0.61, 1.49, 2.42$.

Thus, $\cos 3x \geq \frac{1}{2}x - \sin x$ on $[-\pi, -1.63] \cup [-0.45, 0.61] \cup [1.49, 2.42]$.

$[-\pi, \pi]$ by $[-2.09, 2.09]$

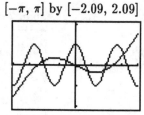   Xscl $= \pi/4$
                        Yscl $= 1$

Figure 69

Note: If $y = a\tan(bx + c)$ or $y = a\cot(bx + c)$,

then the periods for the tangent and cotangent graphs are $\pi/|b|$.

If $y = a\sec(bx + c)$ or $y = a\csc(bx + c)$,

then the periods for the secant and cosecant graphs are $2\pi/|b|$.

$\boxed{1}$ $y = 4\tan x$ • Vertically stretch $\tan x$ by a factor of 4. The *x-intercepts* of $\tan x$ and $\cot x$ are not affected by vertically stretching or compressing their graphs. The *vertical asymptotes* of $\tan x$, $\cot x$, $\sec x$, and $\csc x$ are not affected by vertically stretching or compressing their graphs. ★ π

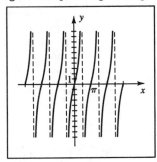

Figure 1

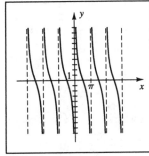

Figure 3

$\boxed{3}$ $y = 3\cot x$ • Vertically stretch $\cot x$ by a factor of 3. ★ π

$\boxed{5}$ $y = 2\csc x$ • Vertically stretch $\csc x$ by a factor of 2.

Note that there is now a minimum value of 2 at $x = \frac{\pi}{2}$.

The range of this function is $(-\infty, -2] \cup [2, \infty)$, or $|y| \geq 2$. ★ 2π

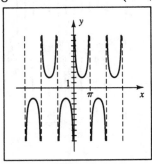

Figure 5

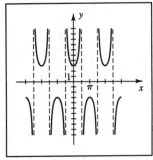

Figure 7

$\boxed{7}$ $y = 3\sec x$ • Vertically stretch $\sec x$ by a factor of 3.

Note that there is now a minimum value of 3 at $x = 0$.

The range of this function is $(-\infty, -3] \cup [3, \infty)$ or $|y| \geq 3$. ★ 2π

Note: The vertical asymptotes of each function are denoted by *VA @ x =* . The work to determine two consecutive vertical asymptotes is shown for each exercise. For the tangent and secant functions, the region from $-\frac{\pi}{2}$ to $\frac{\pi}{2}$ is used. For the cotangent and cosecant functions, the region from 0 to π is used.

☐9☐ $y = \tan\left(x - \frac{\pi}{4}\right)$ • Shift $\tan x$ right $\frac{\pi}{4}$ units, *VA @* $x = -\frac{\pi}{4} + \pi n$. Note that the asymptotes remain π units apart. $-\frac{\pi}{2} \le x - \frac{\pi}{4} \le \frac{\pi}{2} \Rightarrow -\frac{\pi}{4} \le x \le \frac{3\pi}{4}$ ★ π

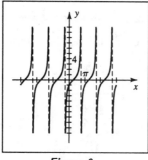

Figure 9

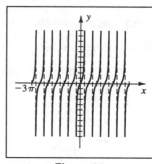

Figure 11

☐11☐ $y = \tan 2x$ • Horizontally compress $\tan x$ by a factor of 2, *VA @* $x = -\frac{\pi}{4} + \frac{\pi}{2}n$. Note that the asymptotes are only $\frac{\pi}{2}$ units apart. $-\frac{\pi}{2} \le 2x \le \frac{\pi}{2} \Rightarrow -\frac{\pi}{4} \le x \le \frac{\pi}{4}$ ★ $\frac{\pi}{2}$

☐13☐ $y = \tan\frac{1}{4}x$ • Horizontally stretch $\tan x$ by a factor of 4, *VA @* $x = -2\pi + 4\pi n$. Note that the asymptotes are 4π units apart. $-\frac{\pi}{2} \le \frac{1}{4}x \le \frac{\pi}{2} \Rightarrow -2\pi \le x \le 2\pi$ ★ 4π

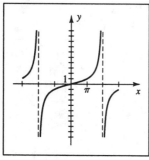

Figure 13

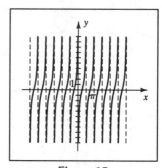

Figure 15

☐15☐ $y = 2\tan\left(2x + \frac{\pi}{2}\right) = 2\tan\left[2\left(x + \frac{\pi}{4}\right)\right]$ •

The phase shift is $-\frac{\pi}{4}$, the period is $\frac{\pi}{2}$, and we have a vertical stretching factor of 2. $-\frac{\pi}{2} \le 2x + \frac{\pi}{2} \le \frac{\pi}{2} \Rightarrow -\pi \le 2x \le 0 \Rightarrow -\frac{\pi}{2} \le x \le 0$, *VA @* $x = \frac{\pi}{2}n$ ★ $\frac{\pi}{2}$

$\boxed{17}$ $y = -\frac{1}{4}\tan\left(\frac{1}{2}x + \frac{\pi}{3}\right) = -\frac{1}{4}\tan\left[\frac{1}{2}\left(x + \frac{2\pi}{3}\right)\right]$. • Note that the "−" in front of the $\frac{1}{4}$ reflects the graph through the x-axis. This changes the appearance of a tangent graph to that of a cotangent graph { increasing to decreasing }.

$-\frac{\pi}{2} \le \frac{1}{2}x + \frac{\pi}{3} \le \frac{\pi}{2} \Rightarrow -\frac{5\pi}{6} \le \frac{1}{2}x \le \frac{\pi}{6} \Rightarrow -\frac{5\pi}{3} \le x \le \frac{\pi}{3}$, VA @ $x = -\frac{5\pi}{3} + 2\pi n$ ★ 2π

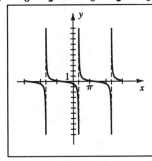

Figure 17

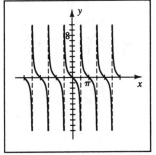

Figure 19

$\boxed{19}$ $y = \cot\left(x - \frac{\pi}{2}\right)$ • $0 \le x - \frac{\pi}{2} \le \pi \Rightarrow \frac{\pi}{2} \le x \le \frac{3\pi}{2}$, VA @ $x = \frac{\pi}{2} + \pi n$ ★ π

$\boxed{21}$ $y = \cot 2x$ • $0 \le 2x \le \pi \Rightarrow 0 \le x \le \frac{\pi}{2}$, VA @ $x = \frac{\pi}{2}n$ ★ $\frac{\pi}{2}$

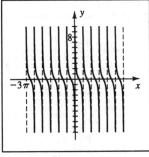

Figure 21

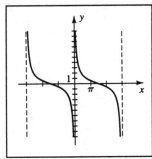

Figure 23

$\boxed{23}$ $y = \cot\frac{1}{3}x$ • $0 \le \frac{1}{3}x \le \pi \Rightarrow 0 \le x \le 3\pi$, VA @ $x = 3\pi n$ ★ 3π

$\boxed{25}$ $y = 2\cot\left(2x + \frac{\pi}{2}\right) = 2\cot\left[2\left(x + \frac{\pi}{4}\right)\right]$. •

$0 \le 2x + \frac{\pi}{2} \le \pi \Rightarrow -\frac{\pi}{2} \le 2x \le \frac{\pi}{2} \Rightarrow -\frac{\pi}{4} \le x \le \frac{\pi}{4}$, VA @ $x = -\frac{\pi}{4} + \frac{\pi}{2}n$ ★ $\frac{\pi}{2}$

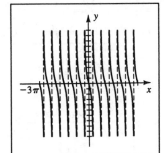

Figure 25

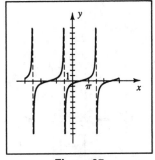

Figure 27

$\boxed{27}$ $y = -\frac{1}{2}\cot\left(\frac{1}{2}x + \frac{\pi}{4}\right) = -\frac{1}{2}\cot\left[\frac{1}{2}\left(x + \frac{\pi}{2}\right)\right]$. •

$0 \le \frac{1}{2}x + \frac{\pi}{4} \le \pi \Rightarrow -\frac{\pi}{4} \le \frac{1}{2}x \le \frac{3\pi}{4} \Rightarrow -\frac{\pi}{2} \le x \le \frac{3\pi}{2}$, VA @ $x = -\frac{\pi}{2} + 2\pi n$ ★ 2π

$\boxed{29}$ $y = \sec\left(x - \frac{\pi}{2}\right)$ • Note that this is the same graph as the graph of $y = \csc x$.

$-\frac{\pi}{2} \le x - \frac{\pi}{2} \le \frac{\pi}{2} \Rightarrow 0 \le x \le \pi$, *VA* @ $x = \pi n$ ★ 2π

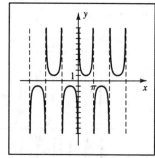

Figure 29

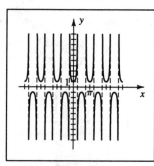

Figure 31

$\boxed{31}$ $y = \sec 2x$ • Note that the asymptotes move closer together by a factor of 2.

$-\frac{\pi}{2} \le 2x \le \frac{\pi}{2} \Rightarrow -\frac{\pi}{4} \le x \le \frac{\pi}{4}$, *VA* @ $x = -\frac{\pi}{4} + \frac{\pi}{2}n$ ★ π

$\boxed{33}$ $y = \sec\frac{1}{3}x$ • Note that the asymptotes move farther apart by a factor of 3,

just as they did with the graphs of tan and cot.

$-\frac{\pi}{2} \le \frac{1}{3}x \le \frac{\pi}{2} \Rightarrow -\frac{3\pi}{2} \le x \le \frac{3\pi}{2}$, *VA* @ $x = -\frac{3\pi}{2} + 3\pi n$ ★ 6π

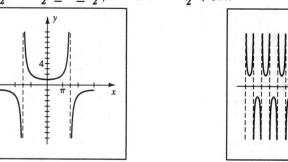

Figure 33

Figure 35

$\boxed{35}$ $y = 2\sec\left(2x - \frac{\pi}{2}\right) = 2\sec\left[2\left(x - \frac{\pi}{4}\right)\right]$. •

$-\frac{\pi}{2} \le 2x - \frac{\pi}{2} \le \frac{\pi}{2} \Rightarrow 0 \le 2x \le \pi \Rightarrow 0 \le x \le \frac{\pi}{2}$, *VA* @ $x = \frac{\pi}{2}n$ ★ π

$\boxed{37}$ $y = -\frac{1}{3}\sec\left(\frac{1}{2}x + \frac{\pi}{4}\right) = -\frac{1}{3}\sec\left[\frac{1}{2}\left(x + \frac{\pi}{2}\right)\right]$. •

$-\frac{\pi}{2} \le \frac{1}{2}x + \frac{\pi}{4} \le \frac{\pi}{2} \Rightarrow -\frac{3\pi}{4} \le \frac{1}{2}x \le \frac{\pi}{4} \Rightarrow -\frac{3\pi}{2} \le x \le \frac{\pi}{2}$, *VA* @ $x = -\frac{3\pi}{2} + 2\pi n$ ★ 4π

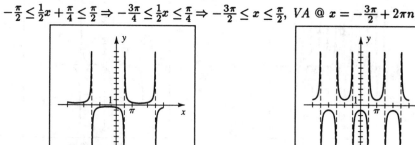

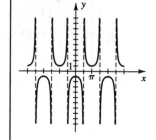

Figure 37

Figure 39

$\boxed{39}$ $y = \csc\left(x - \frac{\pi}{2}\right)$ • $0 \le x - \frac{\pi}{2} \le \pi \Rightarrow \frac{\pi}{2} \le x \le \frac{3\pi}{2}$, *VA* @ $x = \frac{\pi}{2} + \pi n$ ★ 2π

41 $y = \csc 2x$ • $0 \le 2x \le \pi \Rightarrow 0 \le x \le \frac{\pi}{2}$, $VA @ x = \frac{\pi}{2}n$ ★ π

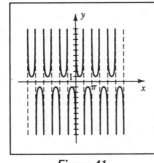

Figure 41

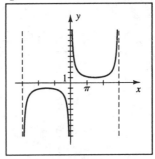

Figure 43

43 $y = \csc \frac{1}{3}x$ • $0 \le \frac{1}{3}x \le \pi \Rightarrow 0 \le x \le 3\pi$, $VA @ x = 3\pi n$ ★ 6π

45 $y = 2\csc\left(2x + \frac{\pi}{2}\right) = 2\csc\left[2(x + \frac{\pi}{4})\right]$. •

$0 \le 2x + \frac{\pi}{2} \le \pi \Rightarrow -\frac{\pi}{2} \le 2x \le \frac{\pi}{2} \Rightarrow -\frac{\pi}{4} \le x \le \frac{\pi}{4}$, $VA @ x = -\frac{\pi}{4} + \frac{\pi}{2}n$ ★ π

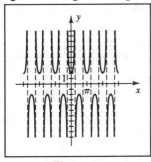

Figure 45

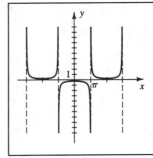

Figure 47

47 $y = -\frac{1}{4}\csc\left(\frac{1}{2}x + \frac{\pi}{2}\right) = -\frac{1}{4}\csc\left[\frac{1}{2}(x + \pi)\right]$. •

$0 \le \frac{1}{2}x + \frac{\pi}{2} \le \pi \Rightarrow -\frac{\pi}{2} \le \frac{1}{2}x \le \frac{\pi}{2} \Rightarrow -\pi \le x \le \pi$, $VA @ x = -\pi + 2\pi n$ ★ 4π

49 $y = \tan \frac{\pi}{2}x$ • Horizontally stretch $\tan x$ by a factor of $2/\pi$, $VA @ x = -1 + 2n$.

$-\frac{\pi}{2} \le \frac{\pi}{2}x \le \frac{\pi}{2} \Rightarrow -1 \le x \le 1$ ★ 2

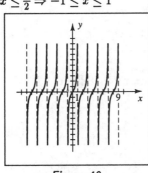

Figure 49

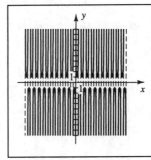

Figure 51

51 $y = \csc 2\pi x$ • $0 \le 2\pi x \le \pi \Rightarrow 0 \le x \le \frac{1}{2}$, $VA @ x = \frac{1}{2}n$ ★ 1

53 Reflecting the graph of $y = \cot x$ about the x-axis, which is $y = -\cot x$, gives us the graph of $y = \tan\left(x + \frac{\pi}{2}\right)$. If we shift this graph to the left (or right), we will obtain the graph of $y = \tan x$. Thus, one equation is $y = -\cot\left(x + \frac{\pi}{2}\right)$.

55 $y = |\sin x|$ ●   Reflect the negative values of $y = \sin x$ through the x-axis. In general, when sketching the graph of $y = |f(x)|$, reflect the negative values of $f(x)$ through the x-axis. The absolute value does not affect the nonnegative values.

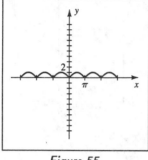

Figure 55

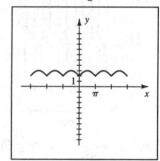

Figure 57

57 $y = |\sin x| + 2$ ●   Shift $y = |\sin x|$ up 2 units.

59 $y = -|\cos x| + 1$ ●   Similar to Exercise 55, we first reflect the negative values of $y = \cos x$ through the x-axis. The "$-$" in front of $|\cos x|$ has the effect of reflecting $y = |\cos x|$ through the x-axis. Finally, we shift that graph 1 unit up to obtain the graph of $y = -|\cos x| + 1$.

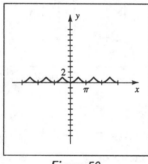

Figure 59

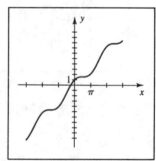

Figure 61

61 $y = x + \cos x$ ●   The value of $\cos x$ is between -1 and 1—adding this relatively small amount to the value of x has the effect of oscillating the graph about the line $y = x$.

63 $y = 2^{-x} \cos x$ ●   This graph is similar to the graph in Example 8.

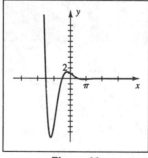

Figure 63

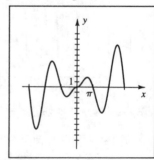

Figure 65

65 $y = |x| \sin x$ • See *Figure 65* on the preceding page. The graph will coincide with the graph of $y = |x|$ if $\sin x = 1$—that is, if $x = \frac{\pi}{2} + 2\pi n$. The graph will coincide with the graph of $y = -|x|$ if $\sin x = -1$—that is, if $x = \frac{3\pi}{2} + 2\pi n$.

67 $f(x) = \tan(0.5x);$ $g(x) = \tan[0.5(x + \pi/2)]$ • Since $g(x) = f(x + \pi/2)$, the graph of g can be obtained by shifting the graph of f left a distance of $\frac{\pi}{2}$.

$[-2\pi, 2\pi]$ by $[-4, 4]$ $[-2\pi, 2\pi]$ by $[-4, 4]$

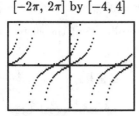   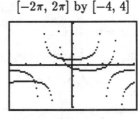

Xscl $= \pi/2$ Xscl $= \pi/2$
Yscl $= 1$ Yscl $= 1$

Figure 67   *Figure 69*

69 $f(x) = 0.5 \sec 0.5x;$ $g(x) = 0.5 \sec[0.5(x - \pi/2)] - 1$ •

Since $g(x) = f(x - \pi/2) - 1$, the graph of g can be obtained by shifting the graph of f horizontally to the right a distance of $\frac{\pi}{2}$ and vertically downward a distance of 1.

73 The damping factor of $y = e^{-x/4} \sin 4x$ is $e^{-x/4}$.

$[-2\pi, 2\pi]$ by $[-4.19, 4.19]$ $[-\pi, \pi]$ by $[-4, 4]$

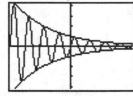   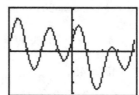

Xscl $= \pi/2$ Xscl $= \pi/4$
Yscl $= 1$ Yscl $= 1$

Figure 73   *Figure 75*

75 From the graph, we see that the maximum occurs at the approximate coordinates $(-2.76, 3.09)$, and the minimum occurs at the approximate coordinates $(1.23, -3.68)$.

77 From the graph, we see that f is increasing and one-to-one between

$a \approx -0.70$ and $b \approx 0.12$. Thus, the interval is approximately $[-0.70, 0.12]$.

$[-2, 2]$ by $[-1.33, 1.33]$ $[-\pi, \pi]$ by $[-2.09, 2.09]$

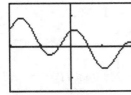   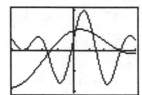

Xscl $= 1$ Xscl $= \pi/4$
Yscl $= 1$ Yscl $= 1$

Figure 77   *Figure 79*

79 Graph $Y_1 = \cos(2x - 1) + \sin 3x$ and $Y_2 = \sin\frac{1}{3}x + \cos x$.

From the graph, Y_1 intersects Y_2 at $x \approx -1.31, 0.11, 0.95, 2.39$.

Thus, $\cos(2x - 1) + \sin 3x \geq \sin\frac{1}{3}x + \cos x$ on $[-\pi, -1.31] \cup [0.11, 0.95] \cup [2.39, \pi]$.

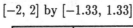

$\boxed{81}$ (a) $\theta = 0 \Rightarrow I = \frac{1}{2}I_0[1 + \cos{(\pi \sin 0)}] = \frac{1}{2}I_0[1 + \cos{(0)}] = \frac{1}{2}I_0(2) = I_0$.

(b) $\theta = \pi/3 \Rightarrow I = \frac{1}{2}I_0[1 + \cos{(\pi \sin{(\pi/3)})}] \approx 0.044I_0$.

(c) $\theta = \pi/7 \Rightarrow I = \frac{1}{2}I_0[1 + \cos{(\pi \sin{(\pi/7)})}] \approx 0.603I_0$.

$\boxed{83}$ (a) The damping factor of $S = A_0 e^{-\alpha z} \sin{(kt - \alpha z)}$ is $A_0 e^{-\alpha z}$.

(b) The phase shift at depth z_0 can be found by solving the equation $kt - \alpha z_0 = 0$ for
t. Doing so gives us $kt = \alpha z_0$, and hence, $t = \frac{\alpha}{k} z_0$.

(c) At the surface, $z = 0$. Hence, $S = A_0 \sin{kt}$ and the amplitude at the surface is
A_0. $\text{Amplitude}_{\text{wave}} = \frac{1}{2}\text{Amplitude}_{\text{surface}} \Rightarrow$

$$A_0 e^{-\alpha z} = \frac{1}{2}A_0 \Rightarrow e^{-\alpha z} = \frac{1}{2} \Rightarrow -\alpha z = \ln{\frac{1}{2}} \Rightarrow z = \frac{-\ln 2}{-\alpha} = \frac{\ln 2}{\alpha}.$$

2.8 Exercises

Note: The missing values are found in terms of the given values.

We could also use proportions to find the remaining parts.

$\boxed{1}$ Since α is given and $\gamma = 90°$, we can easily find β. $\beta = 90° - \alpha = 90° - 30° = 60°$.

To find a, we will relate it to the given parts, α and b, using the tangent function.

$$\tan{\alpha} = \frac{a}{b} \Rightarrow a = b \tan{\alpha} = 20 \tan{30°} = 20(\tfrac{1}{3}\sqrt{3}) = \frac{20}{3}\sqrt{3}.$$

$$\sec{\alpha} = \frac{c}{b} \Rightarrow c = b \sec{\alpha} = 20 \sec{30°} = 20(\tfrac{2}{3}\sqrt{3}) = \frac{40}{3}\sqrt{3}.$$

$\boxed{3}$ $\alpha = 90° - \beta = 90° - 45° = 45°$.

$$\cos{\beta} = \frac{a}{c} \Rightarrow a = c \cos{\beta} = 30 \cos{45°} = 30(\tfrac{1}{2}\sqrt{2}) = 15\sqrt{2}. \quad b = a \text{ in a } 45°\text{-}45°\text{-}90° \triangle.$$

$\boxed{5}$ $\tan{\alpha} = \frac{a}{b} = \frac{5}{5} = 1 \Rightarrow \alpha = 45°$. $\beta = 90° - \alpha = 90° - 45° = 45°$. Using the

Pythagorean theorem, $a^2 + b^2 = c^2 \Rightarrow c = \sqrt{a^2 + b^2} = \sqrt{25 + 25} = \sqrt{50} = 5\sqrt{2}$.

$\boxed{7}$ $\cos{\alpha} = \frac{b}{c} = \dfrac{5\sqrt{3}}{10\sqrt{3}} = \frac{1}{2} \Rightarrow \alpha = 60°$. $\beta = 90° - \alpha = 90° - 60° = 30°$.

$$a = \sqrt{c^2 - b^2} = \sqrt{300 - 75} = \sqrt{225} = 15.$$

$\boxed{9}$ $\beta = 90° - \alpha = 90° - 37° = 53°$. $\tan{\alpha} = \frac{a}{b} \Rightarrow a = b \tan{\alpha} = 24 \tan{37°} \approx 18$.

$$\sec{\alpha} = \frac{c}{b} \Rightarrow c = b \sec{\alpha} = 24 \sec{37°} \approx 30.$$

$\boxed{11}$ $\alpha = 90° - \beta = 90° - 71°51' = 18°9'$. $\cot{\beta} = \frac{a}{b} \Rightarrow a = b \cot{\beta} = 240.0 \cot{71°51'} \approx 78.7$.

$$\csc{\beta} = \frac{c}{b} \Rightarrow c = b \csc{\beta} = 240.0 \csc{71°51'} \approx 252.6.$$

$\boxed{13}$ $\tan{\alpha} = \frac{a}{b} = \frac{25}{45} \Rightarrow \alpha = \tan^{-1}{\frac{25}{45}} \approx 29°$. $\beta = 90° - \alpha \approx 90° - 29° = 61°$.

$$c = \sqrt{a^2 + b^2} = \sqrt{25^2 + 45^2} = \sqrt{625 + 2025} = \sqrt{2650} \approx 51.$$

$\boxed{15}$ $\cos\alpha = \frac{b}{c} = \frac{2.1}{5.8} \Rightarrow \alpha = \cos^{-1}\frac{21}{58} \approx 69°.$ $\beta = 90° - \alpha \approx 90° - 69° = 21°.$

$$a = \sqrt{c^2 - b^2} = \sqrt{(5.8)^2 - (2.1)^2} = \sqrt{33.64 - 4.41} = \sqrt{29.23} \approx 5.4.$$

Note: Refer to Figures 74 and 75 in the text for the labeling of the sides and angles.

$\boxed{17}$ We need to find a relationship involving b, c, and α. We want angle α with its adjacent side b and hypotenuse c. The cosine or secant are the functions of α that involve b and c. We choose the cosine since it is easier to solve for b { b is in the numerator }. $\cos\alpha = \frac{b}{c} \Rightarrow b = c\cos\alpha.$

$\boxed{19}$ We want angle β with its adjacent side a and opposite side b. The tangent or cotangent are the functions of β that involve a and b. $\cot\beta = \frac{a}{b} \Rightarrow a = b\cot\beta.$

$\boxed{21}$ We want angle α with its opposite side a and hypotenuse c. The sine or cosecant are the functions of α that involve a and c. $\csc\alpha = \frac{c}{a} \Rightarrow c = a\csc\alpha.$

$\boxed{23}$ $a^2 + b^2 = c^2 \Rightarrow b^2 = c^2 - a^2 \Rightarrow b = \sqrt{c^2 - a^2}$

$\boxed{25}$ Let h denote the height of the kite and $x = h - 4$. $\sin 60° = \frac{x}{500} \Rightarrow$

$$x = 500\sin 60° = 500(\tfrac{1}{2}\sqrt{3}) = 250\sqrt{3}.\ h = x + 4 = 250\sqrt{3} + 4 \approx 437 \text{ ft.}$$

$\boxed{27}$ $\sin 10° = \frac{5000}{x} \Rightarrow x = \frac{5000}{\sin 10°} \Rightarrow$

$x = 5000\csc 10° \approx 28{,}793.85$, or 28,800 ft.

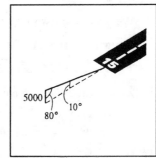

Figure 27

$\boxed{31}$ The 10,000 feet would represent the hypotenuse in a triangle depicting this information. Let h denote the altitude. $\sin 75° = \frac{h}{10{,}000} \Rightarrow h \approx 9659$ ft.

$\boxed{33}$ (a) The bridge section is 75 feet long. Using the right triangle with the 75 foot section as its hypotenuse and $(d - 15)$ as the side opposite the 35° angle, we have

$$\sin 35° = \frac{d - 15}{75} \Rightarrow d = 75\sin 35° + 15 \approx 58 \text{ ft.}$$

(b) Let x be the horizontal distance from the end of a bridge section to a point directly underneath the end of the section. $\cos 35° = \frac{x}{75} \Rightarrow x = 75\cos 35°.$

The distance between the ends of the two sections is (total distance) −

(the 2 horizontal distances under the bridge sections) $= 150 - 2x \approx 27$ ft.

37 Let D denote the position of the duck and t the number of seconds required for a direct hit. The duck will move $(7t)$ cm. and the bullet will travel $(25t)$ cm.

$$\sin \varphi = \frac{\overline{AD}}{\overline{OD}} = \frac{7t}{25t} \Rightarrow \sin \varphi = \frac{7}{25} \Rightarrow \varphi \approx 16.3°.$$

39 Let h denote the height of the tower.

$$\tan 21°20'24'' = \tan 21.34° = \frac{h}{5280} \Rightarrow h = 5280 \tan 21.34° \approx 2063 \text{ ft.}$$

41 The central angle of a section of the Pentagon has measure $\frac{360°}{5} = 72°$.

Bisecting that angle, we have an angle of $36°$ whose opposite side is $\frac{921}{2}$.

The height h is given by $\tan 36° = \dfrac{\frac{921}{2}}{h} \Rightarrow h = \dfrac{921}{2 \tan 36°}$.

$$\text{Area} = 5(\tfrac{1}{2}bh) = 5(\tfrac{1}{2})(921)\left(\frac{921}{2 \tan 36°}\right) \approx 1{,}459{,}379 \text{ ft}^2.$$

43 The diagonal of the base is $\sqrt{8^2 + 6^2} = 10$. $\tan \theta = \frac{4}{10} \Rightarrow \theta \approx 21.8°$

45 $\cot 53°30' = \frac{x}{h} \Rightarrow x = h \cot 53°30'$. $\cot 26°50' = \frac{x + 25}{h} \Rightarrow$

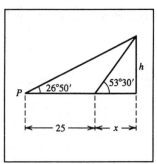

$x + 25 = h \cot 26°50' \Rightarrow x = h \cot 26°50' - 25$.

We now have two expressions for x.

We will set these equal to each other and solve for h.

Thus, $h \cot 53°30' = h \cot 26°50' - 25 \Rightarrow$

$25 = h \cot 26°50' - h \cot 53°30' \Rightarrow$

$h = \dfrac{25}{\cot 26°50' - \cot 53°30'} \approx 20.2 \text{ m.}$

Figure 45

47 When the angle of elevation is $19°20'$, $\tan 19°20' = \frac{h_1}{110} \Rightarrow h_1 = 110 \tan 19°20'$.

When the angle of elevation is $31°50'$, $\tan 31°50' = \frac{h_2}{110} \Rightarrow h_2 = 110 \tan 31°50'$.

The change in elevation is $h_2 - h_1 \approx 68.29 - 38.59 = 29.7 \text{ km.}$

49 The distance from the spacelab to the center of the earth is $(380 + r)$ miles.

$$\sin 65.8° = \frac{r}{r + 380} \Rightarrow r \sin 65.8° + 380 \sin 65.8° = r \Rightarrow r - r \sin 65.8° = 380 \sin 65.8° \Rightarrow$$

$$r(1 - \sin 65.8°) = 380 \sin 65.8° \Rightarrow r = \frac{380 \sin 65.8°}{1 - \sin 65.8°} \approx 3944 \text{ mi.}$$

51 Let d be the distance traveled. $\tan 42° = \frac{10{,}000}{d} \Rightarrow d = 10{,}000 \cot 42°$.

Converting to mi/hr, we have $\dfrac{10{,}000 \cot 42° \text{ ft}}{1 \text{ minute}} \cdot \dfrac{60 \text{ minutes}}{1 \text{ hour}} \cdot \dfrac{1 \text{ mile}}{5280 \text{ ft}} \approx 126 \text{ mi/hr.}$

[53] (a) As in Exercise 49, there is a right angle formed on the earth's surface.

Bisecting angle θ and forming a right triangle, we have

$$\cos\frac{\theta}{2} = \frac{R}{R+a} = \frac{4000}{26,300}. \text{ Thus, } \frac{\theta}{2} \approx 81.25° \Rightarrow \theta \approx 162.5°.$$

The percentage of the equator that is within signal range is $\frac{162.5°}{360°} \times 100 \approx 45\%$.

(b) Each satellite has a signal range of more than 120°,

and thus all 3 will cover all points on the equator.

[55] Let $x = h - c$. $\sin\alpha = \frac{x}{d} \Rightarrow x = d\sin\alpha$. $h = x + c = d\sin\alpha + c$.

[57] Let x denote the distance from the base of the tower to the closer point.

$$\cot\beta = \frac{x}{h} \Rightarrow x = h\cot\beta. \quad \cot\alpha = \frac{x+d}{h} \Rightarrow x + d = h\cot\alpha \Rightarrow x = h\cot\alpha - d.$$

Thus, $h\cot\beta = h\cot\alpha - d \Rightarrow d = h\cot\alpha - h\cot\beta \Rightarrow d = h(\cot\alpha - \cot\beta) \Rightarrow$

$$h = \frac{d}{\cot\alpha - \cot\beta}.$$

[59] When the angle of elevation is α, $\tan\alpha = \frac{h_1}{d} \Rightarrow h_1 = d\tan\alpha$.

When the angle of elevation is β, $\tan\beta = \frac{h_2}{d} \Rightarrow h_2 = d\tan\beta$.

$$h = h_2 - h_1 = d\tan\beta - d\tan\alpha = d(\tan\beta - \tan\alpha).$$

[61] The bearing from P to A is $90° - 20° = 70°$ east of north and is denoted by N70°E.

The bearing from P to B is $40°$ west of north and is denoted by N40°W.

The bearing from P to C is $90° - 75° = 15°$ west of south and is denoted by S15°W.

The bearing from P to D is $25°$ east of south and is denoted by S25°E.

[63] (a) The first ship travels 2 hours @ 24 mi/hr for a
distance of 48 miles. The second ship travels
$1\frac{1}{2}$ hours @ 18 mi/hr for a distance of 27 miles.
The paths form a right triangle with legs of 48 miles
and 27 miles. The distance between the two ships is
$\sqrt{27^2 + 48^2} \approx 55$ miles.

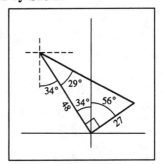

Figure 63

(b) The angle between the side of length 48 and the
hypotenuse is found by solving $\tan\alpha = \frac{27}{48}$ for α. Now
$\alpha \approx 29°$, so the second ship is approximately $29° + 34° = $ S63°E of the first ship.

[65] 30 minutes @ 360 mi/hr = 180 miles. 45 minutes @ 360 mi/hr = 270 miles. $227° - 137° = 90°$, and hence, the plane's flight forms a right triangle with legs 180 miles and 270 miles. The distance from A to the airplane is equal to the hypotenuse of the triangle—that is, $\sqrt{180^2 + 270^2} \approx 324.5$ mi.

[67] Amplitude, 10 cm; period $= \frac{2\pi}{6\pi} = \frac{1}{3}$ sec; frequency $= \frac{6\pi}{2\pi} = 3$ oscillations/sec. The point is at the origin at $t = 0$. It moves upward with decreasing speed, reaching the point with coordinate 10 when $6\pi t = \frac{\pi}{2}$ or $t = \frac{1}{12}$.

It then reverses direction and moves downward, gaining speed until it reaches the origin when $6\pi t = \pi$ or $t = \frac{1}{6}$. It continues downward with decreasing speed, reaching the point with coordinate -10 when $6\pi t = \frac{3\pi}{2}$ or $t = \frac{1}{4}$.

It then reverses direction and moves upward with increasing speed, returning to the origin when $6\pi t = 2\pi$ or $t = \frac{1}{3}$ to complete one oscillation.

Another approach is to simply model this movement in terms of proportions of the sine curve. For one period, the sine increases for $\frac{1}{4}$ period, decreases for $\frac{1}{2}$ period, and increases for its last $\frac{1}{4}$ period.

[69] Amplitude, 4 cm; period $= \frac{2\pi}{3\pi/2} = \frac{4}{3}$ sec; frequency $= \frac{3\pi/2}{2\pi} = \frac{3}{4}$ oscillation/sec. The point is at $d = 4$ when $t = 0$. It then decreases in height until $\frac{3\pi}{2}t = \pi$ or $t = \frac{2}{3}$ where it obtains a minimum of $d = -4$. It then reverses direction and increases to a height of $d = 4$ when $\frac{3\pi}{2}t = 2\pi$ or $t = \frac{4}{3}$ to complete one oscillation.

[71] Period $= 3 \Rightarrow \frac{2\pi}{\omega} = 3 \Rightarrow \omega = \frac{2\pi}{3}$. Amplitude $= 5 \Rightarrow a = 5$. $d = 5\cos\frac{2\pi}{3}t$

[73] (a) period $= 30 \Rightarrow \frac{2\pi}{\omega} = 30 \Rightarrow \omega = \frac{\pi}{15}$. When $t = 0$, the wave is at its highest

point, thus, we use the cosine function. $y = 25\cos\frac{\pi}{15}t$, where t is in minutes.

(b) 180 ft/sec = 10,800 ft/min.

10,800 ft/min for 30 minutes is a distance of 324,000 ft, or $61\frac{4}{11}$ miles.

1. $330° \cdot \frac{\pi}{180} = \frac{11 \cdot 30\pi}{6 \cdot 30} = \frac{11\pi}{6};$ $\qquad\qquad$ $405° \cdot \frac{\pi}{180} = \frac{9 \cdot 45\pi}{4 \cdot 45} = \frac{9\pi}{4}$

2. $\frac{9\pi}{2} \cdot \left(\frac{180}{\pi}\right)° = \left(\frac{9 \cdot 90 \cdot 2\pi}{2\pi}\right)° = 810°;$ $\qquad$ $-\frac{2\pi}{3} \cdot \left(\frac{180}{\pi}\right)° = -\left(\frac{2 \cdot 60 \cdot 3\pi}{3\pi}\right)° = -120°;$

3. (a) $\theta = \frac{s}{r} = \frac{20\,\text{cm}}{2\,\text{m}} = \frac{20\,\text{cm}}{2\,(100)\,\text{cm}} = 0.1\,\text{radian}$

   (b) $A = \frac{1}{2}r^2\theta = \frac{1}{2}(2)^2(0.1) = 0.2\,\text{m}^2$

4. (a) $s = r\theta = (15 \cdot \frac{1}{2})(70 \cdot \frac{\pi}{180}) = \frac{35\pi}{12} \approx 9.16\,\text{cm}$

   (b) $A = \frac{1}{2}r^2\theta = \frac{1}{2}(15 \cdot \frac{1}{2})^2(70 \cdot \frac{\pi}{180}) = \frac{175\pi}{16} \approx 34.4\,\text{cm}^2$

5. 7π is coterminal with π. $P(7\pi) = P(\pi) = (-1, 0)$.

   $\qquad\qquad -\frac{5\pi}{2}$ is coterminal with $-\frac{\pi}{2}$. $P\left(-\frac{5\pi}{2}\right) = P\left(-\frac{\pi}{2}\right) = (0, -1)$.

6. $P(t) = \left(-\frac{3}{5}, -\frac{4}{5}\right)$ is in QIII. $P(t + \pi)$ is in QI and will have the same coordinates as

   $P(t)$, but with opposite (positive) signs. $P(t + 3\pi) = P(t + \pi) = P(t - \pi) = \left(\frac{3}{5}, \frac{4}{5}\right)$.

7. (b) $\cot t = \frac{\cos t}{\sin t} = \frac{\csc t}{\sec t} \Rightarrow -\frac{3}{2} = \frac{\sqrt{13}/2}{\sec t} \Rightarrow \sec t = -\frac{\sqrt{13}}{3};$

   $\qquad\qquad\qquad\qquad\qquad\qquad$ the other values are just the reciprocals.

8. (a) $\sec t < 0 \Rightarrow P$ is in QII or QIII. $\sin t > 0 \Rightarrow P$ is in QI or QII.

   $\qquad\qquad\qquad\qquad\qquad\qquad\qquad\qquad$ Hence, P is in QII.

   (b) $\cot t > 0 \Rightarrow P$ is in QI or QIII. $\csc t < 0 \Rightarrow P$ is in QIII or QIV.

   $\qquad\qquad\qquad\qquad\qquad\qquad\qquad\qquad$ Hence, P is in QIII.

9. $1 + \tan^2 t = \sec^2 t \Rightarrow \tan^2 t = \sec^2 t - 1 \Rightarrow \tan t = \sqrt{\sec^2 t - 1}$

11. $\sin t\,(\csc t - \sin t)\ = \sin t \csc t - \sin^2 t$ $\qquad$ { multiply terms }

    $\qquad\qquad\qquad\qquad = \sin t \cdot \frac{1}{\sin t} - \sin^2 t$ $\qquad$ { reciprocal identity }

    $\qquad\qquad\qquad\qquad = 1 - \sin^2 t$ $\qquad\qquad\qquad$ { simplify }

    $\qquad\qquad\qquad\qquad = \cos^2 t$ $\qquad\qquad\qquad\qquad$ { Pythagorean identity }

13. $(\cos^2 t - 1)(\tan^2 t + 1)\ = (\cos^2 t - 1)(\sec^2 t)$ $\qquad$ { Pythagorean identity }

    $\qquad\qquad\qquad\qquad\qquad = \cos^2 t \sec^2 t - \sec^2 t$ $\qquad$ { multiply terms }

    $\qquad\qquad\qquad\qquad\qquad = 1 - \sec^2 t$ $\qquad\qquad\qquad$ { reciprocal identity }

15. $\frac{1 + \tan^2 t}{\tan^2 t}\ = \frac{1}{\tan^2 t} + \frac{\tan^2 t}{\tan^2 t}$ $\qquad$ { split up the fraction }

    $\qquad\qquad = \cot^2 t + 1$ $\qquad\qquad$ { reciprocal identity, simplify }

    $\qquad\qquad = \csc^2 t$ $\qquad\qquad\qquad$ { Pythagorean identity }

$\boxed{17}$ $\dfrac{\cot t - 1}{1 - \tan t} = \dfrac{\dfrac{\cos t}{\sin t} - 1}{1 - \dfrac{\sin t}{\cos t}}$ { put in terms of sines and cosines }

$\qquad = \dfrac{\dfrac{\cos t - \sin t}{\sin t}}{\dfrac{\cos t - \sin t}{\cos t}}$ $\left\{ \begin{array}{c} \text{make the numerator and the} \\ \text{denominator each a single fraction} \end{array} \right\}$

$\qquad = \dfrac{(\cos t - \sin t)\cos t}{(\cos t - \sin t)\sin t}$ { simplify a complex fraction }

$\qquad = \dfrac{\cos t}{\sin t}$ { cancel like term }

$\qquad = \cot t$ { cotangent identity }

$\boxed{19}$ $\dfrac{\tan(-t) + \cot(-t)}{\tan t} = \dfrac{-\tan t - \cot t}{\tan t}$ { formulas for negatives }

$\qquad = -\dfrac{\tan t}{\tan t} - \dfrac{\cot t}{\tan t}$ { split up fraction }

$\qquad = -1 - \cot^2 t$ { simplify, reciprocal identity }

$\qquad = -(1 + \cot^2 t)$ { factor out -1 }

$\qquad = -\csc^2 t$ { Pythagorean identity }

$\boxed{21}$ (a) $x = 30$ and $y = -40 \Rightarrow r = \sqrt{30^2 + (-40)^2} = 50.$ ★ (a) $-\frac{4}{5}, \frac{3}{5}, -\frac{4}{3}, -\frac{3}{4}, \frac{5}{3}, -\frac{5}{4}$

(b) $2x + 3y + 6 = 0 \Leftrightarrow y = -\frac{2}{3}x - 2$, so the slope of the given line is $-\frac{2}{3}$.

The line through the origin with that slope is $y = -\frac{2}{3}x$.

If $x = -3$, then $y = 2$ and $(-3, 2)$ is a point on the terminal side of θ.

$$x = -3 \text{ and } y = 2 \Rightarrow r = \sqrt{(-3)^2 + 2^2} = \sqrt{13}.$$

★ (b) $\dfrac{2}{\sqrt{13}}, -\dfrac{3}{\sqrt{13}}, -\dfrac{2}{3}, -\dfrac{3}{2}, -\dfrac{\sqrt{13}}{3}, \dfrac{\sqrt{13}}{2}$

(c) For $\theta = -90°$, choose $x = 0$ and $y = -1$. r is 1.   ★ (c) $-1, 0, U, 0, U, -1$

$\boxed{22}$ opp $= \sqrt{\text{hyp}^2 - \text{adj}^2} = \sqrt{7^2 - 4^2} = \sqrt{33}.$ ★ $\dfrac{\sqrt{33}}{7}, \dfrac{4}{7}, \dfrac{\sqrt{33}}{4}, \dfrac{4}{\sqrt{33}}, \dfrac{7}{4}, \dfrac{7}{\sqrt{33}}$

$\boxed{23}$ $\sin 60° = \frac{9}{x} \Rightarrow \frac{\sqrt{3}}{2} = \frac{9}{x} \Rightarrow x = 6\sqrt{3}$; $\tan 60° = \frac{9}{y} \Rightarrow \sqrt{3} = \frac{9}{y} \Rightarrow y = 3\sqrt{3}$

$\boxed{25}$ (a) $t = -\frac{9\pi}{8} \Rightarrow \theta_C = \frac{7\pi}{8}$ and $t_R = \pi - \frac{7\pi}{8} = \frac{\pi}{8}$.

(b) $\theta = 892° \Rightarrow \theta_C = 172°$ and $\theta_R = 180° - 172° = 8°$.

$\boxed{26}$ (b) For $\theta = -\frac{5\pi}{4}$, choose $x = -1$ and $y = 1$. $r = \sqrt{2}$.

★ (b) $\dfrac{\sqrt{2}}{2}, -\dfrac{\sqrt{2}}{2}, -1, -1, -\sqrt{2}, \sqrt{2}$

(d) For $\theta = \frac{11\pi}{6}$, choose $x = \sqrt{3}$ and $y = -1$. $r = 2$.

★ (d) $-\dfrac{1}{2}, \dfrac{\sqrt{3}}{2}, -\dfrac{\sqrt{3}}{3}, -\sqrt{3}, \dfrac{2}{\sqrt{3}}, -2$

$\boxed{27}$ (a) $\cos 225° = -\cos 45° = -\dfrac{\sqrt{2}}{2}$ (b) $\tan 150° = -\tan 30° = -\dfrac{\sqrt{3}}{3}$

(c) $\sin\left(-\dfrac{\pi}{6}\right) = -\sin\dfrac{\pi}{6} = -\dfrac{1}{2}$ (d) $\sec\dfrac{4\pi}{3} = -\sec\dfrac{\pi}{3} = -2$

(e) $\cot\dfrac{7\pi}{4} = -\cot\dfrac{\pi}{4} = -1$ (f) $\csc 300° = -\csc 60° = -\dfrac{2}{\sqrt{3}}$

$\boxed{28}$ $\sin\theta = -0.7604 \Rightarrow \theta = \sin^{-1}(-0.7604) \approx -49.5° \Rightarrow \theta_R \approx 49.5°$.

Since the sine is negative in QIII and QIV, and the secant is positive in QIV,

we want the fourth-quadrant angle having $\theta_R = 49.5°$. $360° - 49.5° = 310.5°$

$\boxed{29}$ $y = 5\cos x$ • Vertically stretch $\cos x$ by a factor of 5.   ★ $5, 2\pi$, x-int. @ $\dfrac{\pi}{2} + \pi n$

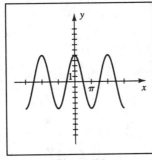

Figure 29

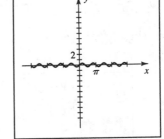

Figure 31

$\boxed{31}$ $y = \frac{1}{3}\sin 3x$ • Horizontally compress $\sin x$ by a factor of 3 and vertically compress

that graph by a factor of 3.   ★ $\dfrac{1}{3}, \dfrac{2\pi}{3}$, x-int. @ $\dfrac{\pi}{3}n$

$\boxed{33}$ $y = -3\cos\frac{1}{2}x$ • Horizontally stretch $\cos x$ by a factor of 2, vertically stretch that

graph by a factor of 3, and reflect that graph through the x-axis.

★ $3, 4\pi$, x-int. @ $\pi + 2\pi n$

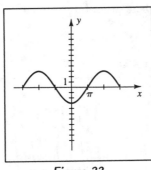

Figure 33

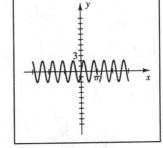

Figure 35

$\boxed{35}$ $y = 2\sin\pi x$ • Horizontally compress $\sin x$ by a factor of π, and then vertically

stretch that graph by a factor of 2.   ★ $2, 2$, x-int. @ n

Note: Let a denote the amplitude and p the period.

37 (a) $a = |-1.43| = 1.43$. We have $\frac{3}{4}$ of a period from $(0, 0)$ to $(1.5, -1.43)$.

$$\tfrac{3}{4}p = 1.5 \Rightarrow p = 2.$$

(b) Since the period p is given by $\frac{2\pi}{b}$, we can solve for b, obtaining $b = \frac{2\pi}{p}$.

$$\text{Hence, } b = \frac{2\pi}{p} = \frac{2\pi}{2} = \pi \text{ and consequently, } y = 1.43 \sin \pi x.$$

39 (a) Since the y-intercept is -3, $a = |-3| = 3$. The second positive x-intercept is π,

so $\frac{3}{4}$ of a period occurs from $(0, -3)$ to $(\pi, 0)$. Thus, $\frac{3}{4}p = \pi \Rightarrow p = \frac{4\pi}{3}$.

(b) $b = \frac{2\pi}{p} = \frac{2\pi}{4\pi/3} = \frac{3}{2}$, $y = -3 \cos \frac{3}{2}x$.

41 $y = 2 \sin\left(x - \frac{2\pi}{3}\right)$ • $0 \le x - \frac{2\pi}{3} \le 2\pi \Rightarrow \frac{2\pi}{3} \le x \le \frac{8\pi}{3}$.

There are x-intercepts at $x = \frac{2\pi}{3} + \pi n$.

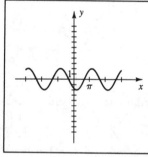

Figure 41

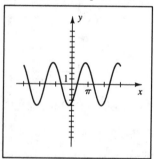

Figure 43

43 $y = -4 \cos\left(x + \frac{\pi}{6}\right)$ • $-\frac{\pi}{2} \le x + \frac{\pi}{6} \le \frac{3\pi}{2} \Rightarrow -\frac{2\pi}{3} \le x \le \frac{4\pi}{3}$.

There are x-intercepts at $x = -\frac{2\pi}{3} + \pi n$.

45 $y = 2 \tan\left(\frac{1}{2}x - \pi\right) = 2 \tan\left[\frac{1}{2}(x - 2\pi)\right]$. •

$-\frac{\pi}{2} \le \frac{1}{2}x - \pi \le \frac{\pi}{2} \Rightarrow \frac{\pi}{2} \le \frac{1}{2}x \le \frac{3\pi}{2} \Rightarrow \pi \le x \le 3\pi$, VA @ $x = \pi + 2\pi n$

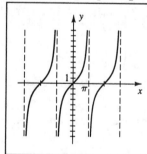

Figure 45

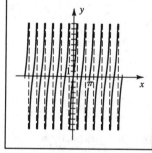

Figure 47

47 $y = -4 \cot\left(2x - \frac{\pi}{2}\right) = -4 \cot\left[2(x - \frac{\pi}{4})\right]$. •

$0 \le 2x - \frac{\pi}{2} \le \pi \Rightarrow \frac{\pi}{2} \le 2x \le \frac{3\pi}{2} \Rightarrow \frac{\pi}{4} \le x \le \frac{3\pi}{4}$, VA @ $x = \frac{\pi}{4} + \frac{\pi}{2}n$

$\boxed{49}$ $y = \sec\left(\frac{1}{2}x + \pi\right) = \sec\left[\frac{1}{2}(x + 2\pi)\right]$. •

$-\frac{\pi}{2} \le \frac{1}{2}x + \pi \le \frac{\pi}{2} \Rightarrow -\frac{3\pi}{2} \le \frac{1}{2}x \le -\frac{\pi}{2} \Rightarrow -3\pi \le x \le -\pi,\ VA\ @\ x = -3\pi + 2\pi n$

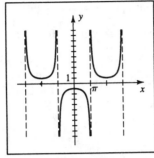

Figure 49

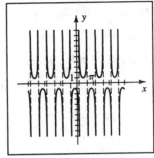

Figure 51

$\boxed{51}$ $y = \csc\left(2x - \frac{\pi}{4}\right) = \csc\left[2\left(x - \frac{\pi}{8}\right)\right]$. •

$0 \le 2x - \frac{\pi}{4} \le \pi \Rightarrow \frac{\pi}{4} \le 2x \le \frac{5\pi}{4} \Rightarrow \frac{\pi}{8} \le x \le \frac{5\pi}{8},\ VA\ @\ x = \frac{\pi}{8} + \frac{\pi}{2}n$

$\boxed{53}$ α and β are complementary, so $\alpha = 90° - \beta = 90° - 60° = 30°$.

$\cot\beta = \frac{a}{b} \Rightarrow a = b\cot\beta = 40\cot 60° = 40(\frac{1}{3}\sqrt{3}) \approx 23$.

$\csc\beta = \frac{c}{b} \Rightarrow c = b\csc\beta = 40\csc 60° = 40(\frac{2}{3}\sqrt{3}) \approx 46$.

$\boxed{55}$ $\tan\alpha = \frac{a}{b} = \frac{62}{25} \Rightarrow \alpha \approx 68°$. $\beta = 90° - \alpha \approx 90° - 68° = 22°$.

$$c = \sqrt{a^2 + b^2} = \sqrt{62^2 + 25^2} = \sqrt{3844 + 625} = \sqrt{4469} \approx 67.$$

$\boxed{57}$ (a) $\left(\frac{545\ \text{rev}}{1\ \text{min}}\right)\left(\frac{2\pi\ \text{rad}}{1\ \text{rev}}\right)\left(\frac{1\ \text{min}}{60\ \text{sec}}\right) = \frac{109\pi}{6}$ rad/sec ≈ 57 rad/sec

(b) $d = 22.625$ ft $\Rightarrow C = \pi d = 22.625\pi$ ft.

$$\left(\frac{22.625\pi\ \text{ft}}{1\ \text{rev}}\right)\left(\frac{545\ \text{rev}}{1\ \text{min}}\right)\left(\frac{1\ \text{mile}}{5280\ \text{ft}}\right)\left(\frac{60\ \text{min}}{1\ \text{hour}}\right) \approx 440.2\ \text{mi/hr}$$

$\boxed{59}$ $\Delta f = \frac{2fv}{c} \Rightarrow v = \frac{c(\Delta f)}{2f} = \frac{186,000 \times 10^8}{2 \times 10^{14}} = 0.093$ mi/sec

$\boxed{60}$ The angle φ has an adjacent side of $\frac{1}{2}(230$ m), or 115 m. $\tan\varphi = \frac{147}{115} \Rightarrow \varphi \approx 52°$.

$\boxed{62}$ The depth of the cone is 4 inches and its slant height is 5 inches. A right triangle is

formed by a cross section of the cone with sides 4, 5, and r. Thus, $4^2 + r^2 = 5^2 \Rightarrow$

$r = 3$ inches. The circumference of the rim of the cone is $2\pi r = 2\pi(3) = 6\pi$.

This circumference is the arc length from A to B on the circle.

Using the formula for arc length, $\theta = \frac{s}{r} = \frac{6\pi}{5}$ radians $= 216°$.

$\boxed{64}$ (a) Let h denote the height of the building and x the distance between the two buildings. $\tan 59° = \dfrac{h-50}{x}$ and $\tan 62° = \dfrac{h}{x}$. We now solve both of these equations for h, giving us $h = x\tan 59° + 50$ and $h = x\tan 62°$. Setting these expressions equal to each other and solving for x, we have $x\tan 62° = x\tan 59° + 50 \Rightarrow x\tan 62° - x\tan 59° = 50 \Rightarrow$

$$x(\tan 62° - \tan 59°) = 50 \Rightarrow x = \frac{50}{\tan 62° - \tan 59°} \approx 231.0 \text{ ft.}$$

(b) From part (a), $h = x\tan 62° \approx 434.5$ ft.

$\boxed{66}$ (a) Extend the two boundary lines for h { call these l_{top} and l_{bottom} } to the right until they intersect a line l extended down from the front edge of the building. Let x denote the distance from the intersection of the incline and l_{bottom} to l and y the distance on l from l_{top} to the lower left corner of the building.

$\cos\alpha = \dfrac{x}{d} \Rightarrow x = d\cos\alpha.$ $\sin\alpha = \dfrac{h+y}{d} \Rightarrow h+y = d\sin\alpha \Rightarrow y = d\sin\alpha - h.$

$\tan\theta = \dfrac{y+T}{x} \Rightarrow y+T = x\tan\theta \Rightarrow$

$$\begin{aligned} T = x\tan\theta - y &= (d\cos\alpha)\tan\theta - (d\sin\alpha - h) \\ &= d\cos\alpha\tan\theta - d\sin\alpha + h \\ &= h + (d\cos\alpha\tan\theta - d\sin\alpha) = h + d(\cos\alpha\tan\theta - \sin\alpha). \end{aligned}$$

(b) $T = 6 + 50\,(\cos 15°\tan 31.4° - \sin 15°) \approx 6 + 50\,(0.3308) \approx 22.54$ ft.

$\boxed{68}$ (a) Let $x = \overline{PT}$ and $y = \overline{QT}$. Now $x^2 + d^2 = y^2$, $h = x\sin\alpha$, and $h = y\sin\beta$.

$$d^2 = y^2 - x^2 = \left(\frac{h}{\sin\beta}\right)^2 - \left(\frac{h}{\sin\alpha}\right)^2 = \frac{h^2}{\sin^2\beta} - \frac{h^2}{\sin^2\alpha} = \frac{h^2\sin^2\alpha - h^2\sin^2\beta}{\sin^2\beta\,\sin^2\alpha} =$$

$$\frac{h^2(\sin^2\alpha - \sin^2\beta)}{\sin^2\alpha\,\sin^2\beta} \Rightarrow h^2 = \frac{d^2\sin^2\alpha\,\sin^2\beta}{\sin^2\alpha - \sin^2\beta} \Rightarrow h = \frac{d\sin\alpha\,\sin\beta}{\sqrt{\sin^2\alpha - \sin^2\beta}}.$$

(b) $\alpha = 30°$, $\beta = 20°$, and $d = 10 \Rightarrow h = \dfrac{10\sin 30°\sin 20°}{\sqrt{\sin^2 30° - \sin^2 20°}} \approx 4.69$ miles.

$\boxed{70}$ (a) The side opposite angle θ is one-half the length of the base, $\frac{1}{2}x$. $\sin\theta = \dfrac{\frac{1}{2}x}{a} \Rightarrow$

$x = 2a\sin\theta$. The area of each face is the area of a triangle, and S is the total

area of the four faces. The area of one face is $\frac{1}{2}$(base)(height) $= \frac{1}{2}xa$.

$$S = 4(\tfrac{1}{2}ax) = 2ax = 2a(2a\sin\theta) = 4a^2\sin\theta.$$

(b) $\cos\theta = \frac{y}{a} \Rightarrow y = a\cos\theta$. The volume of a pyramid is one-third times the area

of its base times its height. Hence,

$$V = \tfrac{1}{3}(\text{base area})(\text{height}) = \tfrac{1}{3}x^2y = \tfrac{1}{3}(2a\sin\theta)^2(a\cos\theta) = \tfrac{4}{3}a^3\sin^2\theta\cos\theta.$$

$\boxed{72}$ $y = 1 - 1\cos(\frac{1}{2}\pi x/10) = -\cos(\frac{\pi}{20}x) + 1$. To obtain the range for y, we will start with

the range for x, perform operations on these values, and try to obtain the expression

for y. For $0 \le x \le 10$, $0 \le \frac{\pi}{20}x \le \frac{\pi}{2}$ { multiply by $\frac{\pi}{20}$ }, $1 \ge \cos(\frac{\pi}{20}x) \ge 0$ { take the

cosine of all 3 parts }, $-1 \le -\cos(\frac{\pi}{20}x) \le 0$ { multiply by -1 }, $0 \le -\cos(\frac{\pi}{20}x) + 1 \le 1$

{ add 1 }, which is equivalent to $0 \le y \le 1$.

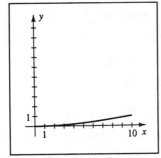

Figure 72

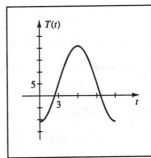

Figure 74

$\boxed{74}$ (a) $p = \frac{2\pi}{\pi/6} = 12$ months; range $= \underline{5 - 15.8}$ to $\underline{5 + 15.8}$, or, equivalently,

-10.8 to 20.8; phase shift $= 3$

(b) The highest temperature will occur when the argument of the sine is $\frac{\pi}{2}$ since this

is when the sine function has a maximum. $\frac{\pi}{6}(t - 3) = \frac{\pi}{2} \Rightarrow t - 3 = 3 \Rightarrow$

$t = 6$ months. This is July 1st and the temperature is $20.8\,°C$, or $69.44\,°F$.

$\boxed{76}$ (a) The cork is in simple harmonic motion. At $t = 0$, its height is

$s(0) = 12 + \cos 0 = 12 + 1 = 13$ ft. It decreases until $t = 1$, reaching a minimum

of $s(1) = 12 + \cos\pi = 12 + (-1) = 11$ ft. It then increases, reaching a maximum

of 13 ft at $t = 2$.

(b) From part (a), the cork is rising for $1 \le t \le 2$.

$\boxed{1}$ On the TI-82/83 with $a = 15$, there is an indication that there are 15 sine waves on each side of the y-axis, but the minimums and maximums do not get to -1 and 1, respectively. With $a = 30$, the number of sine waves is undetectable—there simply aren't enough pixels for any degree of clarity. With $a = 45$, there are 2 sine waves on each side of the y-axis—there should be 45!

$\boxed{3}$ The sum on the left-side of the equation can never be greater than 3—no solutions.

$\boxed{5}$ The graph of $y_1 = x$, $y_2 = \sin x$, and $y_3 = \tan x$ in the suggested viewing rectangle, $[-0.1,\ 0.1]$ by $[-0.1,\ 0.1]$, indicates that their values are very close to each other near $x = 0$—in fact, the graphs of the functions are indistinguishable. Creating a table of values on the order of 10^{-10} also shows that all three functions are nearly equal.

$\boxed{7}$ (a) S is at $(0,\ -1)$ on the rectangular coordinate system. Starting at S, subtract 1 from the 2 km to get to the circular portion of the track. Now consider $t = (1 + \frac{3\pi}{2})$ as the radian measurement (starting from S) of the circular track in Discussion Exercise 6. $\{\frac{3\pi}{2}$ to get to the bottom of the circle and 1 to make the second km.$\}$ $x = \cos t + 1 \approx 1.8415$ and $y = \sin t \approx -0.5403$.

(b) The perimeter of the track is $(4 + 2\pi)$ km. $\frac{500}{4 + 2\pi} \approx 48.623066$ laps. Subtracting the 48 whole laps, we have a portion of a lap left, 0.623066 lap. Multiply by $4 + 2\pi$ to see how many km we are from S. $(4 + 2\pi)(0.623066) \approx 6.4071052$. To determine where this places us on the track, start at S and subtract 1 $\{$to get to $(1,\ -1)\}$, subtract π $\{$to get to $(1,\ 1)\}$, and subtract 2 $\{$to get to $(-1,\ 1)\}$. $6.4071052 - (3 + \pi) \approx 0.2655126$. Now consider $t = (0.2655126 + \frac{\pi}{2})$ as the radian measurement of the circular track in Discussion Exercise 6 (but with S at the bottom). $x \approx \cos t - 1 \approx -1.2624$ and $y \approx \sin t \approx 0.9650$.

Chapter 3: Analytic Trigonometry

$\boxed{1}$ $\csc\theta - \sin\theta = \dfrac{1}{\sin\theta} - \sin\theta = \dfrac{1-\sin^2\theta}{\sin\theta} = \dfrac{\cos^2\theta}{\sin\theta} = \dfrac{\cos\theta}{\sin\theta}\cdot\cos\theta = \cot\theta\,\cos\theta$

$\boxed{3}$ $\dfrac{\sec^2 u - 1}{\sec^2 u} = 1 - \dfrac{1}{\sec^2 u}$ { split up fraction } $= 1 - \cos^2 u = \sin^2 u$

$\boxed{5}$ $\dfrac{\csc^2\theta}{1+\tan^2\theta} = \dfrac{\csc^2\theta}{\sec^2\theta} = \dfrac{1/\sin^2\theta}{1/\cos^2\theta} = \dfrac{\cos^2\theta}{\sin^2\theta} = \left(\dfrac{\cos\theta}{\sin\theta}\right)^2 = \cot^2\theta$

$\boxed{7}$ $\dfrac{1+\cos t}{\sin t} + \dfrac{\sin t}{1+\cos t} = \dfrac{(1+\cos t)^2 + \sin^2 t}{\sin t\,(1+\cos t)}$ { combine fractions }

$\qquad\qquad\quad = \dfrac{1 + 2\cos t + \cos^2 t + \sin^2 t}{\sin t\,(1+\cos t)}$ { expand }

$\qquad\qquad\quad = \dfrac{2 + 2\cos t}{\sin t\,(1+\cos t)}$ { $\cos^2 t + \sin^2 t = 1$ }

$\qquad\qquad\quad = \dfrac{2(1+\cos t)}{\sin t\,(1+\cos t)}$ { factor out 2 }

$\qquad\qquad\quad = 2\csc t$ { cancel like term }

$\boxed{9}$ $\dfrac{1}{1-\cos\gamma} + \dfrac{1}{1+\cos\gamma} = \dfrac{1+\cos\gamma + 1 - \cos\gamma}{1-\cos^2\gamma} = \dfrac{2}{\sin^2\gamma} = 2\csc^2\gamma$

$\boxed{11}$ $(\sec u - \tan u)(\csc u + 1) = \left(\dfrac{1}{\cos u} - \dfrac{\sin u}{\cos u}\right)\left(\dfrac{1}{\sin u} + 1\right)$ { change to sines and cosines }

$\qquad\qquad\qquad\qquad\quad = \left(\dfrac{1-\sin u}{\cos u}\right)\left(\dfrac{1+\sin u}{\sin u}\right)$ { combine fractions }

$\qquad\qquad\qquad\qquad\quad = \dfrac{1-\sin^2 u}{\cos u\,\sin u}$ { multiply terms }

$\qquad\qquad\qquad\qquad\quad = \dfrac{\cos^2 u}{\cos u\,\sin u} = \dfrac{\cos u}{\sin u} = \cot u$

$\boxed{13}$ $\csc^4 t - \cot^4 t = (\csc^2 t)^2 - (\cot^2 t)^2$ { recognize as the diff. of 2 squares }

$\qquad\qquad\quad = (\csc^2 t + \cot^2 t)(\csc^2 t - \cot^2 t)$ { factor }

$\qquad\qquad\quad = (\csc^2 t + \cot^2 t)(1)$ { Pythagorean id., $1 + \cot^2 t = \csc^2 t$ }

$\qquad\qquad\quad = \csc^2 t + \cot^2 t$

[15] The first step in the following verification is to multiply the denominator, $1 - \sin\beta$, by its conjugate, $1 + \sin\beta$. This procedure will give us the difference of two squares, $1 - \sin^2\beta$, which is equal to $\cos^2\beta$. This step is often helpful when simplifying trigonometric expressions because manipulation of the Pythagorean identities often allows us to reduce the resulting expression to a single term.

$$\frac{\cos\beta}{1-\sin\beta} = \frac{\cos\beta}{1-\sin\beta} \cdot \frac{1+\sin\beta}{1+\sin\beta} = \frac{\cos\beta\,(1+\sin\beta)}{1-\sin^2\beta} = \frac{\cos\beta\,(1+\sin\beta)}{\cos^2\beta} = \frac{1+\sin\beta}{\cos\beta} =$$

$$\frac{1}{\cos\beta} + \frac{\sin\beta}{\cos\beta} = \sec\beta + \tan\beta$$

[17] $\dfrac{\tan^2 x}{\sec x + 1} = \dfrac{\sec^2 x - 1}{\sec x + 1} = \dfrac{(\sec x + 1)(\sec x - 1)}{\sec x + 1} = \sec x - 1 = \dfrac{1}{\cos x} - 1 = \dfrac{1 - \cos x}{\cos x}$

[19] $\dfrac{\cot u - 1}{\cot u + 1} = \dfrac{\dfrac{1}{\tan u} - 1}{\dfrac{1}{\tan u} + 1} = \dfrac{\dfrac{1-\tan u}{\tan u}}{\dfrac{1+\tan u}{\tan u}} = \dfrac{1 - \tan u}{1 + \tan u}$

[21] $\sin^4 r - \cos^4 r = (\sin^2 r - \cos^2 r)(\sin^2 r + \cos^2 r) = (\sin^2 r - \cos^2 r)(1) = \sin^2 r - \cos^2 r$

[23] $\tan^4 k - \sec^4 k = (\tan^2 k - \sec^2 k)(\tan^2 k + \sec^2 k)$ { factor as the diff. of 2 squares }

$\qquad\qquad = (-1)(\sec^2 k - 1 + \sec^2 k)$ { Pythagorean identity }

$\qquad\qquad = (-1)(2\sec^2 k - 1) = 1 - 2\sec^2 k$

[25] $(\sec t + \tan t)^2 = \left(\dfrac{1}{\cos t} + \dfrac{\sin t}{\cos t}\right)^2 = \left(\dfrac{1+\sin t}{\cos t}\right)^2 = \dfrac{(1+\sin t)^2}{\cos^2 t} =$

$$\frac{(1+\sin t)^2}{1-\sin^2 t} = \frac{(1+\sin t)^2}{(1+\sin t)(1-\sin t)} = \frac{1+\sin t}{1-\sin t}$$

[27] $(\sin^2\theta + \cos^2\theta)^3 = (1)^3 = 1$

[29] $\dfrac{1+\csc\beta}{\cot\beta + \cos\beta} = \dfrac{1 + \dfrac{1}{\sin\beta}}{\dfrac{\cos\beta}{\sin\beta} + \cos\beta} = \dfrac{\dfrac{\sin\beta + 1}{\sin\beta}}{\dfrac{\cos\beta + \cos\beta\sin\beta}{\sin\beta}} = \dfrac{\sin\beta + 1}{\cos\beta\,(1+\sin\beta)} = \dfrac{1}{\cos\beta} = \sec\beta$

[31] As demonstrated in the following verification, it may be beneficial to try to combine expressions rather than expand them. $(\csc t - \cot t)^4 (\csc t + \cot t)^4 =$

$$\big[(\csc t - \cot t)(\csc t + \cot t)\big]^4 = (\csc^2 t - \cot^2 t)^4 = (1)^4 = 1$$

[33] $\mathrm{RS} = \dfrac{\tan\alpha + \tan\beta}{1 - \tan\alpha\tan\beta} = \dfrac{\dfrac{\sin\alpha}{\cos\alpha} + \dfrac{\sin\beta}{\cos\beta}}{1 - \dfrac{\sin\alpha}{\cos\alpha}\cdot\dfrac{\sin\beta}{\cos\beta}} = \dfrac{\dfrac{\sin\alpha\cos\beta + \cos\alpha\sin\beta}{\cos\alpha\cos\beta}}{\dfrac{\cos\alpha\cos\beta - \sin\alpha\sin\beta}{\cos\alpha\cos\beta}} =$

$$\frac{\sin\alpha\cos\beta + \cos\alpha\sin\beta}{\cos\alpha\cos\beta - \sin\alpha\sin\beta} = \mathrm{LS}$$

Note: We could obtain the RS by dividing

the numerator and denominator of the LS by $(\cos\alpha\cos\beta)$.

35 $\dfrac{\tan\alpha}{1+\sec\alpha}+\dfrac{1+\sec\alpha}{\tan\alpha}=\dfrac{\tan^2\alpha+(1+\sec\alpha)^2}{(1+\sec\alpha)\tan\alpha}=\dfrac{(\sec^2\alpha-1)+(1+2\sec\alpha+\sec^2\alpha)}{(1+\sec\alpha)\tan\alpha}=$

$$\dfrac{2\sec^2\alpha+2\sec\alpha}{(1+\sec\alpha)\tan\alpha}=\dfrac{2\sec\alpha\,(\sec\alpha+1)\cot\alpha}{1+\sec\alpha}=\dfrac{2}{\cos\alpha}\cdot\dfrac{\cos\alpha}{\sin\alpha}=\dfrac{2}{\sin\alpha}=2\csc\alpha$$

37 $\dfrac{1}{\tan\beta+\cot\beta}=\dfrac{1}{\dfrac{\sin\beta}{\cos\beta}+\dfrac{\cos\beta}{\sin\beta}}=\dfrac{1}{\dfrac{\sin^2\beta+\cos^2\beta}{\cos\beta\,\sin\beta}}=\dfrac{1}{\dfrac{1}{\cos\beta\,\sin\beta}}=\sin\beta\,\cos\beta$

39 $\sec\theta+\csc\theta-\cos\theta-\sin\theta\ =\dfrac{1}{\cos\theta}+\dfrac{1}{\sin\theta}-\cos\theta-\sin\theta$

$$=\left(\dfrac{1}{\cos\theta}-\cos\theta\right)+\left(\dfrac{1}{\sin\theta}-\sin\theta\right)$$

$$=\dfrac{1-\cos^2\theta}{\cos\theta}+\dfrac{1-\sin^2\theta}{\sin\theta}$$

$$=\dfrac{\sin^2\theta}{\cos\theta}+\dfrac{\cos^2\theta}{\sin\theta}$$

$$=\left(\sin\theta\cdot\dfrac{\sin\theta}{\cos\theta}\right)+\left(\cos\theta\cdot\dfrac{\cos\theta}{\sin\theta}\right)=\sin\theta\,\tan\theta+\cos\theta\,\cot\theta$$

41 $\mathrm{RS}=\sec^4\phi-4\tan^2\phi=(\sec^2\phi)^2-4\tan^2\phi=(1+\tan^2\phi)^2-4\tan^2\phi=$

$$(1+2\tan^2\phi+\tan^4\phi)-4\tan^2\phi=1-2\tan^2\phi+\tan^4\phi=(1-\tan^2\phi)^2=\mathrm{LS}$$

43 $\dfrac{\cot(-t)+\tan(-t)}{\cot t}=\dfrac{-\cot t-\tan t}{\cot t}=-\dfrac{\cot t}{\cot t}-\dfrac{\tan t}{\cot t}=-(1+\tan^2 t)=-\sec^2 t$

45 $\log 10^{\tan t}=\log_{10}10^{\tan t}=\tan t$, since $\log_a a^x=x$

47 $\ln\cot x=\ln(\cot x)=\ln(\tan x)^{-1}=-\ln(\tan x)\ \{\text{since }\ln a^x=x\ln a\}=-\ln\tan x$

49 $\ln|\sec\theta+\tan\theta|\ =\ln\left|\dfrac{(\sec\theta+\tan\theta)(\sec\theta-\tan\theta)}{\sec\theta-\tan\theta}\right|$ $\{\text{multiply by the conjugate}\}$

$$=\ln\left|\dfrac{\sec^2\theta-\tan^2\theta}{\sec\theta-\tan\theta}\right|\qquad\qquad\{\text{simplify}\}$$

$$=\ln\left|\dfrac{1}{\sec\theta-\tan\theta}\right|\qquad\qquad\{\text{Pythagorean identity}\}$$

$$=\ln\dfrac{|1|}{|\sec\theta-\tan\theta|}\qquad\qquad\left\{\left|\dfrac{a}{b}\right|=\dfrac{|a|}{|b|}\right\}$$

$$=\ln|1|-\ln|\sec\theta-\tan\theta|\qquad\left\{\ln\dfrac{a}{b}=\ln a-\ln b\right\}$$

$$=-\ln|\sec\theta-\tan\theta|\qquad\qquad\{\ln 1=0\}$$

51 $\cos^2 t=1-\sin^2 t\Rightarrow\cos t=\pm\sqrt{1-\sin^2 t}$. Since the given equation is

$\cos t=+\sqrt{1-\sin^2 t}$, we may choose any t such that $\cos t<0$.

Using $t=\pi$, $\mathrm{LS}=\cos\pi=-1$. $\mathrm{RS}=\sqrt{1-\sin^2\pi}=1$. Since $-1\neq 1$, $\mathrm{LS}\neq\mathrm{RS}$.

53 $\sqrt{\sin^2 t} = |\sin t| = \pm \sin t$. Hence, choose any t such that $\sin t < 0$.

Using $t = \frac{3\pi}{2}$, LS $= \sqrt{(-1)^2} = 1$. RS $= \sin \frac{3\pi}{2} = -1$. Since $1 \neq -1$, LS $\neq$ RS.

55 $(\sin \theta + \cos \theta)^2 = \sin^2 \theta + 2 \sin \theta \cos \theta + \cos^2 \theta$. Since the right side of the given equation is only $\sin^2 \theta + \cos^2 \theta$, we may choose any θ such that $2 \sin \theta \cos \theta \neq 0$.

Using $\theta = \frac{\pi}{4}$, LS $= (\frac{1}{2}\sqrt{2} + \frac{1}{2}\sqrt{2})^2 = (\sqrt{2})^2 = 2$.

RS $= (\frac{1}{2}\sqrt{2})^2 + (\frac{1}{2}\sqrt{2})^2 = \frac{1}{2} + \frac{1}{2} = 1$. Since $2 \neq 1$, LS $\neq$ RS.

57 $\cos(-t) = -\cos t$ • Since $\cos(-t) = \cos t$, we may choose any t such that $\cos t \neq -\cos t$—that is, any t such that $\cos t \neq 0$. Using $t = \pi$, LS $= \cos(-\pi) = -1$.

RS $= -\cos \pi = -(-1) = 1$. Since $-1 \neq 1$, LS $\neq$ RS.

59 Don't confuse $\cos(\sec t) = 1$ with $\cos t \cdot \sec t = 1$. The former is true if $\sec t = 2\pi$ or an integer multiple of 2π. The latter is true for any value of t as long as $\sec t$ is defined. Choose any t such that $\sec t \neq 2\pi n$.

Using $t = \frac{\pi}{4}$, LS $= \cos(\sec \frac{\pi}{4}) = \cos \sqrt{2} \neq 1 =$ RS.

61 $\sin^2 t - 4 \sin t - 5 = 0 \Rightarrow (\sin t - 5)(\sin t + 1) = 0 \Rightarrow \sin t = 5$ or $\sin t = -1$.

Since $\sin t$ cannot equal 5, we may choose any t such that $\sin t \neq -1$.

Using $t = \pi$, LS $= -5 \neq 0 =$ RS.

Note: Exer. 63–66: Use $\sqrt{a^2 - x^2} = a \cos \theta$ because

$$\sqrt{a^2 - x^2} = \sqrt{a^2 - a^2 \sin^2 \theta} = \sqrt{a^2(1 - \sin^2 \theta)} = \sqrt{a^2 \cos^2 \theta} = |a| \, |\cos \theta| = a \cos \theta$$

since $\cos \theta > 0$ if $-\frac{\pi}{2} < \theta < \frac{\pi}{2}$ and $a > 0$.

63 $(a^2 - x^2)^{3/2} = (\sqrt{a^2 - x^2})^3 = (a \cos \theta)^3$ {see above note} $= a^3 \cos^3 \theta$

65 $\dfrac{x^2}{\sqrt{a^2 - x^2}} = \dfrac{a^2 \sin^2 \theta}{a \cos \theta} = a \cdot \dfrac{\sin \theta}{\cos \theta} \cdot \sin \theta = a \tan \theta \sin \theta$

Note: Exer. 67–70: Use $\sqrt{a^2 + x^2} = a \sec \theta$ because

$$\sqrt{a^2 + x^2} = \sqrt{a^2 + a^2 \tan^2 \theta} = \sqrt{a^2(1 + \tan^2 \theta)} = \sqrt{a^2 \sec^2 \theta} = |a| \, |\sec \theta| =$$

$a \sec \theta$ since $\sec \theta > 0$ if $-\frac{\pi}{2} < \theta < \frac{\pi}{2}$ and $a > 0$.

67 $\sqrt{a^2 + x^2} = a \sec \theta$ {see above note}

69 $\dfrac{1}{x^2 + a^2} = \dfrac{1}{(\sqrt{a^2 + x^2})^2} = \dfrac{1}{(a \sec \theta)^2} = \dfrac{1}{a^2 \sec^2 \theta} = \dfrac{1}{a^2} \cos^2 \theta$

Note: Exer. 71–74: Use $\sqrt{x^2 - a^2} = a \tan \theta$ because

$$\sqrt{x^2 - a^2} = \sqrt{a^2 \sec^2 \theta - a^2} = \sqrt{a^2(\sec^2 \theta - 1)} = \sqrt{a^2 \tan^2 \theta} = |a| \, |\tan \theta| =$$

$a \tan \theta$ since $\tan \theta > 0$ if $0 < \theta < \frac{\pi}{2}$ and $a > 0$.

71 $\sqrt{x^2 - a^2} = a \tan \theta$ {see above note}

73 $x^3 \sqrt{x^2 - a^2} = (a^3 \sec^3 \theta)(a \tan \theta) = a^4 \sec^3 \theta \tan \theta$

75 The graph of $f(x) = \dfrac{\sin^2 x - \sin^4 x}{(1 - \sec^2 x)\cos^4 x}$ appears to be that of the horizontal line

$y = g(x) = -1$. Verifying this identity, we have

$$\frac{\sin^2 x - \sin^4 x}{(1 - \sec^2 x)\cos^4 x} = \frac{\sin^2 x(1 - \sin^2 x)}{-\tan^2 x \cos^4 x} = \frac{\sin^2 x \cos^2 x}{-(\sin^2 x/\cos^2 x)\cos^4 x} = \frac{\sin^2 x \cos^2 x}{-\sin^2 x \cos^2 x} = -1.$$

77 The graph of $f(x) = \sec x(\sin x \cos x + \cos^2 x) - \sin x$ appears to be that of

$y = g(x) = \cos x$. Verifying this identity, we have

$$\sec x(\sin x \cos x + \cos^2 x) - \sin x = \sec x \cos x(\sin x + \cos x) - \sin x$$

$$= (\sin x + \cos x) - \sin x = \cos x.$$

3.2 Exercises

1 In $[0, 2\pi)$, $\sin x = -\dfrac{\sqrt{2}}{2}$ only if $x = \dfrac{5\pi}{4}, \dfrac{7\pi}{4}$. All solutions would include these angles plus all angles coterminal with them. Hence, $x = \dfrac{5\pi}{4} + 2\pi n, \dfrac{7\pi}{4} + 2\pi n$.

3 $\tan\theta = \sqrt{3} \Rightarrow \theta = \dfrac{\pi}{3} + \pi n$. *Note:* This solution could be written as $\dfrac{\pi}{3} + 2\pi n$ and $\dfrac{4\pi}{3} + 2\pi n$. Since the period of the tangent (and the cotangent) is π, we use the abbreviated form $\dfrac{\pi}{3} + \pi n$ to describe all solutions. We will use "πn" for solutions of exercises involving the tangent and cotangent functions.

7 $\sin x = \dfrac{\pi}{2}$ has no solution since $\dfrac{\pi}{2} > 1$, which is not in the range $[-1, 1]$. Your calculator will give some kind of error message if you attempt to find a solution to this problem or any similar type problem.

9 $\cos\theta = \dfrac{1}{\sec\theta}$ is true for all values *for which the equation is defined.*

★ All θ except $\theta = \dfrac{\pi}{2} + \pi n$

11 $2\cos 2\theta - \sqrt{3} = 0 \Rightarrow \cos 2\theta = \dfrac{\sqrt{3}}{2}$ { 2θ is just an angle — so we solve this equation for 2θ and then divide those solutions by 2 } $\Rightarrow$

$$2\theta = \tfrac{\pi}{6} + 2\pi n, \tfrac{11\pi}{6} + 2\pi n \Rightarrow \theta = \tfrac{\pi}{12} + \pi n, \tfrac{11\pi}{12} + \pi n$$

13 $\sqrt{3}\tan\tfrac{1}{3}t = 1 \Rightarrow \tan\tfrac{1}{3}t = \dfrac{1}{\sqrt{3}} \Rightarrow \tfrac{1}{3}t = \tfrac{\pi}{6} + \pi n \Rightarrow t = \tfrac{\pi}{2} + 3\pi n$

15 $\sin\left(\theta + \tfrac{\pi}{4}\right) = \tfrac{1}{2} \Rightarrow \theta + \tfrac{\pi}{4} = \tfrac{\pi}{6} + 2\pi n, \tfrac{5\pi}{6} + 2\pi n \Rightarrow \theta = -\tfrac{\pi}{12} + 2\pi n, \tfrac{7\pi}{12} + 2\pi n$

17 $\sin\left(2x - \tfrac{\pi}{3}\right) = \tfrac{1}{2} \Rightarrow 2x - \tfrac{\pi}{3} = \tfrac{\pi}{6} + 2\pi n, \tfrac{5\pi}{6} + 2\pi n \Rightarrow 2x = \tfrac{\pi}{2} + 2\pi n, \tfrac{7\pi}{6} + 2\pi n \Rightarrow$

$$x = \tfrac{\pi}{4} + \pi n, \tfrac{7\pi}{12} + \pi n$$

21 $\tan^2 x = 1 \Rightarrow \tan x = \pm 1 \Rightarrow x = \tfrac{\pi}{4} + \pi n, \tfrac{3\pi}{4} + \pi n$, or simply $\tfrac{\pi}{4} + \tfrac{\pi}{2}n$

23 $(\cos\theta - 1)(\sin\theta + 1) = 0 \Rightarrow (\cos\theta - 1) = 0$ or $(\sin\theta + 1) = 0 \Rightarrow$

$$\cos\theta = 1 \text{ or } \sin\theta = -1 \Rightarrow \theta = 2\pi n \text{ or } \theta = \tfrac{3\pi}{2} + 2\pi n$$

[25] $\sec^2\alpha - 4 = 0 \Rightarrow \sec^2\alpha = 4 \Rightarrow \sec\alpha = \pm 2 \Rightarrow$

$\alpha = \frac{\pi}{3} + 2\pi n, \frac{5\pi}{3} + 2\pi n, \frac{2\pi}{3} + 2\pi n, \frac{4\pi}{3} + 2\pi n$, or simply $\frac{\pi}{3} + \pi n, \frac{2\pi}{3} + \pi n$

[29] $\cot^2 x - 3 = 0 \Rightarrow \cot^2 x = 3 \Rightarrow \cot x = \pm\sqrt{3} \Rightarrow x = \frac{\pi}{6} + \pi n, \frac{5\pi}{6} + \pi n$

[31] $(2\sin\theta + 1)(2\cos\theta + 3) = 0 \Rightarrow \sin\theta = -\frac{1}{2}$ or $\sin\theta = -\frac{3}{2} \Rightarrow$

$\theta = \frac{7\pi}{6} + 2\pi n, \frac{11\pi}{6} + 2\pi n$ $\left\{ \sin\theta = -\frac{3}{2} \text{ has no solutions} \right\}$

[33] $\sin 2x (\csc 2x - 2) = 0 \Rightarrow 1 - 2\sin 2x = 0 \Rightarrow \sin 2x = \frac{1}{2} \Rightarrow$

$2x = \frac{\pi}{6} + 2\pi n, \frac{5\pi}{6} + 2\pi n \Rightarrow x = \frac{\pi}{12} + \pi n, \frac{5\pi}{12} + \pi n$

[35] $\cos(\ln x) = 0 \Rightarrow \ln x = \frac{\pi}{2} + \pi n \Rightarrow x = e^{(\pi/2) + \pi n}$ $\left\{ \text{since } \ln x = y \Leftrightarrow x = e^y \right\}$

[37] $\cos\left(2x - \frac{\pi}{4}\right) = 0 \Rightarrow 2x - \frac{\pi}{4} = \frac{\pi}{2} + \pi n \Rightarrow 2x = \frac{3\pi}{4} + \pi n \Rightarrow x = \frac{3\pi}{8} + \frac{\pi}{2}n.$

x will be in the interval $[0, 2\pi)$ if $n = 0, 1, 2,$ or 3. Thus, $x = \frac{3\pi}{8}, \frac{7\pi}{8}, \frac{11\pi}{8}, \frac{15\pi}{8}$.

[39] $2 - 8\cos^2 t = 0 \Rightarrow \cos^2 t = \frac{1}{4} \Rightarrow \cos t = \pm\frac{1}{2} \Rightarrow t = \frac{\pi}{3}, \frac{2\pi}{3}, \frac{4\pi}{3}, \frac{5\pi}{3}$

[41] $2\sin^2 u = 1 - \sin u \Rightarrow 2\sin^2 u + \sin u - 1 = 0 \Rightarrow (2\sin u - 1)(\sin u + 1) = 0 \Rightarrow$

$\sin u = \frac{1}{2}, -1 \Rightarrow u = \frac{\pi}{6}, \frac{5\pi}{6}, \frac{3\pi}{2}$

[43] $\tan^2 x \sin x = \sin x \Rightarrow \tan^2 x \sin x - \sin x = 0 \Rightarrow \sin x (\tan^2 x - 1) = 0 \Rightarrow \sin x = 0$ or

$\tan x = \pm 1 \Rightarrow x = 0, \pi, \frac{\pi}{4}, \frac{3\pi}{4}, \frac{5\pi}{4}, \frac{7\pi}{4}$. *Note:* A common mistake is to divide both sides of the given equation by $\sin x$—doing so results in losing the solutions for $\sin x = 0$.

[45] $2\cos^2\gamma + \cos\gamma = 0 \Rightarrow \cos\gamma (2\cos\gamma + 1) = 0 \Rightarrow \cos\gamma = 0, -\frac{1}{2} \Rightarrow \gamma = \frac{\pi}{2}, \frac{3\pi}{2}, \frac{2\pi}{3}, \frac{4\pi}{3}$

[47] $\sin^2\theta + \sin\theta - 6 = 0 \Rightarrow (\sin\theta + 3)(\sin\theta - 2) = 0 \Rightarrow \sin\theta = -3, 2.$

There are *no solutions* for either equation.

[49] $1 - \sin t = \sqrt{3}\cos t$ •   Square both sides to obtain an equation in either sin or cos.

$(1 - \sin t)^2 = (\sqrt{3}\cos t)^2 \Rightarrow 1 - 2\sin t + \sin^2 t = 3\cos^2 t \Rightarrow$

$\sin^2 t - 2\sin t + 1 = 3(1 - \sin^2 t) \Rightarrow 4\sin^2 t - 2\sin t - 2 = 0 \Rightarrow 2\sin^2 t - \sin t - 1 = 0 \Rightarrow$

$(2\sin t + 1)(\sin t - 1) = 0 \Rightarrow \sin t = -\frac{1}{2}, 1 \Rightarrow t = \frac{7\pi}{6}, \frac{11\pi}{6}, \frac{\pi}{2}$. Since each side of the equation was squared, the solutions must be checked in the original equation. Checking $\frac{7\pi}{6}$, we have $\text{LS} = 1 - \sin\frac{7\pi}{6} = 1 - (-\frac{1}{2}) = \frac{3}{2}$ and $\text{RS} = \sqrt{3}\cos\frac{7\pi}{6} = \sqrt{3}\left(-\frac{\sqrt{3}}{2}\right) = -\frac{3}{2}$. Since $\text{LS} \neq \text{RS}$, $\frac{7\pi}{6}$ is an extraneous solution. Checking $\frac{11\pi}{6}$, we have $\text{LS} = \frac{3}{2}$ and $\text{RS} = \frac{3}{2}$. Since $\text{LS} = \text{RS}$, $\frac{11\pi}{6}$ is a valid solution. Similarly, $\frac{\pi}{2}$ is a valid solution and our solution is $\frac{\pi}{2}$ and $\frac{11\pi}{6}$.

[51] $\cos\alpha + \sin\alpha = 1 \Rightarrow \cos\alpha = 1 - \sin\alpha$ {square both sides} $\Rightarrow$

$\cos^2\alpha = 1 - 2\sin\alpha + \sin^2\alpha$

{change $\cos^2\alpha$ to $1 - \sin^2\alpha$ to obtain an equation involving only $\sin\alpha$} $\Rightarrow$

$1 - \sin^2\alpha = 1 - 2\sin\alpha + \sin^2\alpha \Rightarrow 2\sin^2\alpha - 2\sin\alpha = 0 \Rightarrow$

$2\sin\alpha (\sin\alpha - 1) = 0 \Rightarrow \sin\alpha = 0, 1 \Rightarrow \alpha = 0, \pi, \frac{\pi}{2}$. π is an extraneous solution.

[53] $2\tan t - \sec^2 t = 0 \Rightarrow 2\tan t - (1 + \tan^2 t) = 0 \Rightarrow \tan^2 t - 2\tan t + 1 = 0 \Rightarrow$

$$(\tan t - 1)^2 = 0 \Rightarrow \tan t = 1 \Rightarrow t = \tfrac{\pi}{4}, \tfrac{5\pi}{4}$$

[55] $\cot \alpha + \tan \alpha = \csc \alpha \sec \alpha \Rightarrow \dfrac{\cos \alpha}{\sin \alpha} + \dfrac{\sin \alpha}{\cos \alpha} = \dfrac{1}{\sin \alpha \cos \alpha} \Rightarrow$

$\dfrac{\cos^2 \alpha + \sin^2 \alpha}{\sin \alpha \cos \alpha} = \dfrac{1}{\sin \alpha \cos \alpha}$. This is an identity and is true for *all numbers in* $[0, 2\pi)$

except $0, \tfrac{\pi}{2}, \pi$, and $\tfrac{3\pi}{2}$ since these values make the original equation undefined.

[57] $2\sin^3 x + \sin^2 x - 2\sin x - 1 = 0$ { factor by grouping since there are four terms } $\Rightarrow$

$\sin^2 x \, (2\sin x + 1) - 1(2\sin x + 1) = 0 \Rightarrow$

$$(\sin^2 x - 1)(2\sin x + 1) = 0 \Rightarrow \sin x = \pm 1, -\tfrac{1}{2} \Rightarrow x = \tfrac{\pi}{2}, \tfrac{3\pi}{2}, \tfrac{7\pi}{6}, \tfrac{11\pi}{6}$$

[59] $2\tan t \csc t + 2\csc t + \tan t + 1 = 0 \Rightarrow 2\csc t \, (\tan t + 1) + 1(\tan t + 1) \Rightarrow$

$(2\csc t + 1)(\tan t + 1) = 0 \Rightarrow \csc t = -\tfrac{1}{2}$ or $\tan t = -1 \Rightarrow t = \tfrac{3\pi}{4}, \tfrac{7\pi}{4}$ { since $\csc t \neq -\tfrac{1}{2}$ }

[61] $\sin^2 t - 4\sin t + 1 = 0 \Rightarrow \sin t = \dfrac{4 \pm \sqrt{12}}{2} = 2 \pm \sqrt{3}$.

$(2 + \sqrt{3}) > 1$ is not in the range of the sine, so $\sin t = 2 - \sqrt{3} \Rightarrow$

$$t = 15°30' \text{ or } 164°30' \, \{ \text{to the nearest ten minutes} \}$$

[63] $\tan^2 \theta + 3\tan \theta + 2 = 0 \Rightarrow (\tan \theta + 1)(\tan \theta + 2) = 0 \Rightarrow$

$$\tan \theta = -1, -2 \Rightarrow \theta = 135°, 315°, 116°30', 296°30'$$

[65] $12\sin^2 u - 5\sin u - 2 = 0 \Rightarrow (3\sin u - 2)(4\sin u + 1) = 0 \Rightarrow \sin u = \tfrac{2}{3}, -\tfrac{1}{4} \Rightarrow$

$$u = 41°50', 138°10', 194°30', 345°30'$$

[67] The top of the wave will be above the sea wall when its height is greater than 12.5.

$y > 12.5 \Rightarrow 25\cos \tfrac{\pi}{15}t > 12.5 \Rightarrow \cos \tfrac{\pi}{15}t > \tfrac{1}{2} \Rightarrow$

{ To visualize this step, it may help to look at a unit circle and draw a vertical line

through 0.5 on the x-axis — $x > 0.5$ is the same as $\cos \tfrac{\pi}{15}t > 0.5$. }

$-\tfrac{\pi}{3} < \tfrac{\pi}{15}t < \tfrac{\pi}{3} \Rightarrow -5 < t < 5$ { multiply by $\tfrac{15}{\pi}$ } $\Rightarrow$

$$y > 12.5 \text{ for about } 5 - (-5) = 10 \text{ minutes of each 30-minute period.}$$

[69] (a)   [1, 25] by [0, 100]

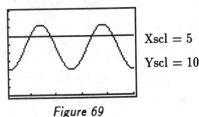

Xscl = 5

Yscl = 10

Figure 69

(b) July: $T(7) = 83°F$; October: $T(10) = 56.5°F$.

(c) Graph $Y_1 = 26.5\sin\left(\tfrac{\pi}{6}x - \tfrac{2\pi}{3}\right) + 56.5$ and $Y_2 = 69$. Their graphs intersect at $t \approx 4.94, 9.06$ on $[1, 13]$. The average high temperature is above $69°F$ approximately May through September.

(continued)

(d) A sine function is periodic and varies between a maximum and minimum value. Average monthly high temperatures are also seasonal with a 12-month period. Therefore, a sine function is a reasonable function to model these temperatures.

$\boxed{71}$ $I = \frac{1}{2}I_M$ and $D = 12 \Rightarrow \frac{1}{2}I_M = I_M \sin^3\frac{\pi}{12}t \Rightarrow \sin^3\frac{\pi}{12}t = \frac{1}{2} \Rightarrow \sin\frac{\pi}{12}t = \sqrt[3]{\frac{1}{2}} \Rightarrow$

$\frac{\pi}{12}t \approx 0.9169$ and 2.2247 { $\pi - 0.9169 \approx 2.2247$ is the reference angle for 0.9169 in

QII. } $\Rightarrow t \approx 3.50$ and $t \approx 8.50$

$\boxed{73}$ 75% of the maximum intensity $= 0.75\,I_M$

(a) $I > 0.75\,I_M \Rightarrow I_M \sin^3\frac{\pi}{12}t > 0.75\,I_M \Rightarrow \sin^3\frac{\pi}{12}t > \frac{3}{4} \Rightarrow \sin\frac{\pi}{12}t > \sqrt[3]{\frac{3}{4}} \Rightarrow$

$1.1398 < \frac{\pi}{12}t < 2.0018 \Rightarrow 4.3538 < t < 7.6462$, or approximately 3.29 hours.

(b) $I > 0.75\,I_M \Rightarrow I_M \sin^2\frac{\pi}{12}t > 0.75\,I_M \Rightarrow \sin^2\frac{\pi}{12}t > \frac{3}{4} \Rightarrow \sin\frac{\pi}{12}t > \frac{1}{2}\sqrt{3} \Rightarrow$

$\frac{\pi}{3} < \frac{\pi}{12}t < \frac{2\pi}{3} \Rightarrow 4 < t < 8$, or 4 hours.

$\boxed{75}$ (a) $N(t) = 1000\cos\frac{\pi}{5}t + 4000$, amplitude $= 1000$,

period $= \frac{2\pi}{\pi/5} = 10$ years

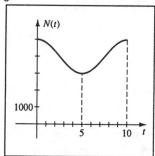

Figure 75

(b) $N > 4500 \Rightarrow 1000\cos\frac{\pi}{5}t + 4000 > 4500 \Rightarrow$

$\cos\frac{\pi}{5}t > \frac{1}{2} \Rightarrow$ { The cosine function is greater

than $\frac{1}{2}$ on $[0, \frac{\pi}{3})$ and $(\frac{5\pi}{3}, 2\pi]$. }

$0 \le \frac{\pi}{5}t < \frac{\pi}{3}$ and $\frac{5\pi}{3} < \frac{\pi}{5}t \le 2\pi \Rightarrow$

$0 \le t < \frac{5}{3}$ and $\frac{25}{3} < t \le 10$

$\boxed{77}$ $\frac{1}{2} + \cos x = 0 \Rightarrow \cos x = -\frac{1}{2} \Rightarrow x = -\frac{4\pi}{3}, -\frac{2\pi}{3}, \frac{2\pi}{3}$, and $\frac{4\pi}{3}$ { for x in $[-2\pi, 2\pi]$ }

for A, B, C, and D, respectively. The corresponding y values are found by using

$y = \frac{1}{2}x + \sin x$ with each of the above values. The points are:

$A(-\frac{4\pi}{3}, -\frac{2\pi}{3} + \frac{1}{2}\sqrt{3})$, $B(-\frac{2\pi}{3}, -\frac{\pi}{3} - \frac{1}{2}\sqrt{3})$, $C(\frac{2\pi}{3}, \frac{\pi}{3} + \frac{1}{2}\sqrt{3})$, and $D(\frac{4\pi}{3}, \frac{2\pi}{3} - \frac{1}{2}\sqrt{3})$

$\boxed{79}$ $I(t) = k \Rightarrow 20\sin(60\pi t - 6\pi) = -10 \Rightarrow \sin(60\pi t - 6\pi) = -\frac{1}{2} \Rightarrow 60\pi t_1 - 6\pi = \frac{7\pi}{6} + 2\pi n$

or $60\pi t_2 - 6\pi = \frac{11\pi}{6} + 2\pi n$ { We use t_1 and t_2 to distinguish between angles coterminal with $\frac{7\pi}{6}$ and those coterminal with $\frac{11\pi}{6}$. } $\Rightarrow 60t_1 = \frac{43}{6} + 2n$ or $60t_2 = \frac{47}{6} + 2n \Rightarrow t_1 = \frac{43}{360} + \frac{1}{30}n$ or $t_2 = \frac{47}{360} + \frac{1}{30}n$. We must now find the smallest positive value of t_1 or t_2. $t_1 > 0 \Rightarrow \frac{1}{30}n > -\frac{43}{360} \Rightarrow n > -\frac{43}{12} \approx -3.58$. The last result indicates that n must be an integer in the set $\{-3, -2, -1, \ldots\}$. If $n = -3$, then $t_1 = \frac{7}{360}$. Greater values of n yield greater values of t_1. Similarly, for t_2, $t_2 > 0 \Rightarrow \frac{1}{30}n > -\frac{47}{360} \Rightarrow n > -\frac{47}{12}$. If $n = -3$, then $t_1 = \frac{11}{360}$. Thus, the smallest exact value of t for which $I(t) = -10$ is $t = \frac{7}{360}$ sec.

$\boxed{81}$ Graph $y = \cos x$ and $y = 0.3$ on the same coordinate plane. See *Figure 81* on the next page. The points of intersection are located at $x \approx 1.27$, 5.02, and $\cos x$ is less than 0.3 between these values. Therefore, $\cos x \ge 0.3$ on $[0, 1.27] \cup [5.02, 2\pi]$.

$[0, 2\pi]$ by $[-2.09, 2.09]$

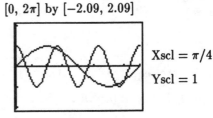

$\text{Xscl} = \pi/4$
$\text{Yscl} = 1$

$[0, 2\pi]$ by $[-2.09, 2.09]$

$\text{Xscl} = \pi/4$
$\text{Yscl} = 1$

Figure 81   Figure 83

83 Graph $y = \cos 3x$ and $y = \sin x$ on the same coordinate plane.

The points of intersection are located at $x \approx 0.39, 1.96, 2.36, 3.53, 5.11, 5.50$.

From the graph, we see that $\cos 3x$ is less than $\sin x$ on

$$(0.39, 1.96) \cup (2.36, 3.53) \cup (5.11, 5.50).$$

85 (a) The largest zero occurs when $x \approx 0.6366$.

(b) As x becomes large, the graph of $f(x) = \cos(1/x)$ approaches the horizontal asymptote $y = 1$.

(c) There appears to be an infinite number of zeros on $[0, c]$ for any $c > 0$.

$[0, 3]$ by $[-1.5, 1.5]$

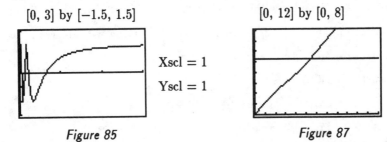

$\text{Xscl} = 1$
$\text{Yscl} = 1$

$[0, 12]$ by $[0, 8]$

$\text{Xscl} = 1$
$\text{Yscl} = 1$

Figure 85   Figure 87

Note: Exer. 87–90: Graph $Y_1 = M$ and $Y_2 = \theta + e\sin\theta$ and approximate the value of θ such that $Y_1 = Y_2$.

87 Mercury: $Y_1 = 5.241$ and $Y_2 = \theta + 0.206\sin\theta$ intersect when $\theta \approx 5.400$ (radians).

91 Graph $y = \sin 2x$ and $y = 2 - x^2$. From the graph, we see that there are two points of intersection. The x-coordinates of these points are $x \approx -1.48, 1.08$.

$[-\pi, \pi]$ by $[-2.09, 2.09]$

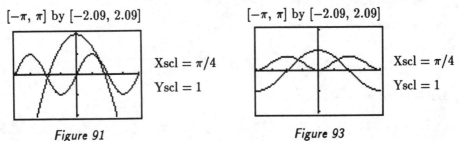

$\text{Xscl} = \pi/4$
$\text{Yscl} = 1$

$[-\pi, \pi]$ by $[-2.09, 2.09]$

$\text{Xscl} = \pi/4$
$\text{Yscl} = 1$

Figure 91   Figure 93

93 Graph $y = \ln(1 + \sin^2 x)$ and $y = \cos x$. From the graph, we see that there are two points of intersection. The x-coordinates of these points are $x \approx \pm 1.00$.

97 (a) $g = 9.8 \Rightarrow 9.8 = 9.8066(1 - 0.00264\cos 2\theta) \Rightarrow 0.00264\cos 2\theta = 1 - \frac{9.8}{9.8066} \Rightarrow$

$$\cos 2\theta = \frac{0.0066}{(9.8066)(0.00264)} \Rightarrow 2\theta \approx 75.2° \Rightarrow \theta \approx 37.6°$$

(b) At the equator, $g_0 = 9.8066(1 - 0.00264\cos 0°) = 9.8066(0.99736)$.   Since the weight W of a person on the earth's surface is directly proportional to the force of gravity, we have $W = kg$.

At $\theta = 0°$, $W = kg \Rightarrow 150 = kg_0 \Rightarrow k = \frac{150}{g_0} \Rightarrow W = \frac{150}{g_0}g$.

$W = 150.5 \Rightarrow 150.5 = \frac{150}{g_0}g \Rightarrow 150.5 = \frac{150 \cdot 9.8066(1 - 0.00264\cos 2\theta)}{9.8066(0.99736)} \Rightarrow$

$150.5 = \frac{150}{0.99736}(1 - 0.00264\cos 2\theta) \Rightarrow 0.00264\cos 2\theta = 1 - \frac{150.5(0.99736)}{150} \Rightarrow$

$\cos 2\theta \approx -0.2593 \Rightarrow 2\theta \approx 105.0° \Rightarrow \theta \approx 52.5°$.

3.3 Exercises

1 *Note:* Use the cofunction formulas with $\left(\frac{\pi}{2} - u\right)$ for the argument if you're working with radian measure, $(90° - u)$ if you're working with degree measure.

(a) $\sin 46°37' = \cos(90° - 46°37') = \cos 43°23'$

(b) $\cos 73°12' = \sin(90° - 73°12') = \sin 16°48'$

(c) $\tan\frac{\pi}{6} = \cot\left(\frac{\pi}{2} - \frac{\pi}{6}\right) = \cot\left(\frac{3\pi}{6} - \frac{\pi}{6}\right) = \cot\frac{2\pi}{6} = \cot\frac{\pi}{3}$

(d) $\sec 17.28° = \csc(90° - 17.28°) = \csc 72.72°$

3 (a) $\cos\frac{7\pi}{20} = \sin\left(\frac{\pi}{2} - \frac{7\pi}{20}\right) = \sin\frac{3\pi}{20}$ (b) $\sin\frac{1}{4} = \cos\left(\frac{\pi}{2} - \frac{1}{4}\right) = \cos\left(\frac{2\pi - 1}{4}\right)$

(c) $\tan 1 = \cot\left(\frac{\pi}{2} - 1\right) = \cot\left(\frac{\pi - 2}{2}\right)$ (d) $\csc 0.53 = \sec\left(\frac{\pi}{2} - 0.53\right)$

5 (a) $\cos\frac{\pi}{4} + \cos\frac{\pi}{6} = \frac{\sqrt{2}}{2} + \frac{\sqrt{3}}{2} = \frac{\sqrt{2} + \sqrt{3}}{2}$

(b) $\cos\frac{5\pi}{12} = \cos\left(\frac{\pi}{4} + \frac{\pi}{6}\right) = \cos\frac{\pi}{4}\cos\frac{\pi}{6} - \sin\frac{\pi}{4}\sin\frac{\pi}{6} = \frac{\sqrt{2}}{2} \cdot \frac{\sqrt{3}}{2} - \frac{\sqrt{2}}{2} \cdot \frac{1}{2} = \frac{\sqrt{6} - \sqrt{2}}{4}$

7 (a) $\tan 60° + \tan 225° = \sqrt{3} + 1$

(b) $\tan 285° = \tan(60° + 225°) =$

$$\frac{\tan 60° + \tan 225°}{1 - \tan 60° \tan 225°} = \frac{\sqrt{3} + 1}{1 - (\sqrt{3})(1)} \cdot \frac{1 + \sqrt{3}}{1 + \sqrt{3}} = \frac{4 + 2\sqrt{3}}{-2} = -2 - \sqrt{3}$$

9 (a) $\sin\frac{3\pi}{4} - \sin\frac{\pi}{6} = \frac{\sqrt{2}}{2} - \frac{1}{2} = \frac{\sqrt{2} - 1}{2}$

(b) $\sin\frac{7\pi}{12} = \sin\left(\frac{3\pi}{4} - \frac{\pi}{6}\right) = \sin\frac{3\pi}{4}\cos\frac{\pi}{6} - \cos\frac{3\pi}{4}\sin\frac{\pi}{6} = \frac{\sqrt{2}}{2} \cdot \frac{\sqrt{3}}{2} - \left(-\frac{\sqrt{2}}{2}\right) \cdot \frac{1}{2} = \frac{\sqrt{6} + \sqrt{2}}{4}$

11 Since the expression is of the form "cos cos plus sin sin", we recognize it as the subtraction formula for the cosine.

$$\cos 48° \cos 23° + \sin 48° \sin 23° = \cos(48° - 23°) = \cos 25°$$

$\boxed{13}$ $\cos 10° \sin 5° - \sin 10° \cos 5° = \sin(5° - 10°) = \sin(-5°)$

$\boxed{15}$ Since we have angle arguments of 2, 3, and -2, we want to change one of them so that we can apply one of the formulas. We recognize that $\cos 2 = \cos(-2)$ and this is probably the simplest change. $\cos 3 \sin(-2) - \cos 2 \sin 3 =$

$$\sin(-2) \cos 3 - \cos(-2) \sin 3 = \sin(-2 - 3) = \sin(-5)$$

$\boxed{17}$ See *Figure 17* for a drawing of angles α and β.

    (a) $\sin(\alpha + \beta) = \sin\alpha \cos\beta + \cos\alpha \sin\beta = \frac{3}{5} \cdot \frac{15}{17} + \frac{4}{5} \cdot \frac{8}{17} = \frac{77}{85}$

    (b) $\cos(\alpha + \beta) = \cos\alpha \cos\beta - \sin\alpha \sin\beta = \frac{4}{5} \cdot \frac{15}{17} - \frac{3}{5} \cdot \frac{8}{17} = \frac{36}{85}$

    (c) Since the sine and cosine of $(\alpha + \beta)$ are positive, $(\alpha + \beta)$ is in QI.

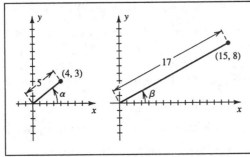

Figure 17   *Figure 19*

$\boxed{19}$ See *Figure 19* for a drawing of angles α and β.

    (a) $\sin(\alpha + \beta) = \sin\alpha \cos\beta + \cos\alpha \sin\beta = \left(-\frac{4}{5}\right) \cdot \frac{3}{5} + \left(-\frac{3}{5}\right) \cdot \frac{4}{5} = -\frac{24}{25}$

    (b) $\tan(\alpha + \beta) = \dfrac{\tan\alpha + \tan\beta}{1 - \tan\alpha \tan\beta} = \dfrac{\frac{4}{3} + \frac{4}{3}}{1 - \frac{4}{3} \cdot \frac{4}{3}} \cdot \frac{9}{9} = \dfrac{12 + 12}{9 - 16} = -\frac{24}{7}$

    (c) Since the sine and tangent of $(\alpha + \beta)$ are negative, $(\alpha + \beta)$ is in QIV.

$\boxed{21}$ (a) $\sin(\alpha - \beta) = \sin\alpha \cos\beta - \cos\alpha \sin\beta =$

$$\left(-\frac{\sqrt{21}}{5}\right) \cdot \left(-\frac{3}{5}\right) - \left(-\frac{2}{5}\right) \cdot \left(-\frac{4}{5}\right) = \frac{3\sqrt{21} - 8}{25} \approx 0.23$$

    (b) $\cos(\alpha - \beta) = \cos\alpha \cos\beta + \sin\alpha \sin\beta =$

$$\left(-\frac{2}{5}\right) \cdot \left(-\frac{3}{5}\right) + \left(-\frac{\sqrt{21}}{5}\right) \cdot \left(-\frac{4}{5}\right) = \frac{4\sqrt{21} + 6}{25} \approx 0.97$$

    (c) Since the sine and cosine of $(\alpha - \beta)$ are positive, $(\alpha - \beta)$ is in QI.

$\boxed{23}$ $\sin(\theta + \pi) = \sin\theta \cos\pi + \cos\theta \sin\pi = \sin\theta(-1) + \cos\theta(0) = -\sin\theta$

$\boxed{25}$ $\sin\left(x - \frac{5\pi}{2}\right) = \sin x \cos\frac{5\pi}{2} - \cos x \sin\frac{5\pi}{2} = \sin x(0) - \cos x(1) = -\cos x$

$\boxed{27}$ $\cos(\theta - \pi) = \cos\theta \cos\pi + \sin\theta \sin\pi = \cos\theta(-1) + \sin\theta(0) = -\cos\theta$

$\boxed{29}$ $\cos\left(x + \frac{3\pi}{2}\right) = \cos x \cos\frac{3\pi}{2} - \sin x \sin\frac{3\pi}{2} = \cos x(0) - \sin x(-1) = \sin x$

$\boxed{31}$ $\tan\left(x - \frac{\pi}{2}\right) = \dfrac{\sin\left(x - \frac{\pi}{2}\right)}{\cos\left(x - \frac{\pi}{2}\right)} = \dfrac{\sin x \cos\frac{\pi}{2} - \cos x \sin\frac{\pi}{2}}{\cos x \cos\frac{\pi}{2} + \sin x \sin\frac{\pi}{2}} = \dfrac{-\cos x}{\sin x} = -\cot x$

33 The tangent of a sum formula won't work here since $\tan\frac{\pi}{2}$ is undefined.

We will use a cofunction identity to verify the identity.

$$\tan\left(\theta+\tfrac{\pi}{2}\right)=\cot\left[\tfrac{\pi}{2}-\left(\theta+\tfrac{\pi}{2}\right)\right]=\cot\left(-\theta\right)=-\cot\theta$$

Alternatively, we could also write $\tan\left(\theta+\frac{\pi}{2}\right)$ as $\dfrac{\sin\left(\theta+\frac{\pi}{2}\right)}{\cos\left(\theta+\frac{\pi}{2}\right)}$ and then simplify.

37 $\tan\left(u+\tfrac{\pi}{4}\right)=\dfrac{\tan u+\tan\frac{\pi}{4}}{1-\tan u\tan\frac{\pi}{4}}=\dfrac{\tan u+1}{1-\tan u\,(1)}=\dfrac{1+\tan u}{1-\tan u}$

39 $\cos\left(u+v\right)+\cos\left(u-v\right)=\left(\cos u\cos v-\sin u\sin v\right)+\left(\cos u\cos v+\sin u\sin v\right)=$

$$2\cos u\cos v$$

41 $\sin\left(u+v\right)\cdot\sin\left(u-v\right)=\left(\sin u\cos v+\cos u\sin v\right)\cdot\left(\sin u\cos v-\cos u\sin v\right)$

$$\{\,\text{addition and subtraction formulas for the sine}\,\}$$

$$=\sin^2 u\cos^2 v-\cos^2 u\sin^2 v$$

$$\{\,\text{recognize as the difference of two squares}\,\}$$

$$=\sin^2 u\left(1-\sin^2 v\right)-\left(1-\sin^2 u\right)\sin^2 v$$

$$\{\,\text{change to terms only involving sine}\,\}$$

$$=\sin^2 u-\sin^2 u\sin^2 v-\sin^2 v+\sin^2 u\sin^2 v=\sin^2 u-\sin^2 v$$

43 $\dfrac{1}{\cot\alpha-\cot\beta}=\dfrac{1}{\dfrac{\cos\alpha}{\sin\alpha}-\dfrac{\cos\beta}{\sin\beta}}=\dfrac{1}{\dfrac{\cos\alpha\sin\beta-\cos\beta\sin\alpha}{\sin\alpha\sin\beta}}=\dfrac{\sin\alpha\sin\beta}{\sin\left(\beta-\alpha\right)}$

45 $\sin\left(u+v+w\right)=\sin\left[\left(u+v\right)+w\right]$

$$=\sin\left(u+v\right)\cos w+\cos\left(u+v\right)\sin w$$

$$=\left(\sin u\cos v+\cos u\sin v\right)\cos w+\left(\cos u\cos v-\sin u\sin v\right)\sin w$$

$$=\sin u\cos v\cos w+\cos u\sin v\cos w+\cos u\cos v\sin w-\sin u\sin v\sin w$$

47 The question usually asked here is "why divide by $\sin u\sin v$?" Since we know the form we want to end up with, we need to "force" the term "$-\sin u\sin v$" to equal "-1", hence divide all terms by "$\sin u\sin v$."

$$\cot\left(u+v\right)=\dfrac{\cos\left(u+v\right)}{\sin\left(u+v\right)}=\dfrac{\left(\cos u\cos v-\sin u\sin v\right)\left(1/\sin u\sin v\right)}{\left(\sin u\cos v+\cos u\sin v\right)\left(1/\sin u\sin v\right)}=\dfrac{\cot u\cot v-1}{\cot v+\cot u}$$

49 $\sin\left(u-v\right)=\sin\left[u+\left(-v\right)\right]=\sin u\cos\left(-v\right)+\cos u\sin\left(-v\right)=\sin u\cos v-\cos u\sin v$

51 $\dfrac{f(x+h)-f(x)}{h}=\dfrac{\cos\left(x+h\right)-\cos x}{h}=\dfrac{\cos x\cos h-\sin x\sin h-\cos x}{h}=$

$$\dfrac{\cos x\cos h-\cos x}{h}-\dfrac{\sin x\sin h}{h}=\cos x\left(\dfrac{\cos h-1}{h}\right)-\sin x\left(\dfrac{\sin h}{h}\right)$$

53 $\sin 4t\cos t=\sin t\cos 4t\;\Rightarrow\;\sin 4t\cos t-\sin t\cos 4t=0\;\Rightarrow\;\sin\left(4t-t\right)=0\;\Rightarrow$

$$\sin 3t=0\;\Rightarrow\;3t=\pi n\;\Rightarrow\;t=\tfrac{\pi}{3}n.\ \text{In}\ [0,\,\pi),\ t=0,\,\tfrac{\pi}{3},\,\tfrac{2\pi}{3}.$$

$\boxed{55}$ $\cos 5t \cos 2t = -\sin 5t \sin 2t \Rightarrow \cos 5t \cos 2t + \sin 5t \sin 2t = 0 \Rightarrow$

$\cos(5t - 2t) = 0 \Rightarrow \cos 3t = 0 \Rightarrow 3t = \frac{\pi}{2} + \pi n \Rightarrow t = \frac{\pi}{6} + \frac{\pi}{3}n$. In $[0, \pi)$, $t = \frac{\pi}{6}, \frac{\pi}{2}, \frac{5\pi}{6}$.

$\boxed{57}$ $\tan 2t + \tan t = 1 - \tan 2t \tan t \Rightarrow \dfrac{\tan 2t + \tan t}{1 - \tan 2t \tan t} = 1 \Rightarrow \tan(2t + t) = 1 \Rightarrow$

$\tan 3t = 1 \Rightarrow 3t = \frac{\pi}{4} + \pi n \Rightarrow t = \frac{\pi}{12} + \frac{\pi}{3}n$. In $[0, \pi)$, $t = \frac{\pi}{12}, \frac{5\pi}{12}, \frac{3\pi}{4}$.

However, $\tan 2t$ is undefined if $t = \frac{3\pi}{4}$, so exclude this value of t.

$\boxed{59}$ (a) $f(x) = \sqrt{3}\cos 2x + \sin 2x$ • $A = \sqrt{(\sqrt{3})^2 + 1^2} = 2$. $\tan C = \frac{1}{\sqrt{3}} \Rightarrow C = \frac{\pi}{6}$.

$$f(x) = 2\cos\left(2x - \frac{\pi}{6}\right) = 2\cos\left[2\left(x - \frac{\pi}{12}\right)\right]$$

(b) amplitude $= 2$, period $= \frac{2\pi}{2} = \pi$, phase shift $= \frac{\pi}{12}$

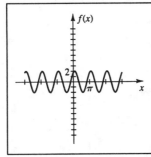

Figure 59

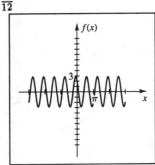

Figure 61

$\boxed{61}$ (a) $f(x) = 2\cos 3x - 2\sin 3x$ • $A = \sqrt{2^2 + 2^2} = 2\sqrt{2}$. $\tan C = \frac{-2}{2} = -1 \Rightarrow$

$C = -\frac{\pi}{4}$. $f(x) = 2\sqrt{2}\cos\left(3x + \frac{\pi}{4}\right) = 2\sqrt{2}\cos\left[3\left(x + \frac{\pi}{12}\right)\right]$

(b) amplitude $= 2\sqrt{2}$, period $= \frac{2\pi}{3}$, phase shift $= -\frac{\pi}{12}$

$\boxed{63}$ $y = 50\sin 60\pi t + 40\cos 60\pi t$ • $A = \sqrt{50^2 + 40^2} = 10\sqrt{41}$. $\tan C = \frac{50}{40} \Rightarrow$

$C = \tan^{-1}\frac{5}{4} \approx 0.8961$. $y = 10\sqrt{41}\cos\left(60\pi t - \tan^{-1}\frac{5}{4}\right) \approx 10\sqrt{41}\cos(60\pi t - 0.8961)$.

$\boxed{65}$ (a) $y = 2\cos t + 3\sin t$; $A = \sqrt{2^2 + 3^2} = \sqrt{13}$; $\tan C = \frac{3}{2} \Rightarrow C \approx 0.98$;

$$y = \sqrt{13}\cos(t - C); \text{ amplitude} = \sqrt{13}, \text{ period} = \frac{2\pi}{1} = 2\pi$$

(b) $y = 0 \Rightarrow \sqrt{13}\cos(t - C) = 0 \Rightarrow t - C = \frac{\pi}{2} + \pi n \Rightarrow$

$t = C + \frac{\pi}{2} + \pi n \approx 2.5536 + \pi n$ for every nonnegative integer n.

$\boxed{67}$ (a) $p(t) = A\sin \omega t + B\sin(\omega t + \tau)$

$= A\sin \omega t + B(\sin \omega t \cos \tau + \cos \omega t \sin \tau)$

$= A\underline{\sin \omega t} + B\cos \tau \underline{\sin \omega t} + B\sin \tau \underline{\cos \omega t}$

$= (B\sin \tau)\cos \omega t + (A + B\cos \tau)\sin \omega t$

$= a\cos \omega t + b\sin \omega t$ with $a = B\sin \tau$ and $b = A + B\cos \tau$

(b) $C^2 = (B\sin \tau)^2 + (A + B\cos \tau)^2$

$= B^2\sin^2\tau + A^2 + 2AB\cos \tau + B^2\cos^2\tau$

$= A^2 + B^2(\sin^2\tau + \cos^2\tau) + 2AB\cos \tau$

$= A^2 + B^2 + 2AB\cos \tau$

[69] (a) $C^2 = A^2 + B^2 + 2AB\cos\tau \le A^2 + B^2 + 2AB$, since $\cos\tau \le 1$ and

$$A > 0,\ B > 0.\ \text{Thus, } C^2 \le (A + B)^2, \text{ and hence } C \le A + B.$$

(b) $C = A + B$ if $\cos\tau = 1$, or $\tau = 0,\ 2\pi$.

(c) Constructive interference will occur if $C > A$. $C > A \Rightarrow C^2 > A^2 \Rightarrow$

$$A^2 + B^2 + 2AB\cos\tau > A^2 \Rightarrow B^2 + 2AB\cos\tau > 0 \Rightarrow B(B + 2A\cos\tau) > 0.$$

Since $B > 0$, the product will be positive if $B + 2A\cos\tau > 0$, i.e., $\cos\tau > -\dfrac{B}{2A}$.

[71] Graph $y = 3\sin 2t + 2\sin(4t + 1)$. Constructive interference will occur when $y > 3$ or

$y < -3$. From the graph, we see that this occurs on the intervals

$$(-2.97,\ -2.69),\ (-1.00,\ -0.37),\ (0.17,\ 0.46),\ \text{and}\ (2.14,\ 2.77).$$

$$[-\pi,\ \pi]\ \text{by}\ [-5,\ 5]$$

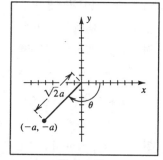

$$\text{Xscl} = \pi/4$$
$$\text{Yscl} = 1$$

Figure 71

3.4 Exercises

[1] From *Figure 1*, $\sin\theta = \frac{4}{5}$ and $\cos\theta = \frac{3}{5}$. Thus, $\sin 2\theta = 2\sin\theta\cos\theta = 2(\frac{4}{5})(\frac{3}{5}) = \frac{24}{25}$.

$$\cos 2\theta = \cos^2\theta - \sin^2\theta = (\tfrac{3}{5})^2 - (\tfrac{4}{5})^2 = -\tfrac{7}{25}.\quad \tan 2\theta = \frac{\sin 2\theta}{\cos 2\theta} = \frac{24/25}{-7/25} = -\frac{24}{7}.$$

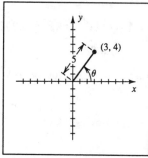

Figure 1

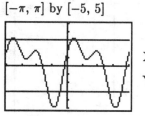

Figure 3

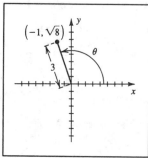

Figure 7

[3] From *Figure 3*, $\sin\theta = \sqrt{8}/3 = \frac{2}{3}\sqrt{2}$ and $\cos\theta = -\frac{1}{3}$.

Thus, $\sin 2\theta = 2\sin\theta\cos\theta = 2(\frac{2}{3}\sqrt{2})(-\frac{1}{3}) = -\frac{4}{9}\sqrt{2}$.

$$\cos 2\theta = \cos^2\theta - \sin^2\theta = (-\tfrac{1}{3})^2 - (\tfrac{2}{3}\sqrt{2})^2 = \tfrac{1}{9} - \tfrac{8}{9} = -\tfrac{7}{9}.$$

$$\tan 2\theta = \frac{\sin 2\theta}{\cos 2\theta} = \frac{-4\sqrt{2}/9}{-7/9} = \frac{4\sqrt{2}}{7}.$$

5 $\sec\theta = \frac{5}{4} \Rightarrow \cos\theta = \frac{4}{5}$. θ acute implies that $\frac{\theta}{2}$ is acute, so all 6 trigonometric functions of $\frac{\theta}{2}$ are positive and we use the " $+$ " sign for the sine and cosine.

$$\sin\frac{\theta}{2} = \sqrt{\frac{1-\cos\theta}{2}} = \sqrt{\frac{1-\frac{4}{5}}{2}} = \sqrt{\frac{\frac{1}{5}}{2}} = \sqrt{\frac{1}{10}\cdot\frac{10}{10}} = \frac{\sqrt{10}}{10}.$$

$$\cos\frac{\theta}{2} = \sqrt{\frac{1+\cos\theta}{2}} = \sqrt{\frac{1+\frac{4}{5}}{2}} = \sqrt{\frac{\frac{9}{5}}{2}} = \sqrt{\frac{9}{10}\cdot\frac{10}{10}} = \frac{3\sqrt{10}}{10}.$$

$$\tan\frac{\theta}{2} = \frac{\sin\frac{\theta}{2}}{\cos\frac{\theta}{2}} = \frac{\sqrt{10}/10}{3\sqrt{10}/10} = \frac{1}{3}.$$

7 From *Figure 7* on the preceding page $(a > 0)$, $\cos\theta = -\dfrac{a}{\sqrt{2}\,a} = -\dfrac{\sqrt{2}}{2}$ and $\frac{\theta}{2}$ is in QIV.

$$\sin\frac{\theta}{2} = -\sqrt{\frac{1-\cos\theta}{2}} = -\sqrt{\frac{1+\sqrt{2}/2}{2}} = -\sqrt{\frac{2+\sqrt{2}}{4}} = -\tfrac{1}{2}\sqrt{2+\sqrt{2}}.$$

$$\cos\frac{\theta}{2} = \sqrt{\frac{1+\cos\theta}{2}} = \sqrt{\frac{1-\sqrt{2}/2}{2}} = \sqrt{\frac{2-\sqrt{2}}{4}} = \tfrac{1}{2}\sqrt{2-\sqrt{2}}.$$

$$\tan\frac{\theta}{2} = \frac{1-\cos\theta}{\sin\theta} = \frac{1+\sqrt{2}/2}{-\sqrt{2}/2}\cdot\frac{2}{2} = \frac{2+\sqrt{2}}{-\sqrt{2}} = -\sqrt{2}-1.$$

9 (a) $\cos 67°30' = \sqrt{\dfrac{1+\cos 135°}{2}} = \sqrt{\dfrac{1-\sqrt{2}/2}{2}} = \sqrt{\dfrac{2-\sqrt{2}}{4}} = \tfrac{1}{2}\sqrt{2-\sqrt{2}}.$

(b) $\sin 15° = \sqrt{\dfrac{1-\cos 30°}{2}} = \sqrt{\dfrac{1-\sqrt{3}/2}{2}} = \sqrt{\dfrac{2-\sqrt{3}}{4}} = \tfrac{1}{2}\sqrt{2-\sqrt{3}}.$

(c) $\tan\dfrac{3\pi}{8} = \dfrac{1-\cos\frac{3\pi}{4}}{\sin\frac{3\pi}{4}} = \dfrac{1+\sqrt{2}/2}{\sqrt{2}/2}\cdot\dfrac{2}{2} = \dfrac{2+\sqrt{2}}{\sqrt{2}} = \sqrt{2}+1.$

13 We recognize the product of the sine and the cosine as being of the form of the right side of the double-angle formula for the sine, and apply that formula "in reverse."

$$4\sin\tfrac{x}{2}\cos\tfrac{x}{2} = 2\cdot 2\sin\tfrac{x}{2}\cos\tfrac{x}{2} = 2\sin\left(2\cdot\tfrac{x}{2}\right) = 2\sin x$$

15 $(\sin t + \cos t)^2 = \sin^2 t + 2\sin t \cos t + \cos^2 t = (\sin^2 t + \cos^2 t) + (2\sin t \cos t) =$

$$1 + \sin 2t$$

17 $\sin 3u = \sin(2u + u)$

$\qquad = \sin 2u \cos u + \cos 2u \sin u$

$\qquad = (2\sin u \cos u)\cos u + (1 - 2\sin^2 u)\sin u$

$\qquad = 2\sin u \cos^2 u + \sin u - 2\sin^3 u$

$\qquad = 2\sin u(1 - \sin^2 u) + \sin u - 2\sin^3 u$

$\qquad = 2\sin u - 2\sin^3 u + \sin u - 2\sin^3 u$

$\qquad = 3\sin u - 4\sin^3 u$

$\qquad = \sin u(3 - 4\sin^2 u)$

$\boxed{19}$ $\cos 4\theta = \cos{(2 \cdot 2\theta)} = 2\cos^2 2\theta - 1 = 2(2\cos^2\theta - 1)^2 - 1 =$

$$2(4\cos^4\theta - 4\cos^2\theta + 1) - 1 = 8\cos^4\theta - 8\cos^2\theta + 1$$

$\boxed{21}$ $\sin^4 t = (\sin^2 t)^2 \quad = \left(\dfrac{1 - \cos 2t}{2}\right)^2$

$$= \tfrac{1}{4}(1 - 2\cos 2t + \cos^2 2t)$$

$$= \tfrac{1}{4} - \tfrac{1}{2}\cos 2t + \tfrac{1}{4}\left(\dfrac{1 + \cos 4t}{2}\right)$$

$$= \tfrac{1}{4} - \tfrac{1}{2}\cos 2t + \tfrac{1}{8} + \tfrac{1}{8}\cos 4t$$

$$= \tfrac{3}{8} - \tfrac{1}{2}\cos 2t + \tfrac{1}{8}\cos 4t$$

$\boxed{23}$ We do not have a formula for $\sec 2\theta$, so we will write $\sec 2\theta$ in terms of $\cos 2\theta$ in order to apply the double-angle formula for the cosine.

$$\sec 2\theta = \frac{1}{\cos 2\theta} = \frac{1}{2\cos^2\theta - 1} = \frac{1}{2\left(\dfrac{1}{\sec^2\theta}\right) - 1} = \frac{1}{\dfrac{2 - \sec^2\theta}{\sec^2\theta}} = \frac{\sec^2\theta}{2 - \sec^2\theta}$$

$\boxed{25}$ We need to match the arguments of the trigonometric functions involved—that is, either write both of them in terms of $2t$ or in terms of $4t$. Converting $\cos 4t$ to an expression with $2t$ as the angle argument gives us

$$2\sin^2 2t + \cos 4t = 2\sin^2 2t + \cos{(2 \cdot 2t)} = 2\sin^2 2t + (1 - 2\sin^2 2t) = 1.$$

Alternatively, if we write both arguments in terms of $4t$, we have

$$2\sin^2 2t + \cos 4t = 2 \cdot \frac{1 - \cos 4t}{2} + \cos 4t = 1 - \cos 4t + \cos 4t = 1.$$

$\boxed{27}$ $\tan 3u = \tan{(2u + u)} = \dfrac{\tan 2u + \tan u}{1 - \tan 2u \tan u}$

$$= \frac{\dfrac{2\tan u}{1 - \tan^2 u} + \tan u}{1 - \dfrac{2\tan u}{1 - \tan^2 u} \cdot \tan u}$$

$$= \frac{\dfrac{2\tan u + \tan u - \tan^3 u}{1 - \tan^2 u}}{\dfrac{1 - \tan^2 u - 2\tan^2 u}{1 - \tan^2 u}}$$

$$= \frac{3\tan u - \tan^3 u}{1 - 3\tan^2 u}$$

$$= \frac{\tan u\,(3 - \tan^2 u)}{1 - 3\tan^2 u}$$

$\boxed{29}$ $\cos^4\dfrac{\theta}{2} = \left(\cos^2\dfrac{\theta}{2}\right)^2 = \left(\dfrac{1 + \cos\theta}{2}\right)^2 = \dfrac{1 + 2\cos\theta + \cos^2\theta}{4} = \tfrac{1}{4} + \tfrac{1}{2}\cos\theta + \tfrac{1}{4}\left(\dfrac{1 + \cos 2\theta}{2}\right) =$

$$\tfrac{1}{4} + \tfrac{1}{2}\cos\theta + \tfrac{1}{8} + \tfrac{1}{8}\cos 2\theta = \tfrac{3}{8} + \tfrac{1}{2}\cos\theta + \tfrac{1}{8}\cos 2\theta$$

33 $\sin 2t + \sin t = 0 \Rightarrow 2\sin t \cos t + \sin t = 0 \Rightarrow \sin t \,(2\cos t + 1) = 0 \Rightarrow$

$$\sin t = 0 \text{ or } \cos t = -\tfrac{1}{2} \Rightarrow t = 0, \ \pi \text{ or } \tfrac{2\pi}{3}, \tfrac{4\pi}{3}$$

35 $\cos u + \cos 2u = 0 \Rightarrow \cos u + 2\cos^2 u - 1 = 0 \Rightarrow 2\cos^2 u + \cos u - 1 = 0 \Rightarrow$

$$(2\cos u - 1)(\cos u + 1) = 0 \Rightarrow \cos u = \tfrac{1}{2}, \ -1 \Rightarrow u = \tfrac{\pi}{3}, \tfrac{5\pi}{3}, \pi$$

37 A first approach uses the concept that if $\tan \alpha = \tan \beta$, then $\alpha = \beta + \pi n$.

$\tan 2x = \tan x \Rightarrow 2x = x + \pi\mathrm{n} \Rightarrow x = \pi\mathrm{n} \Rightarrow x = 0, \ \pi$.

Another approach is: $\tan 2x = \tan x \Rightarrow \dfrac{\sin 2x}{\cos 2x} = \dfrac{\sin x}{\cos x} \Rightarrow \sin 2x \cos x = \sin x \cos 2x \Rightarrow$

$$\sin 2x \cos x - \sin x \cos 2x = 0 \Rightarrow \sin(2x - x) = 0 \Rightarrow \sin x = 0 \Rightarrow x = 0, \ \pi.$$

39 $\sin \tfrac{1}{2}u + \cos u = 1 \Rightarrow \sin \tfrac{1}{2}u + \cos\left[2 \cdot (\tfrac{1}{2}u)\right] = 1 \Rightarrow \sin \tfrac{1}{2}u + (1 - 2\sin^2 \tfrac{1}{2}u) = 1 \Rightarrow$

$\sin \tfrac{1}{2}u - 2\sin^2 \tfrac{1}{2}u = 0 \Rightarrow \sin \tfrac{1}{2}u\,(1 - 2\sin \tfrac{1}{2}u) = 0 \Rightarrow \sin \tfrac{1}{2}u = 0, \ \tfrac{1}{2} \Rightarrow$

$$\tfrac{1}{2}u = 0, \ \tfrac{\pi}{6}, \tfrac{5\pi}{6} \Rightarrow u = 0, \ \tfrac{\pi}{3}, \tfrac{5\pi}{3}$$

41 $\sqrt{a^2 + b^2}\,\sin(u + v) = \sqrt{a^2 + b^2}\,\sin u \cos v + \sqrt{a^2 + b^2}\,\cos u \sin v = a\sin u + b\cos u$

$\{$ equate coefficients of $\sin u$ and $\cos u \} \Rightarrow a = \sqrt{a^2 + b^2}\,\cos v$ and $b = \sqrt{a^2 + b^2}\,\sin v$

$\Rightarrow \cos v = \dfrac{a}{\sqrt{a^2 + b^2}}$ and $\sin v = \dfrac{b}{\sqrt{a^2 + b^2}}$. Since $0 < u < \tfrac{\pi}{2}$, $\sin u > 0$ and $\cos v > 0$.

Now $a > 0$ and $b > 0$ combine with the above to imply that $\cos v > 0$ and $\sin v > 0$.

Thus, $0 < v < \tfrac{\pi}{2}$.

43 (a) $\cos 2x + 2\cos x = 0 \Rightarrow 2\cos^2 x + 2\cos x - 1 = 0 \Rightarrow$

$$\cos x = \frac{-2 \pm \sqrt{12}}{4} = \frac{-1 \pm \sqrt{3}}{2} \approx 0.366 \left\{ \cos x \neq \frac{-1 - \sqrt{3}}{2} < -1 \right\}.$$

Thus, $x \approx 1.20$ and 5.09.

(b) $\sin 2x + \sin x = 0 \Rightarrow 2\sin x \cos x + \sin x = 0 \Rightarrow \sin x\,(2\cos x + 1) = 0 \Rightarrow \sin x = 0$ or

$$\cos x = -\tfrac{1}{2} \Rightarrow x = 0, \ \pi, \ 2\pi \text{ or } \tfrac{2\pi}{3}, \tfrac{4\pi}{3}. \quad P(\tfrac{2\pi}{3}, -1.5), \ Q(\pi, -1), \ R(\tfrac{4\pi}{3}, -1.5)$$

45 (a) $\cos 3x - 3\cos x = 0 \Rightarrow 4\cos^3 x - 3\cos x - 3\cos x = 0 \Rightarrow$

$4\cos^3 x - 6\cos x = 0 \Rightarrow 2\cos x\,(2\cos^2 x - 3) = 0 \Rightarrow \cos x = 0, \ \pm\sqrt{3/2} \Rightarrow$

$$x = -\tfrac{3\pi}{2}, \ -\tfrac{\pi}{2}, \tfrac{\pi}{2}, \tfrac{3\pi}{2} \left\{ \cos x \neq \pm\sqrt{3/2} \text{ since } \sqrt{3/2} > 1 \right\}$$

(b) $\sin 3x - \sin x = 0 \Rightarrow 3\sin x - 4\sin^3 x - \sin x = 0 \Rightarrow 4\sin^3 x - 2\sin x = 0 \Rightarrow$

$2\sin x\,(2\sin^2 x - 1) = 0 \Rightarrow \sin x = 0, \ \pm 1/\sqrt{2} \Rightarrow$

$$x = 0, \ \pm\pi, \ \pm 2\pi, \ \pm\tfrac{\pi}{4}, \ \pm\tfrac{3\pi}{4}, \ \pm\tfrac{5\pi}{4}, \ \pm\tfrac{7\pi}{4}$$

47 (a) Let $y = \overline{BC}$. Form a right triangle with hypotenuse y, side opposite θ, 20, and

side adjacent θ, x. $\sin\theta = \dfrac{20}{y} \Rightarrow y = \dfrac{20}{\sin\theta}$. $\cos\theta = \dfrac{x}{y} \Rightarrow x = y\cos\theta = \dfrac{20\cos\theta}{\sin\theta}$.

　　Now $d = (40 - x) + y = 40 - \dfrac{20\cos\theta}{\sin\theta} + \dfrac{20}{\sin\theta} = 20\left(\dfrac{1 - \cos\theta}{\sin\theta}\right) + 40 = 20\tan\dfrac{\theta}{2} + 40$.

(b) $50 = 20\tan\dfrac{\theta}{2} + 40 \Rightarrow \tan\dfrac{\theta}{2} = \dfrac{1}{2} \Rightarrow \dfrac{1 - \cos\theta}{\sin\theta} = \dfrac{1}{2} \Rightarrow 2 - 2\cos\theta = \sin\theta \Rightarrow$

$4 - 8\cos\theta + 4\cos^2\theta = \sin^2\theta = 1 - \cos^2\theta \Rightarrow 5\cos^2\theta - 8\cos\theta + 3 = 0 \Rightarrow$

$(5\cos\theta - 3)(\cos\theta - 1) = 0 \Rightarrow \cos\theta = \dfrac{3}{5}, 1.$ { $\cos\theta = 1 \Rightarrow \theta = 0$ and 0 is

extraneous }. $\cos\theta = \dfrac{3}{5} \Rightarrow \sin\theta = \dfrac{4}{5}$ and $y = \dfrac{20}{4/5} = 25$. $\cos\theta = \dfrac{x}{y}$ and

$\cos\theta = \dfrac{3}{5} \Rightarrow \dfrac{x}{25} = \dfrac{3}{5} \Rightarrow x = 15$, which means that B would be 25 miles from A.

49 (a) From Example 8, the area A of a cross section is

$A = \dfrac{1}{2}(\text{side})^2(\text{sine of included angle}) = \dfrac{1}{2}\left(\dfrac{1}{2}\right)^2\sin\theta = \dfrac{1}{8}\sin\theta$.

　　The volume $V = (\text{length of gutter})(\text{area of cross section}) = 20(\dfrac{1}{8}\sin\theta) = \dfrac{5}{2}\sin\theta$.

(b) $V = 2 \Rightarrow \dfrac{5}{2}\sin\theta = 2 \Rightarrow \sin\theta = \dfrac{4}{5} \Rightarrow \theta \approx 53.13°$.

51 (a) Let $y = \overline{DB}$ and x denote the distance from D to the midpoint of $\overline{BC}$.

$\sin\dfrac{\theta}{2} = \dfrac{b/2}{y} \Rightarrow y = \dfrac{b}{2} \cdot \dfrac{1}{\sin(\theta/2)}$ and $\tan\dfrac{\theta}{2} = \dfrac{b/2}{x} \Rightarrow x = \dfrac{b}{2} \cdot \dfrac{\cos(\theta/2)}{\sin(\theta/2)}$.

$l = (a - x) + y = a - \dfrac{b}{2} \cdot \dfrac{\cos(\theta/2)}{\sin(\theta/2)} + \dfrac{b}{2} \cdot \dfrac{1}{\sin(\theta/2)} = a + \dfrac{b}{2} \cdot \dfrac{1 - \cos(\theta/2)}{\sin(\theta/2)} =$

$$a + \dfrac{b}{2}\tan\left(\dfrac{\theta/2}{2}\right) = a + \dfrac{b}{2}\tan\dfrac{\theta}{4}.$$

(b) $a = 10$ mm, $b = 6$ mm, and $\theta = 156° \Rightarrow l = 10 + 3\tan 39° \approx 12.43$ mm.

53 The graph of f appears to be that of $y = g(x) = \tan x$.

$$\dfrac{\sin 2x + \sin x}{\cos 2x + \cos x + 1} = \dfrac{2\sin x\cos x + \sin x}{(2\cos^2 x - 1) + \cos x + 1} = \dfrac{\sin x(2\cos x + 1)}{\cos x(2\cos x + 1)} = \dfrac{\sin x}{\cos x} = \tan x$$

57 Graph $Y_1 = 1/\sin(0.25x + 1)$ and $Y_2 = 1.5 - \cos 2x$ on $[-\pi, \pi]$. There are four points

of intersection. They occur at $x \approx -2.03, -0.72, 0.58, 2.62$.

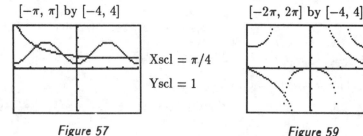

$[-\pi, \pi]$ by $[-4, 4]$　　　　　　　　$[-2\pi, 2\pi]$ by $[-4, 4]$

Xscl $= \pi/4$　　　　　　　Xscl $= \pi/2$

Yscl $= 1$　　　　　　　　Yscl $= 1$

Figure 57　　　　　　　　*Figure 59*

59 Graph $Y_1 = 2/\tan(.25x)$ and $Y_2 = 1 - 1/\cos(.5x)$ on $[-2\pi, 2\pi]$ in Dot mode. There

is one point of intersection. It occurs at $x \approx -2.59$.

Note: We will reference the product-to-sum formulas as [P1]–[P4] and the sum-to-product formulas as [S1]–[S4] in the order they appear in the text. The formulas $\cos(-kx) = \cos kx$ and $\sin(-kx) = -\sin kx$ will be used without mention.

$\boxed{1}$ $\sin 7t \sin 3t = \text{[P4]} \frac{1}{2}\big[\cos(7t-3t) - \cos(7t+3t)\big] = \frac{1}{2}\cos 4t - \frac{1}{2}\cos 10t$

$\boxed{3}$ $\cos 6u \cos(-4u) = \text{[P3]} \frac{1}{2}\Big\{\cos[6u+(-4u)] + \cos[6u-(-4u)]\Big\} = \frac{1}{2}\cos 2u + \frac{1}{2}\cos 10u$

$\boxed{5}$ $2\sin 9\theta \cos 3\theta = \text{[P1]} 2 \cdot \frac{1}{2}\big[\sin(9\theta+3\theta) + \sin(9\theta-3\theta)\big] = \sin 12\theta + \sin 6\theta$

$\boxed{7}$ $3\cos x \sin 2x = \text{[P2]} 3 \cdot \frac{1}{2}\big[\sin(x+2x) - \sin(x-2x)\big] = \frac{3}{2}\sin 3x - \frac{3}{2}\sin(-x) =$
$$\frac{3}{2}\sin 3x + \frac{3}{2}\sin x$$

$\boxed{9}$ $\sin 6\theta + \sin 2\theta = \text{[S1]} 2\sin\dfrac{6\theta+2\theta}{2} \cos\dfrac{6\theta-2\theta}{2} = 2\sin 4\theta \cos 2\theta$

$\boxed{11}$ $\cos 5x - \cos 3x = \text{[S4]} -2\sin\dfrac{5x+3x}{2} \sin\dfrac{5x-3x}{2} = -2\sin 4x \sin x$

$\boxed{13}$ $\sin 3t - \sin 7t = \text{[S2]} 2\cos\dfrac{3t+7t}{2} \sin\dfrac{3t-7t}{2} = 2\cos 5t \sin(-2t) = -2\cos 5t \sin 2t$

$\boxed{15}$ $\cos x + \cos 2x = \text{[S3]} 2\cos\dfrac{x+2x}{2} \cos\dfrac{x-2x}{2} = 2\cos\frac{3}{2}x \cos(-\frac{1}{2}x) = 2\cos\frac{3}{2}x \cos\frac{1}{2}x$

$\boxed{17}$ $\dfrac{\sin 4t + \sin 6t}{\cos 4t - \cos 6t} = \dfrac{\text{[S1]} \ 2\sin 5t \cos(-t)}{\text{[S4]} -2\sin 5t \sin(-t)} = \dfrac{\cos t}{\sin t} = \cot t$

$\boxed{19}$ $\dfrac{\sin u + \sin v}{\cos u + \cos v} = \dfrac{\text{[S1]} \ 2\sin\frac{1}{2}(u+v) \cos\frac{1}{2}(u-v)}{\text{[S3]} \ 2\cos\frac{1}{2}(u+v) \cos\frac{1}{2}(u-v)} = \dfrac{\sin\frac{1}{2}(u+v)}{\cos\frac{1}{2}(u+v)} = \tan\frac{1}{2}(u+v)$

$\boxed{21}$ $\dfrac{\sin u - \sin v}{\sin u + \sin v} = \dfrac{\text{[S2]} \ 2\cos\frac{1}{2}(u+v) \sin\frac{1}{2}(u-v)}{\text{[S1]} \ 2\sin\frac{1}{2}(u+v) \cos\frac{1}{2}(u-v)} = \cot\frac{1}{2}(u+v) \tan\frac{1}{2}(u-v) = \dfrac{\tan\frac{1}{2}(u-v)}{\tan\frac{1}{2}(u+v)}$

$\boxed{23}$ Since the arguments on the right side are all even multiples of x ($2x$, $4x$, and $6x$), we begin by grouping the terms with the odd multiples of x together, and operate on them with a product-to-sum formula, which will convert these expressions to expressions with even multiples of x.

$4\cos x \cos 2x \sin 3x = 2\cos 2x (2\sin 3x \cos x)$
$$= 2\cos 2x (\text{[P1]} \sin 4x + \sin 2x)$$
$$= (2\cos 2x \sin 4x) + (2\cos 2x \sin 2x)$$
$$= \big[\text{[P2]} \sin 6x - \sin(-2x)\big] + (\text{[P2]} \sin 4x - \sin 0)$$
$$= \sin 2x + \sin 4x + \sin 6x$$

25 $(\sin ax)(\cos bx) = $ [P1] $\frac{1}{2}[\sin(ax+bx) + \sin(ax-bx)] = \frac{1}{2}\sin[(a+b)x] + \frac{1}{2}\sin[(a-b)x]$

27 $\sin 5t + \sin 3t = 0 \Rightarrow$ [S1] $2\sin 4t \cos t = 0 \Rightarrow \sin 4t = 0$ or $\cos t = 0 \Rightarrow$

$$4t = \pi n \text{ or } t = \tfrac{\pi}{2} + \pi n \Rightarrow t = \tfrac{\pi}{4}n \text{ \{ which includes } t = \tfrac{\pi}{2} + \pi n \}$$

29 $\cos x = \cos 3x \Rightarrow \cos x - \cos 3x = 0 \Rightarrow$ [S4] $-2\sin 2x \sin(-x) = 0 \Rightarrow$

$$\sin 2x = 0 \text{ or } \sin x = 0 \Rightarrow 2x = \pi n \text{ or } x = \pi n \Rightarrow x = \tfrac{\pi}{2}n \text{ \{ which includes } x = \pi n \}$$

31 $\cos 3x + \cos 5x = \cos x \Rightarrow$ [S3] $2\cos 4x \cos(-x) - \cos x = 0 \Rightarrow$

$$\cos x (2\cos 4x - 1) = 0 \Rightarrow \cos x = 0 \text{ or } \cos 4x = \tfrac{1}{2} \Rightarrow$$

$$x = \tfrac{\pi}{2} + \pi n \text{ or } 4x = \tfrac{\pi}{3} + 2\pi n, \tfrac{5\pi}{3} + 2\pi n \Rightarrow x = \tfrac{\pi}{2} + \pi n, \tfrac{\pi}{12} + \tfrac{\pi}{2}n, \tfrac{5\pi}{12} + \tfrac{\pi}{2}n$$

33 $\sin 2x - \sin 5x = 0 \Rightarrow$ [S2] $2\cos \tfrac{7}{2}x \sin(-\tfrac{3}{2}x) = 0 \Rightarrow \tfrac{7}{2}x = \tfrac{\pi}{2} + \pi n$ or $\tfrac{3}{2}x = \pi n \Rightarrow$

$$x = \tfrac{\pi}{7} + \tfrac{2\pi}{7}n \text{ or } x = \tfrac{2\pi}{3}n$$

35 $\cos x + \cos 3x = 0 \Rightarrow$ [S3] $2\cos 2x \cos(-x) = 0 \Rightarrow \cos 2x = 0$ or $\cos x = 0 \Rightarrow$

$$2x = \tfrac{\pi}{2} + \pi n \text{ or } x = \tfrac{\pi}{2} + \pi n \Rightarrow x = \tfrac{\pi}{4} + \tfrac{\pi}{2}n \text{ or } x = \tfrac{\pi}{2} + \pi n \Rightarrow$$

$$x = \tfrac{\pi}{4}, \tfrac{3\pi}{4}, \tfrac{5\pi}{4}, \tfrac{7\pi}{4}, \tfrac{\pi}{2}, \tfrac{3\pi}{2} \text{ for } 0 \le x \le 2\pi$$

37 $\sin 3x - \sin x = 0 \Rightarrow$ [S2] $2\cos 2x \sin x = 0 \Rightarrow \cos 2x = 0$ or $\sin x = 0 \Rightarrow$

$$2x = \tfrac{\pi}{2} + \pi n \text{ or } x = \pi n \Rightarrow x = \tfrac{\pi}{4} + \tfrac{\pi}{2}n \text{ or } x = \pi n \Rightarrow$$

$$x = 0, \pm\pi, \pm 2\pi, \pm\tfrac{\pi}{4}, \pm\tfrac{3\pi}{4}, \pm\tfrac{5\pi}{4}, \pm\tfrac{7\pi}{4} \text{ for } -2\pi \le x \le 2\pi.$$

39 $f(x) = \sin\left(\tfrac{\pi n}{l}x\right) \cos\left(\tfrac{k\pi n}{l}t\right)$

$$= \text{[P1]} \ \tfrac{1}{2}\left[\sin\left(\tfrac{\pi n}{l}x + \tfrac{k\pi n}{l}t\right) + \sin\left(\tfrac{\pi n}{l}x - \tfrac{k\pi n}{l}t\right)\right]$$

$$= \tfrac{1}{2}\left[\sin\tfrac{\pi n}{l}(x + kt) + \sin\tfrac{\pi n}{l}(x - kt)\right]$$

$$= \tfrac{1}{2}\sin\tfrac{\pi n}{l}(x + kt) + \tfrac{1}{2}\sin\tfrac{\pi n}{l}(x - kt)$$

41 (a) Estimating the x-intercepts, we have $x \approx 0, \pm 1.05, \pm 1.57, \pm 2.09, \pm 3.14$.

(b) $\sin 4x + \sin 2x = 2\sin 3x \cos x = 0 \Rightarrow \sin 3x = 0$ or $\cos x = 0$.

$\sin 3x = 0 \Rightarrow 3x = \pi n \Rightarrow x = \tfrac{\pi}{3}n \Rightarrow x = 0, \pm\tfrac{\pi}{3}, \pm\tfrac{2\pi}{3}, \pm\pi$. $\cos x = 0 \Rightarrow x = \pm\tfrac{\pi}{2}$.

The x-intercepts are $0, \pm\tfrac{\pi}{3}, \pm\tfrac{\pi}{2}, \pm\tfrac{2\pi}{3}, \pm\pi$.

$[-\pi, \pi]$ by $[-2.09, 2.09]$

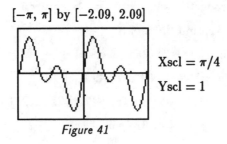

$\text{Xscl} = \pi/4$

$\text{Yscl} = 1$

Figure 41

43 Graphing on an interval of $[-\pi, \pi]$ gives us a figure that resembles a tangent function. It appears that there is a vertical asymptote at about 0.78. Recognizing that this value is about $\frac{\pi}{4}$, we might make the conjecture that we have "halved" the period of the tangent and that the graph of $f(x) = \dfrac{\sin x + \sin 2x + \sin 3x}{\cos x + \cos 2x + \cos 3x}$ appears to be that of $y = g(x) = \tan 2x$. Verifying this identity, we have

$$\frac{\sin x + \sin 2x + \sin 3x}{\cos x + \cos 2x + \cos 3x} = \frac{\sin 2x + (\sin 3x + \sin x)}{\cos 2x + (\cos 3x + \cos x)} = \frac{\sin 2x + 2\sin 2x \cos x}{\cos 2x + 2\cos 2x \cos x} =$$

$$\frac{\sin 2x(1 + 2\cos x)}{\cos 2x(1 + 2\cos x)} = \frac{\sin 2x}{\cos 2x} = \tan 2x.$$

3.6 Exercises

1 (a) $\sin^{-1}\left(-\frac{\sqrt{2}}{2}\right) = -\frac{\pi}{4}$ since $\sin\left(-\frac{\pi}{4}\right) = -\frac{\sqrt{2}}{2}$ and $-\frac{\pi}{2} \le -\frac{\pi}{4} \le \frac{\pi}{2}$

   (b) $\cos^{-1}\left(-\frac{1}{2}\right) = \frac{2\pi}{3}$ since $\cos\frac{2\pi}{3} = -\frac{1}{2}$ and $0 \le \frac{2\pi}{3} \le \pi$

   (c) $\tan^{-1}\left(-\sqrt{3}\right) = -\frac{\pi}{3}$ since $\tan\left(-\frac{\pi}{3}\right) = -\sqrt{3}$ and $-\frac{\pi}{2} < -\frac{\pi}{3} < \frac{\pi}{2}$

3 (a) $\arcsin\frac{\sqrt{3}}{2} = \frac{\pi}{3}$ since $\sin\frac{\pi}{3} = \frac{\sqrt{3}}{2}$ and $-\frac{\pi}{2} \le \frac{\pi}{3} \le \frac{\pi}{2}$

   (b) $\arccos\frac{\sqrt{2}}{2} = \frac{\pi}{4}$ since $\cos\frac{\pi}{4} = \frac{\sqrt{2}}{2}$ and $0 \le \frac{\pi}{4} \le \pi$

   (c) $\arctan\frac{1}{\sqrt{3}} = \frac{\pi}{6}$ since $\tan\frac{\pi}{6} = \frac{1}{\sqrt{3}}$ and $-\frac{\pi}{2} < \frac{\pi}{6} < \frac{\pi}{2}$

5 (a) $\sin^{-1}\frac{\pi}{3}$ is <u>not defined</u> since $\frac{\pi}{3} > 1$, i.e., $\frac{\pi}{3} \notin [-1, 1]$

   (b) $\cos^{-1}\frac{\pi}{2}$ is <u>not defined</u> since $\frac{\pi}{2} > 1$, i.e., $\frac{\pi}{2} \notin [-1, 1]$

   (c) $\tan^{-1}1 = \frac{\pi}{4}$ since $\tan\frac{\pi}{4} = 1$ and $-\frac{\pi}{2} < \frac{\pi}{4} < \frac{\pi}{2}$

Note: Exercises 7–10 refer to the boxed properties of $\sin^{-1}$, $\cos^{-1}$, and $\tan^{-1}$.

7 (a) $\sin\left[\arcsin\left(-\frac{3}{10}\right)\right] = -\frac{3}{10}$ since $-1 \le -\frac{3}{10} \le 1$

   (b) $\cos\left(\arccos\frac{1}{2}\right) = \frac{1}{2}$ since $-1 \le \frac{1}{2} \le 1$

   (c) $\tan\left(\arctan 14\right) = 14$ since $\tan\left(\arctan x\right) = x$ for every x

9 (a) $\sin^{-1}\left(\sin\frac{\pi}{3}\right) = \frac{\pi}{3}$ since $-\frac{\pi}{2} \le \frac{\pi}{3} \le \frac{\pi}{2}$ (b) $\cos^{-1}\left[\cos\left(\frac{5\pi}{6}\right)\right] = \frac{5\pi}{6}$ since $0 \le \frac{5\pi}{6} \le \pi$

   (c) $\tan^{-1}\left[\tan\left(-\frac{\pi}{6}\right)\right] = -\frac{\pi}{6}$ since $-\frac{\pi}{2} < -\frac{\pi}{6} < \frac{\pi}{2}$

11 (a) $\arcsin\left(\sin\frac{5\pi}{4}\right) = \arcsin\left(-\frac{\sqrt{2}}{2}\right) = -\frac{\pi}{4}$

   (b) $\arccos\left(\cos\frac{5\pi}{4}\right) = \arccos\left(-\frac{\sqrt{2}}{2}\right) = \frac{3\pi}{4}$ (c) $\arctan\left(\tan\frac{7\pi}{4}\right) = \arctan(-1) = -\frac{\pi}{4}$

$\boxed{13}$ (a) $\sin\left[\cos^{-1}\left(-\frac{1}{2}\right)\right] = \sin\frac{2\pi}{3} = \frac{\sqrt{3}}{2}$ (b) $\cos\left(\tan^{-1}1\right) = \cos\frac{\pi}{4} = \frac{\sqrt{2}}{2}$

    (c) $\tan\left[\sin^{-1}\left(-1\right)\right] = \tan\left(-\frac{\pi}{2}\right)$, which is <u>not defined</u>.

$\boxed{15}$ (a) Let $\theta = \sin^{-1}\frac{2}{3}$. From *Figure 15(a)*, $\cot\left(\sin^{-1}\frac{2}{3}\right) = \cot\theta = \frac{x}{y} = \frac{\sqrt{5}}{2}$.

    (b) Let $\theta = \tan^{-1}\left(-\frac{3}{5}\right)$. From *Figure 15(b)*, $\sec\left[\tan^{-1}\left(-\frac{3}{5}\right)\right] = \sec\theta = \frac{r}{x} = \frac{\sqrt{34}}{5}$.

    (c) Let $\theta = \cos^{-1}\left(-\frac{1}{4}\right)$. From *Figure 15(c)*, $\csc\left[\cos^{-1}\left(-\frac{1}{4}\right)\right] = \csc\theta = \frac{r}{y} = \frac{4}{\sqrt{15}}$.

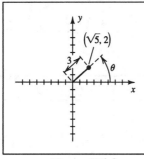   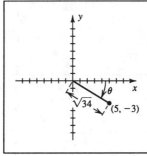   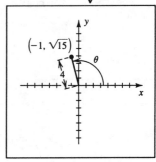

     Figure 15(a)   *Figure 15(b)*   *Figure 15(c)*

$\boxed{17}$ (a) $\sin\left(\arcsin\frac{1}{2} + \arccos 0\right) = \sin\left(\frac{\pi}{6} + \frac{\pi}{2}\right) = \sin\frac{2\pi}{3} = \frac{\sqrt{3}}{2}$.

    (b) Remember, $\arctan\left(-\frac{3}{4}\right)$ and $\arcsin\frac{4}{5}$ are <u>angles</u>. To abbreviate this solution,

       we let $\alpha = \arctan\left(-\frac{3}{4}\right)$ and $\beta = \arcsin\frac{4}{5}$. Using the difference identity for the

       cosine and figures as in Exercises 15 and 16, we have $\cos\left[\arctan\left(-\frac{3}{4}\right) - \arcsin\frac{4}{5}\right]$

$$= \cos\left(\alpha - \beta\right) = \cos\alpha\cos\beta + \sin\alpha\sin\beta = \frac{4}{5}\cdot\frac{3}{5} + \left(-\frac{3}{5}\right)\cdot\frac{4}{5} = 0.$$

    (c) Let $\alpha = \arctan\frac{4}{3}$ and $\beta = \arccos\frac{8}{17}$. $\tan\left(\arctan\frac{4}{3} + \arccos\frac{8}{17}\right) =$

$$\tan\left(\alpha + \beta\right) = \frac{\tan\alpha + \tan\beta}{1 - \tan\alpha\tan\beta} = \frac{\frac{4}{3} + \frac{15}{8}}{1 - \frac{4}{3}\cdot\frac{15}{8}}\cdot\frac{24}{24} = \frac{32 + 45}{24 - 60} = -\frac{77}{36}.$$

$\boxed{19}$ (a) We first recognize this expression as being of the form $\sin\left(\textit{twice an angle}\right)$, where

       $\arccos\left(-\frac{3}{5}\right)$ is the angle. Let $\alpha = \arccos\left(-\frac{3}{5}\right)$. It may help to draw a figure as in

       Exercise 15 to determine that α is a second quadrant angle and that $\sin\alpha = \frac{4}{5}$.

       Applying the double-angle formula for the sine gives us

$$\sin\left[2\arccos\left(-\frac{3}{5}\right)\right] = \sin 2\alpha = 2\sin\alpha\cos\alpha = 2\left(\frac{4}{5}\right)\left(-\frac{3}{5}\right) = -\frac{24}{25}.$$

    (b) Let $\alpha = \sin^{-1}\frac{15}{17}$. Thus, $\cos\alpha = \frac{8}{17}$ and we apply the double-angle formula for the

       cosine. $\cos\left(2\sin^{-1}\frac{15}{17}\right) = \cos 2\alpha = \cos^2\alpha - \sin^2\alpha = \left(\frac{8}{17}\right)^2 - \left(\frac{15}{17}\right)^2 = -\frac{161}{289}$.

    (c) Let $\alpha = \tan^{-1}\frac{3}{4}$. Thus, $\tan\alpha = \frac{3}{4}$ and we apply the double-angle formula for the

       tangent. $\tan\left(2\tan^{-1}\frac{3}{4}\right) = \tan 2\alpha = \frac{2\tan\alpha}{1 - \tan^2\alpha} = \frac{2\cdot\frac{3}{4}}{1 - \left(\frac{3}{4}\right)^2}\cdot\frac{16}{16} = \frac{24}{16 - 9} = \frac{24}{7}$.

21 (a) We first recognize this expression as being of the form $\sin\,(one\text{-}half\ an\ angle)$, where $\sin^{-1}\left(-\frac{7}{25}\right)$ is the angle. Let $\alpha = \sin^{-1}\left(-\frac{7}{25}\right)$. Hence, α is a fourth quadrant angle and $\cos\alpha = \frac{24}{25}$. $-\frac{\pi}{2} < \alpha < 0 \Rightarrow -\frac{\pi}{4} < \frac{1}{2}\alpha < 0$. We need to know the sign of $\sin\frac{1}{2}\alpha$ in order to correctly apply the half-angle formula for the sine. Since $-\frac{\pi}{4} < \frac{1}{2}\alpha < 0$ { QIV }, $\sin\frac{1}{2}\alpha < 0$.

$$\sin\left[\tfrac{1}{2}\sin^{-1}\left(-\tfrac{7}{25}\right)\right] = \sin\tfrac{1}{2}\alpha = -\sqrt{\frac{1-\cos\alpha}{2}} = -\sqrt{\frac{1-\frac{24}{25}}{2}} = -\sqrt{\frac{1}{50}\cdot\frac{2}{2}} = -\frac{1}{10}\sqrt{2}.$$

(b) Let $\alpha = \tan^{-1}\frac{8}{15}$. $0 < \alpha < \frac{\pi}{2} \Rightarrow 0 < \frac{1}{2}\alpha < \frac{\pi}{4}$ and $\cos\frac{1}{2}\alpha > 0$.

Applying the half-angle formula for the cosine,

$$\cos\left(\tfrac{1}{2}\tan^{-1}\tfrac{8}{15}\right) = \cos\tfrac{1}{2}\alpha = \sqrt{\frac{1+\cos\alpha}{2}} = \sqrt{\frac{1+\frac{15}{17}}{2}} = \sqrt{\frac{16}{17}\cdot\frac{17}{17}} = \frac{4}{17}\sqrt{17}.$$

(c) Let $\alpha = \cos^{-1}\frac{3}{5}$. Applying the half-angle formula for the tangent,

$$\tan\left(\tfrac{1}{2}\cos^{-1}\tfrac{3}{5}\right) = \tan\tfrac{1}{2}\alpha = \frac{1-\cos\alpha}{\sin\alpha} = \frac{1-\frac{3}{5}}{\frac{4}{5}} = \frac{1}{2}.$$

23 We recognize this expression as being of the form $\sin\,(some\ angle)$. The angle is *the angle whose tangent is* α. The easiest way to picture this angle is to consider it as being the angle whose ratio of opposite side to adjacent side is x to 1. Let $\alpha = \tan^{-1}x$. From *Figure 23*, $\sin\,(\tan^{-1}x) = \sin\alpha = \dfrac{x}{\sqrt{x^2+1}}$.

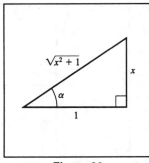

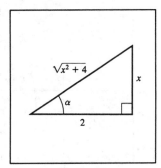

Figure 23   Figure 25

25 Let $\alpha = \sin^{-1}\dfrac{x}{\sqrt{x^2+4}}$. From *Figure 25*, $\sec\left(\sin^{-1}\dfrac{x}{\sqrt{x^2+4}}\right) = \sec\alpha = \dfrac{\sqrt{x^2+4}}{2}$.

$\boxed{27}$ Let $\alpha = \sin^{-1} x$. From *Figure 27*,

$$\sin\left(2\sin^{-1} x\right) = \sin 2\alpha = 2\sin\alpha\cos\alpha = 2 \cdot \frac{x}{1} \cdot \frac{\sqrt{1-x^2}}{1} = 2x\sqrt{1-x^2}.$$

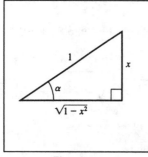

Figure 27

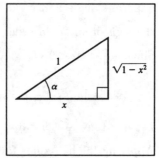

Figure 29

$\boxed{29}$ Let $\alpha = \arccos x$. $0 \le \alpha \le \pi \Rightarrow 0 \le \frac{1}{2}\alpha \le \frac{\pi}{2}$.

Thus $\cos\frac{1}{2}\alpha > 0$ and we use the "+" in the half-angle formula for the cosine.

$$\cos\left(\tfrac{1}{2}\arccos x\right) = \cos\tfrac{1}{2}\alpha = \sqrt{\frac{1 + \cos\alpha}{2}} = \sqrt{\frac{1 + x}{2}}.$$

$\boxed{31}$ (a) See text Figure 19. As $x \to -1^+$, $\sin^{-1} x \to \ \underline{\ -\frac{\pi}{2}\ }$.

(b) See text Figure 21. As $x \to 1^-$, $\cos^{-1} x \to \underline{\ 0\ }$.

(c) See text Figure 23. As $x \to \infty$, $\tan^{-1} x \to \underline{\ \frac{\pi}{2}\ }$.

$\boxed{33}$ $y = \sin^{-1} 2x$ $\quad \bullet \quad$ Horizontally compress $y = \sin^{-1} x$ by a factor of 2.

Note that the domain changes from $[-1,\, 1]$ to $[-\frac{1}{2},\, \frac{1}{2}]$.

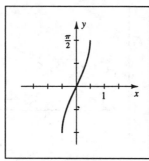

Figure 33

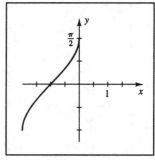

Figure 35

$\boxed{35}$ $y = \sin^{-1}(x + 1)$ $\quad \bullet \quad$ Shift $y = \sin^{-1} x$ left 1 unit.

The domain changes from $[-1,\, 1]$ to $[-2,\, 0]$.

37 $y = \cos^{-1}\frac{1}{2}x$ • Horizontally stretch $y = \cos^{-1}x$ by a factor of 2.

The domain changes from $[-1, 1]$ to $[-2, 2]$.

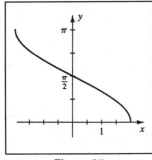

Figure 37

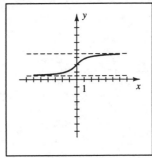

Figure 39

39 $y = 2 + \tan^{-1}x$ • Shift $y = \tan^{-1}x$ up 2 units. The range changes from

$(-\frac{\pi}{2}, \frac{\pi}{2})$ to $(2 - \frac{\pi}{2}, 2 + \frac{\pi}{2})$, which is approximately $(0.43, 3.57)$.

41 If $\alpha = \arccos x$, then $\cos\alpha = x$, where $0 \le \alpha \le \pi$.

Hence, $y = \sin(\arccos x) = \sin\alpha = \sqrt{1 - \cos^2\alpha} = \sqrt{1 - x^2}$.

Thus, we have the graph of the semicircle $y = \sqrt{1 - x^2}$

on the interval $[-1, 1]$.

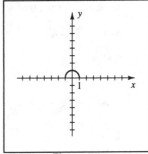

Figure 41

43 (a) Since the domain of the arcsine function is $[-1, 1]$, we know that $x - 3$ must be

in that interval. Thus, $-1 \le x - 3 \le 1 \Rightarrow 2 \le x \le 4$.

(b) The range of the arcsine function is $[-\frac{\pi}{2}, \frac{\pi}{2}]$.

Thus, $-\frac{\pi}{2} \le \sin^{-1}(x - 3) \le \frac{\pi}{2} \Rightarrow -\frac{\pi}{4} \le \frac{1}{2}\sin^{-1}(x - 3) \le \frac{\pi}{4} \Rightarrow -\frac{\pi}{4} \le y \le \frac{\pi}{4}$.

(c) $y = \frac{1}{2}\sin^{-1}(x - 3) \Rightarrow 2y = \sin^{-1}(x - 3) \Rightarrow \sin 2y = x - 3 \Rightarrow x = \sin 2y + 3$

45 (a) $-1 \le \frac{2}{3}x \le 1 \Rightarrow -\frac{3}{2} \le x \le \frac{3}{2}$

(b) $0 \le \cos^{-1}\frac{2}{3}x \le \pi \Rightarrow 0 \le 4\cos^{-1}\frac{2}{3}x \le 4\pi \Rightarrow 0 \le y \le 4\pi$

(c) $y = 4\cos^{-1}\frac{2}{3}x \Rightarrow \frac{1}{4}y = \cos^{-1}\frac{2}{3}x \Rightarrow \cos\frac{1}{4}y = \frac{2}{3}x \Rightarrow x = \frac{3}{2}\cos\frac{1}{4}y$

47 $y = -3 - \sin x \Rightarrow y + 3 = -\sin x \Rightarrow -(y + 3) = \sin x \Rightarrow x = \sin^{-1}(-y - 3)$

49 $y = 15 - 2\cos x \Rightarrow 2\cos x = 15 - y \Rightarrow \cos x = \frac{1}{2}(15 - y) \Rightarrow x = \cos^{-1}\left[\frac{1}{2}(15 - y)\right]$

51 $\frac{\sin x}{3} = \frac{\sin y}{4} \Rightarrow \sin x = \frac{3}{4}\sin y$. Since $0 < y < \pi$, we know that $\sin y > 0$. Thus, solving

$\sin x = \left(\frac{3}{4}\sin y\right)$ is similar to solving $\sin x = a$, where $0 < a < 1$. Remember that when

solving equations of this form, there are two solutions, a first quadrant angle and a

second quadrant angle. The reference angle for x is $x_R = \sin^{-1}(\frac{3}{4}\sin y)$, where

$0 < \frac{3}{4}\sin y \le \frac{3}{4} < 1$. If $0 < x < \frac{\pi}{2}$, then $x = x_R$. If $\frac{\pi}{2} < x < \pi$, then $x = \pi - x_R$.

$\boxed{53}$ $\cos^2 x + 2\cos x - 1 = 0 \Rightarrow \cos x = -1 \pm \sqrt{2} \approx 0.4142, -2.4142$.

Since $-2.4142 < -1$, $x = \cos^{-1}(-1 + \sqrt{2}) \approx 1.1437$ is one answer.

$$x = 2\pi - \cos^{-1}(-1 + \sqrt{2}) \approx 2\pi - 1.1437 \approx 5.1395 \text{ is the other.}$$

$\boxed{57}$ $15\cos^4 x - 14\cos^2 x + 3 = 0 \Rightarrow (5\cos^2 x - 3)(3\cos^2 x - 1) = 0 \Rightarrow \cos^2 x = \frac{3}{5}, \frac{1}{3} \Rightarrow$

$\cos x = \pm\frac{1}{5}\sqrt{15}, \pm\frac{1}{3}\sqrt{3} \Rightarrow x = \cos^{-1}(\pm\frac{1}{5}\sqrt{15}), \cos^{-1}(\pm\frac{1}{3}\sqrt{3})$.

$\cos^{-1}\frac{1}{5}\sqrt{15} \approx 0.6847$, $\cos^{-1}(-\frac{1}{5}\sqrt{15}) \approx 2.4569$,

$$\cos^{-1}\frac{1}{3}\sqrt{3} \approx 0.9553, \cos^{-1}(-\frac{1}{3}\sqrt{3}) \approx 2.1863$$

$\boxed{59}$ $6\sin^3\theta + 18\sin^2\theta - 5\sin\theta - 15 = 0 \Rightarrow 6\sin^2\theta(\sin\theta + 3) - 5(\sin\theta + 3) = 0 \Rightarrow$

$$(6\sin^2\theta - 5)(\sin\theta + 3) = 0 \Rightarrow \sin\theta = \pm\frac{1}{6}\sqrt{30} \Rightarrow \theta = \sin^{-1}(\pm\frac{1}{6}\sqrt{30}) \approx \pm 1.1503$$

$\boxed{61}$ $(\cos x)(15\cos x + 4) = 3 \Rightarrow 15\cos^2 x + 4\cos x - 3 = 0 \Rightarrow (5\cos x + 3)(3\cos x - 1) =$

$0 \Rightarrow \cos x = -\frac{3}{5}, \frac{1}{3} \Rightarrow x = \cos^{-1}(-\frac{3}{5}) \approx 2.2143$, $\cos^{-1}\frac{1}{3} \approx 1.2310$. These angles are in

the second quadrant and the first quadrant. In $[0, 2\pi)$, we must also have a third

quadrant angle and a fourth quadrant angle that satisfy the original equation. These

angles are $2\pi - \cos^{-1}(-\frac{3}{5}) \approx 4.0689$ and $2\pi - \cos^{-1}\frac{1}{3} \approx 5.0522$.

$\boxed{63}$ $3\cos 2x - 7\cos x + 5 = 0 \Rightarrow 3(2\cos^2 x - 1) - 7\cos x + 5 = 0 \Rightarrow$

$6\cos^2 x - 7\cos x + 2 = 0 \Rightarrow (3\cos x - 2)(2\cos x - 1) = 0 \Rightarrow \cos x = \frac{2}{3}, \frac{1}{2} \Rightarrow$

$$x = \cos^{-1}\frac{2}{3} \approx 0.8411, 2\pi - \cos^{-1}\frac{2}{3} \approx 5.4421, \frac{\pi}{3} \approx 1.0472, \frac{5\pi}{3} \approx 5.2360.$$

$\boxed{65}$ (a) $S = 4$, $D = 3.5$, $d = 1 \Rightarrow M = \frac{S}{2}\left(1 - \frac{2}{\pi}\tan^{-1}\frac{d}{D}\right) = \frac{4}{2}\left(1 - \frac{2}{\pi}\tan^{-1}\frac{1}{3.5}\right) \approx 1.65$ m

(b) $d = 4 \Rightarrow M = \frac{S}{2}\left(1 - \frac{2}{\pi}\tan^{-1}\frac{d}{D}\right) = \frac{4}{2}\left(1 - \frac{2}{\pi}\tan^{-1}\frac{4}{3.5}\right) \approx 0.92$ m

(c) $d = 10 \Rightarrow M = \frac{S}{2}\left(1 - \frac{2}{\pi}\tan^{-1}\frac{d}{D}\right) = \frac{4}{2}\left(1 - \frac{2}{\pi}\tan^{-1}\frac{10}{3.5}\right) \approx 0.43$ m

$\boxed{67}$ Form a right triangle with the center line of the fairway and half the width of the

fairway as the legs of the triangle.

$$\text{opp} = \tfrac{1}{2}(30) = 15 \text{ and hyp} = 280 \Rightarrow \sin\theta = \tfrac{15}{280} \Rightarrow \theta = \sin^{-1}\tfrac{15}{280} \approx 3.07°$$

$\boxed{69}$ (a) Let β denote the angle by the sailboat with opposite side d and hypotenuse k.

Now $\sin\beta = \frac{d}{k} \Rightarrow \beta = \sin^{-1}\frac{d}{k}$. Using alternate interior angles,

$$\text{we see that } \alpha + \beta = \theta. \text{ Thus, } \alpha = \theta - \beta = \theta - \sin^{-1}\frac{d}{k}.$$

(b) $d = 50$, $k = 210$, and $\theta = 53.4° \Rightarrow \alpha = 53.4° - \sin^{-1}\frac{50}{210} \approx 39.63°$, or $40°$.

Note: The following is a general outline that can be used for verifying trigonometric identities involving inverse trigonometric functions.

(1) Define angles and their ranges—make sure the range of values for one side of the equation is equal to the range of values for the other side.

(2) Choose a trigonometric function T that is one-to-one on the range of values listed in part (1).

(3) Show that $T(\text{LS}) = T(\text{RS})$. Note that $T(\text{LS}) = T(\text{RS}) \not\Rightarrow \text{LS} = \text{RS}$.

(4) Conclude that since T is one-to-one on the range of values, $\text{LS} = \text{RS}$.

71 Let $\alpha = \sin^{-1} x$ and $\beta = \tan^{-1} \dfrac{x}{\sqrt{1-x^2}}$ with $-\frac{\pi}{2} < \alpha < \frac{\pi}{2}$ and $-\frac{\pi}{2} < \beta < \frac{\pi}{2}$.

Thus, $\sin \alpha = x$ and $\sin \beta = x$. Since the sine function is one-to-one on $\left(-\frac{\pi}{2}, \frac{\pi}{2}\right)$,

we have $\alpha = \beta$—that is, $\sin^{-1} x = \tan^{-1} \dfrac{x}{\sqrt{1-x^2}}$.

73 Let $\alpha = \arcsin(-x)$ and $\beta = \arcsin x$ with $-\frac{\pi}{2} \le \alpha \le \frac{\pi}{2}$ and $-\frac{\pi}{2} \le \beta \le \frac{\pi}{2}$.

Thus, $\sin \alpha = -x$ and $\sin \beta = x$. Consequently, $\sin \alpha = -\sin \beta = \sin(-\beta)$.

Since the sine function is one-to-one on $\left[-\frac{\pi}{2}, \frac{\pi}{2}\right]$,

we have $\alpha = -\beta$—that is, $\arcsin(-x) = -\arcsin x$.

75 Let $\alpha = \arctan x$ and $\beta = \arctan(1/x)$.

Since $x > 0$, we have $0 < \alpha < \frac{\pi}{2}$ and $0 < \beta < \frac{\pi}{2}$, and hence $0 < \alpha + \beta < \pi$.

Thus, $\tan(\alpha + \beta) = \dfrac{\tan \alpha + \tan \beta}{1 - \tan \alpha \tan \beta} = \dfrac{x + (1/x)}{1 - x \cdot (1/x)} = \dfrac{x + (1/x)}{0}$.

Since the denominator is 0, $\tan(\alpha + \beta)$ is undefined and hence $\alpha + \beta = \frac{\pi}{2}$ since $\frac{\pi}{2}$ is the only value between 0 and π for which the tangent is undefined.

77 The domain of $\sin^{-1}(x - 1)$ is $[0, 2]$ and the domain of $\cos^{-1}\frac{1}{2}x$ is $[-2, 2]$.

The domain of f is the intersection of $[0, 2]$ and $[-2, 2]$, i.e., $[0, 2]$.

From the graph, we see that the function is increasing and its range is $\left[-\frac{\pi}{2}, \pi\right]$.

$[-3, 6]$ by $[-2, 4]$

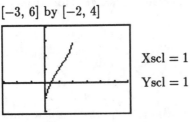

Xscl $= 1$
Yscl $= 1$

Figure 77

$[-3, 3]$ by $[-2, 2]$

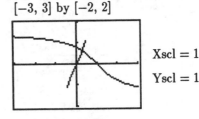

Xscl $= 1$
Yscl $= 1$

Figure 79

79 Graph $y = \sin^{-1} 2x$ and $y = \tan^{-1}(1 - x)$.

From the graph, we see that there is one solution at $x \approx 0.29$.

81 Make the assignments $Y_1 = \sin^{-1}(\sin x / 1.52)$, $Y_2 = x - Y_1$, $Y_3 = x + Y_1$, and $Y_4 = .5((\sin Y_2)^2 / (\sin Y_3)^2 + (\tan Y_2)^2 / (\tan Y_3)^2)$. Now turn *off* Y_1, Y_2, and Y_3—leaving only Y_4 *on* to graph. From the graph, we see that when $f(\theta) = 0.2$, $\theta \approx 1.25$, or approximately $72°$.

$[0, \pi/2]$ by $[0, 1.05]$

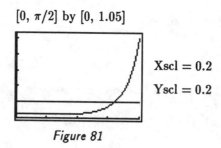

Xscl $= 0.2$

Yscl $= 0.2$

Figure 81

83 Actual distance between x-ticks is equal to $x_A = \frac{3 \text{ units}}{3 \text{ ticks}} = 1$ unit between ticks.

Actual distance between y-ticks is equal to $y_A = \frac{2 \text{ units}}{2 \text{ ticks}} = 1$ unit between ticks.

The ratio is $m_A = \frac{y_A}{x_A} = \frac{1}{1} = 1$. The graph will make an angle of $\theta = \tan^{-1} 1 = 45°$.

$[0, 3]$ by $[0, 2]$ $[0, 3]$ by $[0, 4]$

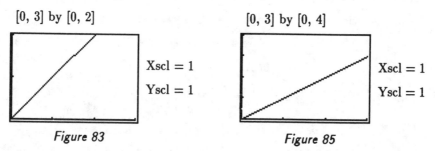

Xscl $= 1$ Xscl $= 1$

Yscl $= 1$ Yscl $= 1$

Figure 83   *Figure 85*

85 $x_A = \frac{3 \text{ units}}{3 \text{ ticks}} = 1$, $y_A = \frac{2 \text{ units}}{4 \text{ ticks}} = \frac{1}{2} \Rightarrow m_A = \frac{1/2}{1} = \frac{1}{2} \Rightarrow \theta = \tan^{-1} \frac{1}{2} \approx 26.6°$.

Chapter 3 Review Exercises

3 $\dfrac{(\sec^2\theta - 1)\cot\theta}{\tan\theta\,\sin\theta + \cos\theta} = \dfrac{(\tan^2\theta)\cot\theta}{\dfrac{\sin\theta}{\cos\theta} \cdot \sin\theta + \cos\theta}$ { Pythagorean and tangent identities }

$\qquad\qquad\qquad\qquad = \dfrac{\tan\theta\,(\tan\theta\,\cot\theta)}{\dfrac{\sin^2\theta}{\cos\theta} + \cos\theta}$ { combine terms }

$\qquad\qquad\qquad\qquad = \dfrac{\tan\theta}{\dfrac{\sin^2\theta + \cos^2\theta}{\cos\theta}}$ { reciprocal identity, common denominator }

$\qquad\qquad\qquad\qquad = \dfrac{\sin\theta / \cos\theta}{1/\cos\theta}$ { Pythagorean and tangent identities }

$\qquad\qquad\qquad\qquad = \sin\theta$ { simplify }

$\boxed{5}$ $\dfrac{1}{1+\sin t} = \dfrac{1}{1+\sin t} \cdot \dfrac{1-\sin t}{1-\sin t}$ $\left\{\begin{array}{l}\text{multiply the numerator and the denominator}\\ \text{by conjugate of the denominator}\end{array}\right\}$

$= \dfrac{1-\sin t}{1-\sin^2 t}$ $\{\text{the difference of two squares}\}$

$= \dfrac{1-\sin t}{\cos^2 t}$ $\{\text{Pythagorean identity}\}$

$= \dfrac{1-\sin t}{\cos t} \cdot \dfrac{1}{\cos t}$ $\{\text{break up since we want } \sec t \text{ on the right side}\}$

$= \left(\dfrac{1}{\cos t} - \dfrac{\sin t}{\cos t}\right) \cdot \sec t$ $\{\text{split up fraction}\}$

$= (\sec t - \tan t)\sec t$ $\{\text{reciprocal and tangent identities}\}$

$\boxed{6}$ $\dfrac{\sin(\alpha-\beta)}{\cos(\alpha+\beta)} = \dfrac{\sin\alpha\cos\beta - \cos\alpha\sin\beta}{\cos\alpha\cos\beta - \sin\alpha\sin\beta}$ $\left\{\begin{array}{l}\underline{\text{subtraction formula for the sine}}\\ \text{addition formula for the cosine}\end{array}\right\}$

The first term in the denominator is $\cos\alpha\cos\beta$, but looking ahead, we see that the first term in the denominator of the expression we want to obtain is 1. Hence, we will divide both the numerator and the denominator by $\cos\alpha\cos\beta$.

$$= \dfrac{(\sin\alpha\cos\beta - \cos\alpha\sin\beta) \,/\, \cos\alpha\cos\beta}{(\cos\alpha\cos\beta - \sin\alpha\sin\beta) \,/\, \cos\alpha\cos\beta}$$

$$= \dfrac{\dfrac{\sin\alpha\cos\beta}{\cos\alpha\cos\beta} - \dfrac{\cos\alpha\sin\beta}{\cos\alpha\cos\beta}}{\dfrac{\cos\alpha\cos\beta}{\cos\alpha\cos\beta} - \dfrac{\sin\alpha\sin\beta}{\cos\alpha\cos\beta}} = \dfrac{\dfrac{\sin\alpha}{\cos\alpha} - \dfrac{\sin\beta}{\cos\beta}}{1 - \dfrac{\sin\alpha}{\cos\alpha}\cdot\dfrac{\sin\beta}{\cos\beta}} = \dfrac{\tan\alpha - \tan\beta}{1 - \tan\alpha\tan\beta}$$

$\boxed{7}$ $\tan 2u = \dfrac{2\tan u}{1-\tan^2 u}$ $\{\text{apply the double-angle formula for the tangent}\}$

$= \dfrac{2\cdot\dfrac{1}{\cot u}}{1-\dfrac{1}{\cot^2 u}}$ $\{\text{put in terms of cot since it appears on the right side}\}$

$= \dfrac{\dfrac{2}{\cot u}}{\dfrac{\cot^2 u - 1}{\cot^2 u}}$ $\{\text{combine into one fraction}\}$

$= \dfrac{2\cot u}{\cot^2 u - 1}$ $\{\text{simplify complex fraction}\}$

$= \dfrac{2\cot u}{(\csc^2 u - 1) - 1}$ $\{\text{Pythagorean identity}\}$

$= \dfrac{2\cot u}{\csc^2 u - 2}$ $\{\text{simplify}\}$

$\boxed{10}$ $LS = \dfrac{\sin u + \sin v}{\csc u + \csc v} = \dfrac{\sin u + \sin v}{\dfrac{1}{\sin u} + \dfrac{1}{\sin v}} = \dfrac{\sin u + \sin v}{\dfrac{\sin v + \sin u}{\sin u \, \sin v}} = \sin u \, \sin v$

At this point, there is no apparent "next step." Thus, we will stop working with the left side and try to simplify the right side to the same expression, $\sin u \, \sin v$.

$RS = \dfrac{1 - \sin u \, \sin v}{-1 + \csc u \, \csc v} = \dfrac{1 - \sin u \, \sin v}{-1 + \dfrac{1}{\sin u \, \sin v}} = \dfrac{1 - \sin u \, \sin v}{\dfrac{1 - \sin u \, \sin v}{\sin u \, \sin v}} = \sin u \, \sin v$

Since the LS and RS equal the same expression and the steps are reversible,

the identity is verified.

$\boxed{12}$ $\dfrac{\cos \gamma}{1 - \tan \gamma} + \dfrac{\sin \gamma}{1 - \cot \gamma} = \dfrac{\cos \gamma}{1 - \dfrac{\sin \gamma}{\cos \gamma}} + \dfrac{\sin \gamma}{1 - \dfrac{\cos \gamma}{\sin \gamma}} = \dfrac{\cos \gamma}{\dfrac{\cos \gamma - \sin \gamma}{\cos \gamma}} + \dfrac{\sin \gamma}{\dfrac{\sin \gamma - \cos \gamma}{\sin \gamma}} =$

$\dfrac{\cos^2 \gamma}{\cos \gamma - \sin \gamma} + \dfrac{\sin^2 \gamma}{\sin \gamma - \cos \gamma} = \dfrac{\cos^2 \gamma}{\cos \gamma - \sin \gamma} - \dfrac{\sin^2 \gamma}{-(\sin \gamma - \cos \gamma)} =$

$\dfrac{\cos^2 \gamma}{\cos \gamma - \sin \gamma} - \dfrac{\sin^2 \gamma}{\cos \gamma - \sin \gamma} = \dfrac{\cos^2 \gamma - \sin^2 \gamma}{\cos \gamma - \sin \gamma} = \dfrac{(\cos \gamma + \sin \gamma)(\cos \gamma - \sin \gamma)}{\cos \gamma - \sin \gamma} =$

$\cos \gamma + \sin \gamma$

$\boxed{13}$ $\dfrac{\cos(-t)}{\sec(-t) + \tan(-t)} = \dfrac{\cos t}{\sec t + (-\tan t)} = \dfrac{\cos t}{\sec t - \tan t} = \dfrac{\cos t}{\dfrac{1}{\cos t} - \dfrac{\sin t}{\cos t}} = \dfrac{\cos t}{\dfrac{1 - \sin t}{\cos t}} =$

$\dfrac{\cos^2 t}{1 - \sin t} = \dfrac{1 - \sin^2 t}{1 - \sin t} = \dfrac{(1 - \sin t)(1 + \sin t)}{1 - \sin t} = 1 + \sin t$

$\boxed{15}$ In the following solution, we could multiply both the numerator and the denominator by *either* the conjugate of the numerator *or* the conjugate of the denominator. Since the numerator on the right side looks like the numerator on the left side, we'll change the denominator.

$\sqrt{\dfrac{1 - \cos t}{1 + \cos t}} = \sqrt{\dfrac{(1 - \cos t)}{(1 + \cos t)} \cdot \dfrac{(1 - \cos t)}{(1 - \cos t)}} = \sqrt{\dfrac{(1 - \cos t)^2}{1 - \cos^2 t}} = \sqrt{\dfrac{(1 - \cos t)^2}{\sin^2 t}} =$

$\dfrac{\sqrt{(1 - \cos t)^2}}{\sqrt{\sin^2 t}} = \dfrac{|1 - \cos t|}{|\sin t|} = \dfrac{1 - \cos t}{|\sin t|}$, since $(1 - \cos t) \geq 0$.

$\boxed{19}$ We need to break down the angle argument of 4β into terms with only β as their argument. We can do this by using either the double-angle formula for the sine or the addition formula for the sine.

$\tfrac{1}{4} \sin 4\beta = \tfrac{1}{4} \sin(2 \cdot 2\beta) = \tfrac{1}{4}(2 \sin 2\beta \cos 2\beta) = \tfrac{1}{2}(2 \sin \beta \cos \beta)(\cos^2 \beta - \sin^2 \beta) =$

$\sin \beta \cos^3 \beta - \cos \beta \sin^3 \beta$

20 $\tan\frac{1}{2}\theta = \frac{1-\cos\theta}{\sin\theta}$ { apply the half-angle formula for the tangent }

$\qquad = \frac{1}{\sin\theta} - \frac{\cos\theta}{\sin\theta}$ { split up the fraction }

$\qquad = \csc\theta - \cot\theta$ { reciprocal and cotangent identities }

22 Let $\alpha = \arctan x$ and $\beta = \arctan\dfrac{2x}{1-x^2}$.

Because $-1 < x < 1$, $\arctan(-1) < \arctan(x) < \arctan(1)$, and hence $-\frac{\pi}{4} < \alpha < \frac{\pi}{4}$.

Thus, $\tan\alpha = x$ and $\tan\beta = \dfrac{2x}{1-x^2} = \dfrac{2\tan\alpha}{1-\tan^2\alpha} = \tan 2\alpha$.

Since the tangent function is one-to-one on $(-\frac{\pi}{2}, \frac{\pi}{2})$, we have $\beta = 2\alpha$ or, equivalently,

$\qquad \alpha = \frac{1}{2}\beta$. In terms of x, we have $\arctan x = \frac{1}{2}\arctan\dfrac{2x}{1-x^2}$.

24 $2\cos\alpha + \tan\alpha = \sec\alpha \Rightarrow 2\cos\alpha + \frac{\sin\alpha}{\cos\alpha} = \frac{1}{\cos\alpha} \Rightarrow$ { multiply by the lcd, $\cos\alpha$ }

$2\cos^2\alpha + \sin\alpha = 1 \Rightarrow 2(1-\sin^2\alpha) + \sin\alpha = 1 \Rightarrow 2 - 2\sin^2\alpha + \sin\alpha = 1 \Rightarrow$

$2\sin^2\alpha - \sin\alpha - 1 = 0 \Rightarrow (2\sin\alpha + 1)(\sin\alpha - 1) = 0 \Rightarrow \sin\alpha = -\frac{1}{2}, 1 \Rightarrow \alpha = \frac{7\pi}{6}, \frac{11\pi}{6}, \frac{\pi}{2}$.

Checking these values in the original equation, we see that $\tan\frac{\pi}{2}$ is undefined so

$\qquad$ exclude $\frac{\pi}{2}$ and our solution is $\alpha = \frac{7\pi}{6}, \frac{11\pi}{6}$.

25 $\sin\theta = \tan\theta \Rightarrow \sin\theta - \frac{\sin\theta}{\cos\theta} = 0 \Rightarrow \sin\theta\left(1 - \frac{1}{\cos\theta}\right) = 0 \Rightarrow \sin\theta = 0$ or $1 = \frac{1}{\cos\theta} \Rightarrow$

$\qquad \sin\theta = 0$ or $\cos\theta = 1 \Rightarrow \theta = 0, \pi$ or $\theta = 0 \Rightarrow \theta = 0, \pi$

27 $2\cos^3 t + \cos^2 t - 2\cos t - 1 = 0 \Rightarrow \cos^2 t(2\cos t + 1) - 1(2\cos t + 1) = 0 \Rightarrow$

$\qquad (\cos^2 t - 1)(2\cos t + 1) = 0 \Rightarrow \cos t = \pm 1, -\frac{1}{2} \Rightarrow t = 0, \pi, \frac{2\pi}{3}, \frac{4\pi}{3}$

29 $\sin\beta + 2\cos^2\beta = 1 \Rightarrow \sin\beta + 2(1-\sin^2\beta) = 1 \Rightarrow 2\sin^2\beta - \sin\beta - 1 = 0 \Rightarrow$

$\qquad (2\sin\beta + 1)(\sin\beta - 1) = 0 \Rightarrow \sin\beta = -\frac{1}{2}, 1 \Rightarrow \beta = \frac{7\pi}{6}, \frac{11\pi}{6}, \frac{\pi}{2}$

32 $\tan 2x \cos 2x = \sin 2x \Rightarrow \sin 2x = \sin 2x$. This is an identity and is true for all values

of x in $[0, 2\pi)$ except those that make $\tan 2x$ undefined, or, equivalently, those that

make $\cos 2x$ equal to 0. $\cos 2x = 0 \Rightarrow 2x = \frac{\pi}{2} + \pi n \Rightarrow x = \frac{\pi}{4} + \frac{\pi}{2}n$.

$\qquad$ Hence, the solutions are all x in $[0, 2\pi)$ except $\frac{\pi}{4}, \frac{3\pi}{4}, \frac{5\pi}{4}, \frac{7\pi}{4}$.

33 $2\cos 3x \cos 2x = 1 - 2\sin 3x \sin 2x \Rightarrow 2\cos 3x \cos 2x + 2\sin 3x \sin 2x = 1 \Rightarrow$

$\qquad 2(\cos 3x \cos 2x + \sin 3x \sin 2x) = 1 \Rightarrow \cos(3x - 2x) = \frac{1}{2} \Rightarrow \cos x = \frac{1}{2} \Rightarrow x = \frac{\pi}{3}, \frac{5\pi}{3}$

35 $\cos\pi x + \sin\pi x = 0 \Rightarrow \sin\pi x = -\cos\pi x \Rightarrow \frac{\sin\pi x}{\cos\pi x} = \frac{-\cos\pi x}{\cos\pi x}$ { divide by $\cos\pi x$ } $\Rightarrow$

$\qquad \tan\pi x = -1 \Rightarrow \pi x = \frac{3\pi}{4} + \pi n \Rightarrow x = \frac{3}{4} + n \Rightarrow x = \frac{3}{4}, \frac{7}{4}, \frac{11}{4}, \frac{15}{4}, \frac{19}{4}, \frac{23}{4}$

$\boxed{37}$ $2\cos^2\frac{1}{2}\theta - 3\cos\theta = 0 \Rightarrow 2\left(\frac{1+\cos\theta}{2}\right) - 3\cos\theta = 0 \Rightarrow$

$$(1+\cos\theta) - 3\cos\theta = 0 \Rightarrow 1 - 2\cos\theta = 0 \Rightarrow \cos\theta = \frac{1}{2} \Rightarrow \theta = \frac{\pi}{3}, \frac{5\pi}{3}$$

$\boxed{39}$ $\sin 5x = \sin 3x \Rightarrow \sin 5x - \sin 3x = 0 \Rightarrow$ [S2] $2\cos\dfrac{5x+3x}{2}\sin\dfrac{5x-3x}{2} = 0 \Rightarrow$

$\cos 4x \sin x = 0 \Rightarrow 4x = \frac{\pi}{2} + \pi n$ or $x = \pi n \Rightarrow x = \frac{\pi}{8} + \frac{\pi}{4}n$ or $x = 0, \pi \Rightarrow$

$$x = 0, \frac{\pi}{8}, \frac{3\pi}{8}, \frac{5\pi}{8}, \frac{7\pi}{8}, \pi, \frac{9\pi}{8}, \frac{11\pi}{8}, \frac{13\pi}{8}, \frac{15\pi}{8}$$

$\boxed{40}$ $\cos 3x = -\cos 2x \Rightarrow \cos 3x + \cos 2x = 0 \Rightarrow$ [S3] $2\cos\dfrac{3x+2x}{2}\cos\dfrac{3x-2x}{2} = 0 \Rightarrow$

$\cos\frac{5}{2}x \cos\frac{1}{2}x = 0 \Rightarrow \frac{5}{2}x = \frac{\pi}{2} + \pi n$ or $\frac{1}{2}x = \frac{\pi}{2} + \pi n \Rightarrow x = \frac{\pi}{5} + \frac{2\pi}{5}n$ or $x = \pi + 2\pi n \Rightarrow$

$$x = \frac{\pi}{5}, \frac{3\pi}{5}, \pi, \frac{7\pi}{5}, \frac{9\pi}{5}$$

$\boxed{42}$ $\tan 285° = \tan(225° + 60°) =$

$$\frac{\tan 225° + \tan 60°}{1 - \tan 225° \tan 60°} = \frac{1 + \sqrt{3}}{1 - 1\cdot\sqrt{3}} = \frac{1+\sqrt{3}}{1-\sqrt{3}} \cdot \frac{1+\sqrt{3}}{1+\sqrt{3}} = \frac{4 + 2\sqrt{3}}{-2} = -2 - \sqrt{3}$$

$\boxed{44}$ $\csc\dfrac{\pi}{8} = \dfrac{1}{\csc\frac{\pi}{8}} = \dfrac{1}{\sin\left(\frac{1}{2}\cdot\frac{\pi}{4}\right)} = \dfrac{1}{\sqrt{\dfrac{1-\cos\frac{\pi}{4}}{2}}} = \dfrac{1}{\sqrt{\dfrac{1-\sqrt{2}/2}{2}}} = \dfrac{1}{\sqrt{\dfrac{2-\sqrt{2}}{4}}} = \dfrac{2}{\sqrt{2-\sqrt{2}}}$

$\boxed{45}$ $\csc\theta = \frac{5}{3}$ and $\cos\phi = \frac{8}{17} \Rightarrow \sin\theta = \frac{3}{5}, \cos\theta = \frac{4}{5}, \tan\theta = \frac{3}{4}$ and $\sin\phi = \frac{15}{17}, \tan\phi = \frac{15}{8}$.

$$\sin(\theta + \phi) = \sin\theta \cos\phi + \cos\theta \sin\phi = \frac{3}{5}\cdot\frac{8}{17} + \frac{4}{5}\cdot\frac{15}{17} = \frac{84}{85}$$

$\boxed{48}$ $\tan(\theta - \phi) = \dfrac{\tan\theta - \tan\phi}{1 + \tan\theta\tan\phi} = \dfrac{\frac{3}{4} - \frac{15}{8}}{1 + \frac{3}{4}\cdot\frac{15}{8}} \cdot \dfrac{32}{32} = \dfrac{24 - 60}{32 + 45} = -\dfrac{36}{77}$

$\boxed{49}$ $\sin(\phi - \theta) = \sin\phi\cos\theta - \cos\phi\sin\theta = \frac{15}{17}\cdot\frac{4}{5} - \frac{8}{17}\cdot\frac{3}{5} = \frac{36}{85}$

$\boxed{50}$ First recognize the relationship to the expression in Exercise 49.

$$\sin(\theta - \phi) = \sin[-(\phi - \theta)] = -\sin(\phi - \theta) = -\frac{36}{85}$$

$\boxed{52}$ $\cos 2\phi = \cos^2\phi - \sin^2\phi = \left(\frac{8}{17}\right)^2 - \left(\frac{15}{17}\right)^2 = -\frac{161}{289}$

$\boxed{54}$ $\sin\frac{1}{2}\theta = \sqrt{\dfrac{1-\cos\theta}{2}} = \sqrt{\dfrac{1-\frac{4}{5}}{2}} = \sqrt{\dfrac{\frac{1}{5}}{2}} = \sqrt{\dfrac{1}{10}\cdot\dfrac{10}{10}} = \dfrac{1}{10}\sqrt{10}$

$\boxed{55}$ $\tan\frac{1}{2}\theta = \dfrac{1-\cos\theta}{\sin\theta} = \dfrac{1-\frac{4}{5}}{\frac{3}{5}} = \dfrac{\frac{1}{5}}{\frac{3}{5}} = \dfrac{1}{3}$

$\boxed{57}$ (a) $\sin 7t \sin 4t = $ [P4] $\frac{1}{2}[\cos(7t - 4t) - \cos(7t + 4t)] = \frac{1}{2}\cos 3t - \frac{1}{2}\cos 11t$

    (b) $\cos\frac{1}{4}u \cos\left(-\frac{1}{6}u\right) = $ [P3] $\frac{1}{2}\left\{\cos\left[\frac{1}{4}u + \left(-\frac{1}{6}u\right)\right] + \cos\left[\frac{1}{4}u - \left(-\frac{1}{6}u\right)\right]\right\} =$

$$\frac{1}{2}\left(\cos\frac{2}{24}u + \cos\frac{10}{24}u\right) = \frac{1}{2}\cos\frac{1}{12}u + \frac{1}{2}\cos\frac{5}{12}u$$

$\boxed{58}$ (b) $\cos 3\theta - \cos 8\theta = [S4]\ -2\sin\dfrac{3\theta + 8\theta}{2}\sin\dfrac{3\theta - 8\theta}{2} = -2\sin\tfrac{11}{2}\theta\,\sin\left(-\tfrac{5}{2}\theta\right) =$

$$2\sin\tfrac{11}{2}\theta\,\sin\tfrac{5}{2}\theta$$

(c) $\sin\tfrac{1}{4}t - \sin\tfrac{1}{5}t = [S2]\ 2\cos\dfrac{\tfrac{1}{4}t + \tfrac{1}{5}t}{2}\sin\dfrac{\tfrac{1}{4}t - \tfrac{1}{5}t}{2} = 2\cos\dfrac{\tfrac{5}{20}t + \tfrac{4}{20}t}{2}\sin\dfrac{\tfrac{5}{20}t - \tfrac{4}{20}t}{2} =$

$$2\cos\tfrac{9}{40}t\,\sin\tfrac{1}{40}t$$

$\boxed{62}$ $\arccos\left(\tan\tfrac{3\pi}{4}\right) = \arccos(-1) = \pi$

$\boxed{63}$ $\arcsin\left(\sin\tfrac{5\pi}{4}\right) = \arcsin\left(-\dfrac{\sqrt{2}}{2}\right) = -\dfrac{\pi}{4}$ $\quad$ $\boxed{64}$ $\cos^{-1}\left(\cos\tfrac{5\pi}{4}\right) = \cos^{-1}\left(-\dfrac{\sqrt{2}}{2}\right) = \dfrac{3\pi}{4}$

$\boxed{69}$ Let $\alpha = \sin^{-1}\tfrac{15}{17}$ and $\beta = \sin^{-1}\tfrac{8}{17}$.

$\cos\left(\sin^{-1}\tfrac{15}{17} - \sin^{-1}\tfrac{8}{17}\right) = \cos(\alpha - \beta) = \cos\alpha\cos\beta + \sin\alpha\sin\beta = \tfrac{8}{17}\cdot\tfrac{15}{17} + \tfrac{15}{17}\cdot\tfrac{8}{17} = \tfrac{240}{289}.$

$\boxed{70}$ Let $\alpha = \sin^{-1}\tfrac{4}{5}$. $\cos\left(2\sin^{-1}\tfrac{4}{5}\right) = \cos(2\alpha) = \cos^2\alpha - \sin^2\alpha = \left(\tfrac{3}{5}\right)^2 - \left(\tfrac{4}{5}\right)^2 = -\tfrac{7}{25}.$

$\boxed{73}$ $y = 1 - \sin^{-1}x = -\sin^{-1}x + 1$ $\quad\bullet$

Reflect $y = \sin^{-1}x$ through the x-axis and shift it up 1 unit.

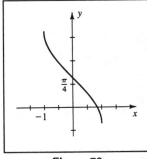 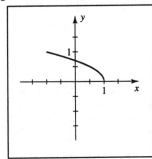

Figure 73 $\qquad\qquad\qquad\qquad$ *Figure 74*

$\boxed{74}$ If $\alpha = \cos^{-1}x$, then $\cos\alpha = x$, where $0 \le \alpha \le \pi$.

Hence, $y = \sin\left(\tfrac{1}{2}\cos^{-1}x\right) = \sin\tfrac{1}{2}\alpha = \sqrt{\dfrac{1 - \cos\alpha}{2}} = \sqrt{\dfrac{1 - x}{2}}.$

Thus, we have the graph of the half-parabola $y = \sqrt{\tfrac{1}{2}(1 - x)}$ on the interval $[-1, 1]$.

$\boxed{75}$ $\cos(\alpha + \beta + \gamma)$

$= \cos\big[(\alpha + \beta) + \gamma\big]$

$= \cos(\alpha + \beta)\cos\gamma - \sin(\alpha + \beta)\sin\gamma$

$= (\cos\alpha\cos\beta - \sin\alpha\sin\beta)\cos\gamma - (\sin\alpha\cos\beta + \cos\alpha\sin\beta)\sin\gamma$

$= \cos\alpha\cos\beta\cos\gamma - \sin\alpha\sin\beta\cos\gamma - \sin\alpha\cos\beta\sin\gamma - \cos\alpha\sin\beta\sin\gamma$

76 (a) $t = -\frac{\pi}{2b} \Rightarrow F = A\left[\cos\left(-\frac{\pi}{2}\right) - a\cos\left(-\frac{3\pi}{2}\right)\right] = A(0 - a \cdot 0) = 0$

$t = \frac{\pi}{2b} \Rightarrow F = A\left(\cos\frac{\pi}{2} - a\cos\frac{3\pi}{2}\right) = A(0 - a \cdot 0) = 0$

(b) $a = \frac{1}{3} \Rightarrow \sin 3bt = \sin bt \Rightarrow \sin 3bt - \sin bt = 0 \Rightarrow$

[S2] $2\cos\dfrac{3bt + bt}{2}\sin\dfrac{3bt - bt}{2} = 0 \Rightarrow \cos 2bt \sin bt = 0 \Rightarrow$

$\cos 2bt = 0$ or $\sin bt = 0 \Rightarrow 2bt = \frac{\pi}{2} + \pi n$ or $bt = \pi n \Rightarrow$

$t = \frac{\pi}{4b} + \frac{\pi}{2b}n$ or $t = \frac{\pi}{b}n$. Since $-\frac{\pi}{2b} < t < \frac{\pi}{2b}$, $t = \pm\frac{\pi}{4b}$, 0.

(c) Using the values from part (b), $t = 0 \Rightarrow F = A\left(\cos 0 - \frac{1}{3}\cos 0\right) = A\left(1 - \frac{1}{3}\right) = \frac{2}{3}A$.

$t = \pm\frac{\pi}{4b} \Rightarrow F = A\left[\cos\left(\pm\frac{\pi}{4}\right) - \frac{1}{3}\cos\left(\pm\frac{3\pi}{4}\right)\right] = A\left(\frac{\sqrt{2}}{2} + \frac{\sqrt{2}}{6}\right) = \frac{4\sqrt{2}}{6}A = \frac{2}{3}\sqrt{2}\,A$.

The second value is $\sqrt{2}$ times the first, hence $\frac{2}{3}\sqrt{2}\,A$ is the maximum force.

78 (a) Bisect θ to form two right triangles. $\tan\frac{1}{2}\theta = \dfrac{\frac{1}{2}x}{d} \Rightarrow x = 2d\tan\frac{1}{2}\theta$.

(b) Using part (a) with $x = 0.5$ ft and $\theta = 0.0005$ radian,

we have $d = \dfrac{x}{2\tan\frac{1}{2}\theta} \approx 1000$ ft, so $d \le 1000$ ft.

79 (a) Bisect θ to form two right triangles.

$\cos\frac{1}{2}\theta = \dfrac{r}{d + r} \Rightarrow d + r = \dfrac{r}{\cos\frac{1}{2}\theta} \Rightarrow d = r\sec\frac{1}{2}\theta - r = r\left(\sec\frac{1}{2}\theta - 1\right)$.

(b) $d = 300$ and $r = 4000 \Rightarrow \cos\frac{1}{2}\theta = \dfrac{r}{d + r} = \frac{4000}{4300} \Rightarrow \frac{1}{2}\theta \approx 21.5° \Rightarrow \theta \approx 43°$.

80 $N = h/w$ and $\tan\theta = N \Rightarrow \tan\theta = \frac{h}{w}$.

(a) $\tan\theta = \frac{h}{w} = \frac{400}{80} = 5 \Rightarrow \theta = \tan^{-1} 5 \approx 78.7°$

(b) $\tan\theta = \frac{h}{w} = \frac{55}{30} = \frac{11}{6} \Rightarrow \theta = \tan^{-1}\frac{11}{6} \approx 61.4°$

1 $\dfrac{\tan x}{1-\cot x}+\dfrac{\cot x}{1-\tan x}=\dfrac{\frac{\sin x}{\cos x}}{1-\frac{\cos x}{\sin x}}+\dfrac{\frac{\cos x}{\sin x}}{1-\frac{\sin x}{\cos x}}=$

$$\dfrac{\sin^2 x}{\cos x\,(\sin x-\cos x)}+\dfrac{\cos^2 x}{\sin x\,(\cos x-\sin x)}=\dfrac{\sin^2 x}{\cos x\,(\sin x-\cos x)}-\dfrac{\cos^2 x}{\sin x\,(\sin x-\cos x)}=$$

$$\dfrac{\sin^3 x-\cos^3 x}{\cos x\,\sin x\,(\sin x-\cos x)}=\dfrac{(\sin x-\cos x)(\sin^2 x+\sin x\,\cos x+\cos^2 x)}{\cos x\,\sin x\,(\sin x-\cos x)}=$$

$$\dfrac{1+\sin x\,\cos x}{\cos x\,\sin x}=\dfrac{1}{\cos x\,\sin x}+1=1+\sec x\,\csc x$$

3 *Note:* Graphing on a TI-82/83 doesn't really help to solve this problem.

$3\cos 45x+4\sin 45x=5\Rightarrow\{$ by Example 6 in Section 3.3 $\}$

$5\cos\left(45x-\tan^{-1}\frac{4}{3}\right)=5\Rightarrow\cos\left(45x-\tan^{-1}\frac{4}{3}\right)=1\Rightarrow45x-\tan^{-1}\frac{4}{3}=2\pi n\Rightarrow$

$45x=2\pi n+\tan^{-1}\frac{4}{3}\Rightarrow x=\dfrac{2\pi n+\tan^{-1}\frac{4}{3}}{45}$. $\;\;n=0,\,1,\,\dots,\,44$ will yield x values in

$[0,\,2\pi)$. *Note:* After using Example 6, you might notice that this is a function with period $2\pi/45$, and it will obtain 45 maximums on an interval of length 2π. The largest value of x is approximately 6.164 {when $n=44$}.

5 Let $\alpha=\tan^{-1}\!\left(\frac{1}{239}\right)$ and $\theta=\tan^{-1}\!\left(\frac{1}{5}\right)$. $\frac{\pi}{4}=4\theta-\alpha\Rightarrow\frac{\pi}{4}+\alpha=4\theta$. Both sides are acute angles, and we will show that the tangent of each side is equal to the same value, hence proving the identity.

$$\text{LS}=\tan\left(\tfrac{\pi}{4}+\alpha\right)=\dfrac{\tan\frac{\pi}{4}+\tan\alpha}{1-\tan\frac{\pi}{4}\tan\alpha}=\dfrac{1+\frac{1}{239}}{1-1\cdot\frac{1}{239}}=\dfrac{\frac{240}{239}}{\frac{238}{239}}=\dfrac{240}{238}=\dfrac{120}{119}.$$

$\text{RS}=\tan 4\theta=$

$$\dfrac{2\tan 2\theta}{1-\tan^2(2\theta)}=\dfrac{2\cdot\frac{2\tan\theta}{1-\tan^2\theta}}{1-\left(\frac{2\tan\theta}{1-\tan^2\theta}\right)^2}=\dfrac{2\cdot\frac{\frac{2}{5}}{1-\frac{1}{25}}}{1-\left(\frac{\frac{2}{5}}{1-\frac{1}{25}}\right)^2}=\dfrac{\frac{\frac{4}{5}}{\frac{24}{25}}}{1-\frac{\frac{4}{25}}{\frac{24\cdot24}{25\cdot25}}}=\dfrac{\frac{5}{6}}{\frac{119}{144}}=\dfrac{120}{119}.$$

Similarly, for the second relationship, we could write $\frac{\pi}{4}=\alpha+\beta+\gamma$ and show that $\tan\left(\frac{\pi}{4}-\alpha\right)=\tan\left(\beta+\gamma\right)=\frac{1}{3}$.

Chapter 4: Applications of Trigonometry

1 $\beta = 180° - \alpha - \gamma = 180° - 41° - 77° = 62°.$

$$\frac{b}{\sin \beta} = \frac{a}{\sin \alpha} \Rightarrow b = \frac{a \sin \beta}{\sin \alpha} = \frac{10.5 \sin 62°}{\sin 41°} \approx 14.1.$$

$$\frac{c}{\sin \gamma} = \frac{a}{\sin \alpha} \Rightarrow c = \frac{a \sin \gamma}{\sin \alpha} = \frac{10.5 \sin 77°}{\sin 41°} \approx 15.6.$$

3 $\gamma = 180° - \alpha - \beta = 180° - 27°40' - 52°10' = 100°10'.$

$$\frac{b}{\sin \beta} = \frac{a}{\sin \alpha} \Rightarrow b = \frac{a \sin \beta}{\sin \alpha} = \frac{32.4 \sin 52°10'}{\sin 27°40'} \approx 55.1.$$

$$\frac{c}{\sin \gamma} = \frac{a}{\sin \alpha} \Rightarrow c = \frac{a \sin \gamma}{\sin \alpha} = \frac{32.4 \sin 100°10'}{\sin 27°40'} \approx 68.7.$$

7 $\dfrac{\sin \beta}{b} = \dfrac{\sin \gamma}{c} \Rightarrow \beta = \sin^{-1}\left(\dfrac{b \sin \gamma}{c}\right) = \sin^{-1}\left(\dfrac{12 \sin 81°}{11}\right) \approx \sin^{-1}(1.0775).$ Since 1.0775 is

not in the domain of the inverse sine function, which is $[-1, 1]$, *no triangle exists.*

9 $\dfrac{\sin \alpha}{a} = \dfrac{\sin \gamma}{c} \Rightarrow \alpha = \sin^{-1}\left(\dfrac{a \sin \gamma}{c}\right) = \sin^{-1}\left(\dfrac{140 \sin 53°20'}{115}\right) \approx \sin^{-1}(0.9765) \approx$

$77°30'$ or $102°30'$ {rounded to the nearest 10 minutes}. When using the inverse sine function to solve for an angle, remember that there are always *two* values if $\theta \neq 90°$. The first angle is the one you obtain using your calculator, and the second is the reference angle for the first angle in the second quadrant. After obtaining these values, we need to check the sum of the angles. If this sum is less than $180°$, a triangle is formed. If this sum is greater than or equal to $180°$, no triangle is formed. For this exercise, there are two triangles possible since in either case $\alpha + \gamma < 180°$.

$\beta = (180° - \gamma) - \alpha \approx (180° - 53°20') - (77°30' \text{ or } 102°30') = 49°10' \text{ or } 24°10'.$

$$\frac{b}{\sin \beta} = \frac{c}{\sin \gamma} \Rightarrow b = \frac{c \sin \beta}{\sin \gamma} \approx \frac{115 \sin(49°10' \text{ or } 24°10')}{\sin 53°20'} \approx 108 \text{ or } 58.7.$$

11 $\dfrac{\sin \alpha}{a} = \dfrac{\sin \gamma}{c} \Rightarrow \alpha = \sin^{-1}\left(\dfrac{a \sin \gamma}{c}\right) = \sin^{-1}\left(\dfrac{131.08 \sin 47.74°}{97.84}\right) \approx \sin^{-1}(0.9915) \approx$

$82.54°$ or $97.46°$. There are two triangles possible since in either case $\alpha + \gamma < 180°$.

$\beta = (180° - \gamma) - \alpha \approx (180° - 47.74°) - (82.54° \text{ or } 97.46°) = 49.72° \text{ or } 34.80°.$

$$\frac{b}{\sin \beta} = \frac{c}{\sin \gamma} \Rightarrow b = \frac{c \sin \beta}{\sin \gamma} \approx \frac{97.84 \sin(49.72° \text{ or } 34.80°)}{\sin 47.74°} \approx 100.85 \text{ or } 75.45.$$

13 $\dfrac{\sin \beta}{b} = \dfrac{\sin \alpha}{a} \Rightarrow \beta = \sin^{-1}\left(\dfrac{b \sin \alpha}{a}\right) = \sin^{-1}\left(\dfrac{18.9 \sin 65°10'}{21.3}\right) \approx \sin^{-1}(0.8053) \approx$

$53°40'$ or $126°20'$ {rounded to the nearest 10 minutes}. Reject $126°20'$ because then $\alpha + \beta \geq 180°.$ $\gamma = 180° - \alpha - \beta \approx 180° - 65°10' - 53°40' = 61°10'.$

$$\frac{c}{\sin \gamma} = \frac{a}{\sin \alpha} \Rightarrow c = \frac{a \sin \gamma}{\sin \alpha} \approx \frac{21.3 \sin 61°10'}{\sin 65°10'} \approx 20.6.$$

15   $\dfrac{\sin \gamma}{c} = \dfrac{\sin \beta}{b} \Rightarrow \gamma = \sin^{-1}\left(\dfrac{c \sin \beta}{b}\right) = \sin^{-1}\left(\dfrac{0.178 \sin 121.624^\circ}{0.283}\right) \approx \sin^{-1}(0.5356) \approx$

32.383° or 147.617°. Reject 147.617° because then $\beta + \gamma \geq 180^\circ$.

$\alpha = 180^\circ - \beta - \gamma \approx 180^\circ - 121.624^\circ - 32.383^\circ = 25.993^\circ.$

$$\dfrac{a}{\sin \alpha} = \dfrac{b}{\sin \beta} \Rightarrow a = \dfrac{b \sin \alpha}{\sin \beta} \approx \dfrac{0.283 \sin 25.993^\circ}{\sin 121.624^\circ} \approx 0.146.$$

17   $\angle ABC = 180^\circ - 54^\circ10' - 63^\circ20' = 62^\circ30'. \quad \dfrac{\overline{AB}}{\sin 54^\circ10'} = \dfrac{240}{\sin 62^\circ30'} \Rightarrow \overline{AB} \approx 219.36$ yd

19   (a) $\angle ABP = 180^\circ - 65^\circ = 115^\circ. \quad \angle APB = 180^\circ - 21^\circ - 115^\circ = 44^\circ.$

$$\dfrac{\overline{AP}}{\sin 115^\circ} = \dfrac{1.2}{\sin 44^\circ} \Rightarrow \overline{AP} \approx 1.57, \text{ or } 1.6 \text{ mi.}$$

(b) $\sin 21^\circ = \dfrac{\text{height of } P}{\overline{AP}} \Rightarrow$ height of $P = \dfrac{1.2 \sin 115^\circ \sin 21^\circ}{\sin 44^\circ}$ { from part (a) } $\approx 0.56,$

or 0.6 mi.

21   Let C denote the base of the balloon and P its projection on the ground.

$\angle ACB = 180^\circ - 24^\circ10' - 47^\circ40' = 108^\circ10'. \quad \dfrac{\overline{AC}}{\sin 47^\circ40'} = \dfrac{8.4}{\sin 108^\circ10'} \Rightarrow \overline{AC} \approx 6.5$ mi.

$$\sin 24^\circ10' = \dfrac{\overline{PC}}{\overline{AC}} \Rightarrow \overline{PC} = \dfrac{8.4 \sin 47^\circ40' \sin 24^\circ10'}{\sin 108^\circ10'} \approx 2.7 \text{ mi.}$$

23   $\angle APQ = 57^\circ - 22^\circ = 35^\circ. \quad \angle AQP = 180^\circ - (63^\circ - 22^\circ) = 139^\circ.$

$$\angle PAQ = 180^\circ - 139^\circ - 35^\circ = 6^\circ. \quad \dfrac{\overline{AP}}{\sin 139^\circ} = \dfrac{100}{\sin 6^\circ} \Rightarrow \overline{AP} = \dfrac{100 \sin 139^\circ}{\sin 6^\circ} \approx 628 \text{ m.}$$

25   $\angle FAB = 90^\circ - 27^\circ10' = 62^\circ50'.$

$\angle FBA = 90^\circ - 52^\circ40' = 37^\circ20'.$

$\angle AFB = 180^\circ - 62^\circ50' - 37^\circ20' = 79^\circ50'.$

$\dfrac{\overline{AF}}{\sin 37^\circ20'} = \dfrac{6}{\sin 79^\circ50'} \Rightarrow \overline{AF} \approx 3.70$ mi.

$\dfrac{\overline{BF}}{\sin 62^\circ50'} = \dfrac{6}{\sin 79^\circ50'} \Rightarrow \overline{BF} \approx 5.42$ mi.

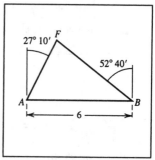

Figure 25

27   Let A denote the base of the hill, B the base of the cathedral, and C the top of the

spire. The angle at the base of the hill is $180^\circ - 48^\circ = 132^\circ$. The angle at the top

of the spire is $180^\circ - 132^\circ - 41^\circ = 7^\circ. \quad \dfrac{\overline{AC}}{\sin 41^\circ} = \dfrac{200}{\sin 7^\circ} \Rightarrow \overline{AC} = \dfrac{200 \sin 41^\circ}{\sin 7^\circ} \approx 1077$ ft.

$\angle BAC = 48^\circ - 32^\circ = 16^\circ. \quad \angle ACB = 90^\circ - 48^\circ = 42^\circ. \quad \angle ABC = 180^\circ - 42^\circ - 16^\circ = 122^\circ.$

$$\dfrac{\overline{BC}}{\sin 16^\circ} = \dfrac{\overline{AC}}{\sin 122^\circ} \Rightarrow \overline{BC} = \dfrac{200 \sin 41^\circ \sin 16^\circ}{\sin 7^\circ \sin 122^\circ} \approx 350 \text{ ft.}$$

29 (a) In the triangle that forms the base, the third angle is $180° - 103° - 52° = 25°$.

Let l denote the length of the dashed line. $\dfrac{l}{\sin 103°} = \dfrac{12.0}{\sin 25°} \Rightarrow l \approx 27.7$ units.

Now $\tan 34° = \dfrac{h}{l} \Rightarrow h \approx 18.7$ units.

(b) Draw a line from the 103° angle that is perpendicular to l and call it d.

$\sin 52° = \dfrac{d}{12} \Rightarrow d \approx 9.5$ units. The area of the triangular base is $B = \frac{1}{2}ld$.

The volume V is $\frac{1}{3}(\frac{1}{2}ld)h = 288 \sin 52° \sin^2 103° \tan 34° \csc^2 25° \approx 814$ cubic units.

31 Draw a line through P perpendicular to the x-axis. Locate points A and B on this line so that $\angle PAQ = \angle PBR = 90°$. $\overline{AP} = 5127.5 - 3452.8 = 1674.7$, $\overline{AQ} = 3145.8 - 1487.7 = 1658.1$, and $\tan \angle APQ = \frac{1658.1}{1674.7} \Rightarrow \angle APQ \approx 44°43'$. Thus, $\angle BPR \approx 180° - 55°50' - 44°43' = 79°27'$.

By the distance formula, $\overline{PQ} \approx \sqrt{(1674.7)^2 + (1658.1)^2} \approx 2356.7$.

Now $\dfrac{\overline{PR}}{\sin 65°22'} = \dfrac{\overline{PQ}}{\sin(180° - 55°50' - 65°22')} \Rightarrow \overline{PR} = \dfrac{2356.7 \sin 65°22'}{\sin 58°48'} \approx 2504.5$.

$\sin \angle BPR = \dfrac{\overline{BR}}{\overline{PR}} \Rightarrow \overline{BR} \approx (2504.5)(\sin 79°27') \approx 2462.2$. $\cos \angle BPR = \dfrac{\overline{BP}}{\overline{PR}} \Rightarrow$

$\overline{BP} \approx (2504.5)(\cos 79°27') \approx 458.6$. Using the coordinates of P, we see that

$$R(x, y) \approx (1487.7 + 2462.2, \ 3452.8 - 458.6) = (3949.9, 2994.2).$$

4.2 Exercises

Note: These formulas will be used to solve problems that involve the law of cosines.

(1) $a^2 = b^2 + c^2 - 2bc \cos \alpha \Rightarrow a = \sqrt{b^2 + c^2 - 2bc \cos \alpha}$

(Similar formulas are used for b and c.)

(2) $a^2 = b^2 + c^2 - 2bc \cos \alpha \Rightarrow 2bc \cos \alpha = b^2 + c^2 - a^2 \Rightarrow$

$\cos \alpha = \left(\dfrac{b^2 + c^2 - a^2}{2bc}\right) \Rightarrow \alpha = \cos^{-1}\left(\dfrac{b^2 + c^2 - a^2}{2bc}\right)$

(Similar formulas are used for β and γ.)

1 $a = \sqrt{b^2 + c^2 - 2bc \cos \alpha} = \sqrt{20^2 + 30^2 - 2(20)(30) \cos 60°} = \sqrt{700} \approx 26$.

$\beta = \cos^{-1}\left(\dfrac{a^2 + c^2 - b^2}{2ac}\right) = \cos^{-1}\left(\dfrac{700 + 30^2 - 20^2}{2(\sqrt{700})(30)}\right) \approx \cos^{-1}(0.7559) \approx 41°$.

$\gamma = 180° - \alpha - \beta \approx 180° - 60° - 41° = 79°$.

$\boxed{3}$ $b = \sqrt{a^2 + c^2 - 2ac\cos\beta} = \sqrt{23,400 + 4500\sqrt{3}} \approx 177$, or 180.

$\alpha = \cos^{-1}\left(\dfrac{b^2 + c^2 - a^2}{2bc}\right) \approx \cos^{-1}(0.9054) \approx 25°10'$, or 25°.

Note: We do not have to worry about an "ambiguous" case of the law of cosines as we did with the law of sines. This is true because the inverse cosine function gives us a unique angle from 0° to 180° for all values in its domain.

$$\gamma = 180° - \alpha - \beta \approx 180° - 25°10' - 150° = 4°50', \text{ or } 5°.$$

$\boxed{7}$ $\alpha = \cos^{-1}\left(\dfrac{b^2 + c^2 - a^2}{2bc}\right) = \cos^{-1}(0.875) \approx 29°.$

$\beta = \cos^{-1}\left(\dfrac{a^2 + c^2 - b^2}{2ac}\right) = \cos^{-1}(0.6875) \approx 47°.$

$$\gamma = 180° - \alpha - \beta \approx 180° - 29° - 47° = 104°.$$

$\boxed{9}$ $\alpha = \cos^{-1}\left(\dfrac{b^2 + c^2 - a^2}{2bc}\right) \approx \cos^{-1}(0.9766) \approx 12°30'.$

$\beta = \cos^{-1}\left(\dfrac{a^2 + c^2 - b^2}{2ac}\right) = \cos^{-1}(-0.725) \approx 136°30'.$

$$\gamma = 180° - \alpha - \beta \approx 180° - 12°30' - 136°30' = 31°00'.$$

Note: When deciding whether to use the law of cosines or the law of sines, it may be helpful to remember that the law of cosines must be used when you have:

(1) 3 sides, or

(2) 2 sides and the angle between them.

$\boxed{11}$ We have two sides and the angle between them. Hence, we apply the law of cosines.

$$\text{Third side} = \sqrt{175^2 + 150^2 - 2(175)(150)\cos 73°40'} \approx 196 \text{ feet.}$$

$\boxed{13}$ 20 minutes $= \frac{1}{3}$ hour $\Rightarrow$ the cars have traveled $60(\frac{1}{3}) = 20$ miles and $45(\frac{1}{3}) = 15$ miles, respectively. The distance d apart is $d = \sqrt{20^2 + 15^2 - 2(20)(15)\cos 84°} \approx 24$ miles.

$\boxed{15}$ The first ship travels $(24)(2) = 48$ miles in two hours. The second ship travels $(18)(1\frac{1}{2}) = 27$ miles in $1\frac{1}{2}$ hours. The angle between the paths is $20° + 35° = 55°.$

$$\overline{AB} = \sqrt{27^2 + 48^2 - 2(27)(48)\cos 55°} \approx 39 \text{ miles.}$$

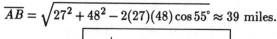

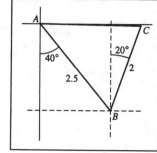

Figure 15   Figure 17

17 $\angle ABC = 40° + 20° = 60°$.

$\overline{AB} = \left(\dfrac{1 \text{ mile}}{8 \text{ min}} \cdot 20 \text{ min}\right) = 2.5$ miles and $\overline{BC} = \left(\dfrac{1 \text{ mile}}{8 \text{ min}} \cdot 16 \text{ min}\right) = 2$ miles.

$$\overline{AC} = \sqrt{2.5^2 + 2^2 - 2(2)(2.5)\cos 60°} = \sqrt{5.25} \approx 2.3 \text{ miles. See } \textit{Figure 17}.$$

19 $\gamma = \cos^{-1}\left(\dfrac{2^2 + 3^2 - 4^2}{2 \cdot 2 \cdot 3}\right) = \cos^{-1}(-0.25) \approx 104°29'. \quad \phi \approx 104°29' - 70° = 34°29'$.

The direction that the third side was traversed is approximately

$$\text{N}(90° - 34°29')\text{E} = \text{N}55°31'\text{E}.$$

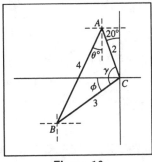

Figure 19

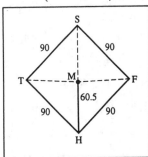

Figure 21

21 Let H denote home plate, M the mound, F first base, S second base, and T third base. $\overline{HS} = \sqrt{90^2 + 90^2} = 90\sqrt{2} \approx 127.3$ ft. $\overline{MS} = 90\sqrt{2} - 60.5 \approx 66.8$ ft.

$\angle MHF = 45°$ so $\overline{MF} = \sqrt{60.5^2 + 90^2 - 2(60.5)(90)\cos 45°} \approx 63.7$ ft.

$$\overline{MT} = \overline{MF} \text{ by the symmetry of the field.}$$

23 $\angle RTP = 21°$ and $\angle RSP = 37°$. $\sin \angle RSP = \dfrac{10,000}{\overline{SP}} \Rightarrow \overline{SP} = 10,000 \csc 37° \approx$

16,616 ft. $\sin \angle RTP = \dfrac{10,000}{\overline{TP}} \Rightarrow \overline{TP} = 10,000 \csc 21° \approx 27,904$ ft.

$$\overline{ST} = \sqrt{\overline{SP}^2 + \overline{TP}^2 - 2(\overline{SP})(\overline{TP})\cos 110°} \approx 37,039 \text{ ft} \approx 7 \text{ miles.}$$

25 Let $d = \overline{ES}$. $d^2 = R^2 + R^2 - 2RR\cos\theta \Rightarrow d^2 = 2R^2(1 - \cos\theta) \Rightarrow$

$d^2 = 4R^2\left(\dfrac{1 - \cos\theta}{2}\right) \Rightarrow d = 2R\sqrt{\dfrac{1 - \cos\theta}{2}} \Rightarrow d = 2R\sin\dfrac{\theta}{2}$.

$$\text{Since } d = vt, \ t = \dfrac{d}{v} = \dfrac{2R}{v}\sin\dfrac{\theta}{2}.$$

27 (a) $\angle BCP = \frac{1}{2}(\angle BCD) = \frac{1}{2}(72°) = 36°$. $\triangle BPC$ is isosceles so

$\angle BPC = \angle PBC$ and $2\angle BPC = 180° - 36° \Rightarrow \underline{\angle BPC = 72°}$.

$\underline{\angle APB} = 180° - \angle BPC = 180° - 72° = \underline{108°}$.

$$\angle ABP = 180° - \angle APB - \angle BAP = 180° - 108° - 36° = \underline{36°}.$$

(b) $\overline{BP} = \sqrt{\overline{BC}^2 + \overline{PC}^2 - 2(\overline{BC})(\overline{PC})\cos 36°} = \sqrt{1^2 + 1^2 - 2(1)(1)\cos 36°} \approx 0.62$.

(c) $\text{Area}_{\text{kite}} = 2(\text{Area of } \triangle BPC) = 2 \cdot \frac{1}{2}(\overline{CB})(\overline{CP})\sin\angle BCP = \sin 36° \approx 0.59$.

$\text{Area}_{\text{dart}} = 2(\text{Area of } \triangle ABP) = 2 \cdot \frac{1}{2}(\overline{AB})(\overline{BP})\sin\angle ABP = \overline{BP}\sin 36° \approx 0.36$.

$$\{\overline{BP} \text{ was found in part (b)}\}$$

Note: Exer. 29–36: $\mathcal{A}$ (the area) is measured in square units.

29 Since α is the angle between sides b and c, we may apply the area of a triangle

formula listed in this section. $\mathcal{A} = \frac{1}{2}bc \sin \alpha = \frac{1}{2}(20)(30)\sin 60° = 300(\sqrt{3}/2) \approx 260$.

31 $\gamma = 180° - \alpha - \beta = 180° - 40.3° - 62.9° = 76.8°$.

$$\frac{a}{\sin \alpha} = \frac{b}{\sin \beta} \Rightarrow a = \frac{b \sin \alpha}{\sin \beta} = \frac{5.63 \sin 40.3°}{\sin 62.9°}. \quad \mathcal{A} = \frac{1}{2}ab \sin \gamma \approx 11.21.$$

33 $\dfrac{\sin \beta}{b} = \dfrac{\sin \alpha}{a} \Rightarrow \sin \beta = \dfrac{b \sin \alpha}{a} = \dfrac{3.4 \sin 80.1°}{8.0} \approx 0.4187 \Rightarrow \beta \approx 24.8°$ or $155.2°$.

Reject $155.2°$ because then $\alpha + \beta = 235.3° \geq 180°$. $\gamma \approx 180° - 80.1° - 24.8° = 75.1°$.

$$\mathcal{A} = \frac{1}{2}ab \sin \gamma = \frac{1}{2}(8.0)(3.4)\sin 75.1° \approx 13.1.$$

35 Given the lengths of the 3 sides of a triangle, we compute the semiperimeter and then

use Heron's formula to find the area of the triangle.

$s = \frac{1}{2}(a + b + c) = \frac{1}{2}(25.0 + 80.0 + 60.0) = 82.5.$

$$\mathcal{A} = \sqrt{s(s-a)(s-b)(s-c)} = \sqrt{(82.5)(57.5)(2.5)(22.5)} \approx 516.56, \text{ or } 517.0.$$

37 $s = \frac{1}{2}(a + b + c) = \frac{1}{2}(115 + 140 + 200) = 227.5. \quad \mathcal{A} = \sqrt{s(s-a)(s-b)(s-c)} =$

$$\sqrt{(227.5)(112.5)(87.5)(27.5)} \approx 7847.6 \text{ yd}^2, \text{ or } \mathcal{A}/4840 \approx 1.62 \text{ acres}.$$

39 The area of the parallelogram is twice the area of the triangle formed by the two

sides and the included angle. $\mathcal{A} = 2(\frac{1}{2})(12.0)(16.0)\sin 40° \approx 123.4 \text{ ft}^2$.

4.3 Exercises

1 $|3 - 4i| = \sqrt{3^2 + (-4)^2} = \sqrt{25} = 5$ 　　　 5 $|8i| = |0 + 8i| = \sqrt{0^2 + 8^2} = \sqrt{64} = 8$

7 From Appendix V or a previous course that covered complex numbers, $i^m = i, -1,$

$-i,$ or 1. Since all of these are 1 unit from the origin, $|i^m| = 1$ for any integer m.

As an alternate solution,

$$\left|i^{500}\right| = \left|(i^4)^{125}\right| = \left|(1)^{125}\right| = |1| = |1 + 0i| = \sqrt{1^2 + 0^2} = \sqrt{1} = 1.$$

9 $|0| = |0 + 0i| = \sqrt{0^2 + 0^2} = \sqrt{0} = 0$

11 $4 + 2i$ 　　　　　 13 $3 - 5i$ 　　　　　 15 $-(3 - 6i) = -3 + 6i$

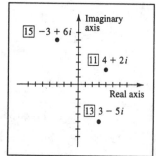

Figure for Exercises 11, 13, 15

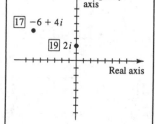

Figure for Exercises 17, 19

$\boxed{17}$ $2i(2+3i) = 4i + 6i^2 = 4i - 6 = -6 + 4i$. See the figure on the previous page.

$\boxed{19}$ $(1+i)^2 = 1 + 2(1)(i) + i^2 = 1 + 2i - 1 = 2i$. See the figure on the previous page.

Note: For each of the following exercises, we need to find r and θ.

If $z = a + bi$, then $r = \sqrt{a^2 + b^2}$. To find θ, we will use the fact that

$\tan\theta = \frac{b}{a}$ and our knowledge of what quadrant the terminal side of θ is in.

$\boxed{21}$ $z = 1 - i \Rightarrow r = \sqrt{1 + (-1)^2} = \sqrt{2}$. $\tan\theta = \frac{-1}{1} = -1$ and θ in QIV $\Rightarrow \theta = \frac{7\pi}{4}$.

Thus, $z = 1 - i = \sqrt{2}\left(\cos\frac{7\pi}{4} + i\sin\frac{7\pi}{4}\right)$, or simply $\sqrt{2}\operatorname{cis}\frac{7\pi}{4}$.

$\boxed{23}$ $z = -4\sqrt{3} + 4i \Rightarrow r = \sqrt{(-4\sqrt{3})^2 + 4^2} = \sqrt{64} = 8$.

$\tan\theta = \frac{4}{-4\sqrt{3}} = -\frac{1}{\sqrt{3}}$ and θ in QII $\Rightarrow \theta = \frac{5\pi}{6}$. $z = 8\operatorname{cis}\frac{5\pi}{6}$.

$\boxed{27}$ $z = -4 - 4i \Rightarrow r = \sqrt{(-4)^2 + (-4)^2} = \sqrt{32} = 4\sqrt{2}$.

$\tan\theta = \frac{-4}{-4} = 1$ and θ in QIII $\Rightarrow \theta = \frac{5\pi}{4}$. $z = 4\sqrt{2}\operatorname{cis}\frac{5\pi}{4}$.

$\boxed{29}$ $z = -20i \Rightarrow r = 20$. θ on the negative y-axis $\Rightarrow \theta = \frac{3\pi}{2}$. $z = 20\operatorname{cis}\frac{3\pi}{2}$.

$\boxed{31}$ $z = 12 \Rightarrow r = 12$. θ on the positive x-axis $\Rightarrow \theta = 0$. $z = 12\operatorname{cis}0$.

$\boxed{33}$ $z = -7 \Rightarrow r = 7$. θ on the negative x-axis $\Rightarrow \theta = \pi$. $z = 7\operatorname{cis}\pi$.

$\boxed{35}$ $z = 6i \Rightarrow r = 6$. θ on the positive y-axis $\Rightarrow \theta = \frac{\pi}{2}$. $z = 6\operatorname{cis}\frac{\pi}{2}$.

$\boxed{39}$ $z = 2 + i \Rightarrow r = \sqrt{2^2 + 1^2} = \sqrt{5}$. $\tan\theta = \frac{1}{2}$ and θ in QI $\Rightarrow \theta = \tan^{-1}\frac{1}{2}$.

$z = \sqrt{5}\operatorname{cis}\left(\tan^{-1}\frac{1}{2}\right)$.

$\boxed{41}$ $z = -3 + i \Rightarrow r = \sqrt{(-3)^2 + 1^2} = \sqrt{10}$.

$\tan\theta = \frac{1}{-3}$ and θ in QII $\Rightarrow \theta = \tan^{-1}\left(-\frac{1}{3}\right) + \pi$. We must add π to $\tan^{-1}\left(-\frac{1}{3}\right)$

because $-\frac{\pi}{2} < \tan^{-1}\left(-\frac{1}{3}\right) < 0$ and we want θ to be in the interval $\left(\frac{\pi}{2}, \pi\right)$.

$z = \sqrt{10}\operatorname{cis}\left[\tan^{-1}\left(-\frac{1}{3}\right) + \pi\right]$.

$\boxed{43}$ $z = -5 - 3i \Rightarrow r = \sqrt{(-5)^2 + (-3)^2} = \sqrt{34}$. $\tan\theta = \frac{-3}{-5} = \frac{3}{5}$ and

θ in QIII $\Rightarrow \theta = \tan^{-1}\frac{3}{5} + \pi$. We must add π to $\tan^{-1}\frac{3}{5}$ because $0 < \tan^{-1}\frac{3}{5} < \frac{\pi}{2}$ and

we want θ to be in the interval $\left(\pi, \frac{3\pi}{2}\right)$. $z = \sqrt{34}\operatorname{cis}\left(\tan^{-1}\frac{3}{5} + \pi\right)$.

$\boxed{45}$ $z = 4 - 3i \Rightarrow r = \sqrt{4^2 + (-3)^2} = \sqrt{25} = 5$.

$\tan\theta = \frac{-3}{4}$ and θ in QIV $\Rightarrow \theta = \tan^{-1}\left(-\frac{3}{4}\right) + 2\pi$. We must add 2π to $\tan^{-1}\left(-\frac{3}{4}\right)$

because $-\frac{\pi}{2} < \tan^{-1}\left(-\frac{3}{4}\right) < 0$ and we want θ to be in the interval $\left(\frac{3\pi}{2}, 2\pi\right)$.

$z = 5\operatorname{cis}\left[\tan^{-1}\left(-\frac{3}{4}\right) + 2\pi\right]$.

$\boxed{47}$ $4\left(\cos\frac{\pi}{4} + i\sin\frac{\pi}{4}\right) = 4\left(\frac{\sqrt{2}}{2} + \frac{\sqrt{2}}{2}i\right) = 2\sqrt{2} + 2\sqrt{2}i$

$\boxed{51}$ $5(\cos\pi + i\sin\pi) = 5(-1 + 0i) = -5$

___

53 For any given angle θ, where $\theta = \tan^{-1} \frac{y}{x}$, we have $r = \sqrt{x^2 + y^2}$. In this case,
$\theta = \tan^{-1} \frac{3}{5}$, so $r = \sqrt{3^2 + 5^2} = \sqrt{34}$. You may want to draw a figure to represent x,
y, and r, as we did in previous chapters. Also, note that $\cos \theta = \frac{x}{r}$ and $\sin \theta = \frac{y}{r}$.

$$\sqrt{34} \operatorname{cis}\left(\tan^{-1} \tfrac{3}{5}\right) = \sqrt{34}\left[\cos\left(\tan^{-1}\tfrac{3}{5}\right) + i\sin\left(\tan^{-1}\tfrac{3}{5}\right)\right] = \sqrt{34}\left(\frac{5}{\sqrt{34}} + \frac{3}{\sqrt{34}}i\right) = 5 + 3i$$

55 $\sqrt{5}\operatorname{cis}\left[\tan^{-1}\left(-\tfrac{1}{2}\right)\right] = \sqrt{5}\left\{\cos\left[\tan^{-1}\left(-\tfrac{1}{2}\right)\right] + i\sin\left[\tan^{-1}\left(-\tfrac{1}{2}\right)\right]\right\} =$

$$\sqrt{5}\left(\frac{2}{\sqrt{5}} - \frac{1}{\sqrt{5}}i\right) = 2 - i$$

Note: For Exercises 57–64, the trigonometric forms for z_1 and z_2 are listed and then used
in the theorem in this section. The trigonometric forms can be found as in
Exercises 21–46.

57 $z_1 = \sqrt{2}\operatorname{cis}\frac{3\pi}{4}$ and $z_2 = \sqrt{2}\operatorname{cis}\frac{\pi}{4}$. $z_1 z_2 = \sqrt{2} \cdot \sqrt{2}\operatorname{cis}\left(\frac{3\pi}{4} + \frac{\pi}{4}\right) = 2\operatorname{cis}\pi = -2 + 0i$.

$$\frac{z_1}{z_2} = \frac{\sqrt{2}}{\sqrt{2}}\operatorname{cis}\left(\frac{3\pi}{4} - \frac{\pi}{4}\right) = 1\operatorname{cis}\frac{\pi}{2} = 0 + i$$

59 $z_1 = 4\operatorname{cis}\frac{4\pi}{3}$ and $z_2 = 5\operatorname{cis}\frac{\pi}{2}$. $z_1 z_2 = 4 \cdot 5\operatorname{cis}\left(\frac{4\pi}{3} + \frac{\pi}{2}\right) = 20\operatorname{cis}\frac{11\pi}{6} = 10\sqrt{3} - 10i$.

$$\frac{z_1}{z_2} = \frac{4}{5}\operatorname{cis}\left(\frac{4\pi}{3} - \frac{\pi}{2}\right) = \frac{4}{5}\operatorname{cis}\frac{5\pi}{6} = -\frac{2}{5}\sqrt{3} + \frac{2}{5}i.$$

61 $z_1 = 10\operatorname{cis}\pi$ and $z_2 = 4\operatorname{cis}\pi$. $z_1 z_2 = 10 \cdot 4\operatorname{cis}\left(\pi + \pi\right) = 40\operatorname{cis}2\pi = 40 + 0i$.

$$\frac{z_1}{z_2} = \frac{10}{4}\operatorname{cis}\left(\pi - \pi\right) = \frac{5}{2}\operatorname{cis}0 = \frac{5}{2} + 0i.$$

63 $z_1 = 4\operatorname{cis}0$ and $z_2 = \sqrt{5}\operatorname{cis}\left[\tan^{-1}\left(-\tfrac{1}{2}\right)\right]$. Let $\theta = \tan^{-1}\left(-\tfrac{1}{2}\right)$.

Thus $\cos\theta = \frac{2}{\sqrt{5}}$ and $\sin\theta = -\frac{1}{\sqrt{5}}$.

$$z_1 z_2 = 4 \cdot \sqrt{5}\operatorname{cis}\left(0 + \theta\right) = 4\sqrt{5}\left(\cos\theta + i\sin\theta\right) = 4\sqrt{5}\left(\frac{2}{\sqrt{5}} + \frac{-1}{\sqrt{5}}i\right) = 8 - 4i.$$

$$\frac{z_1}{z_2} = \frac{4}{\sqrt{5}}\operatorname{cis}\left(0 - \theta\right) = \frac{4}{\sqrt{5}}\left[\cos\left(-\theta\right) + i\sin\left(-\theta\right)\right]$$

$$= \frac{4}{\sqrt{5}}\left(\cos\theta - i\sin\theta\right) = \frac{4}{\sqrt{5}}\left(\frac{2}{\sqrt{5}} + \frac{1}{\sqrt{5}}i\right) = \frac{8}{5} + \frac{4}{5}i.$$

65 Let $z_1 = r_1\operatorname{cis}\theta_1$ and $z_2 = r_2\operatorname{cis}\theta_2$.

$$\frac{z_1}{z_2} = \frac{r_1\operatorname{cis}\theta_1}{r_2\operatorname{cis}\theta_2} = \frac{r_1\left(\cos\theta_1 + i\sin\theta_1\right)\left(\cos\theta_2 - i\sin\theta_2\right)}{r_2\left(\cos\theta_2 + i\sin\theta_2\right)\left(\cos\theta_2 - i\sin\theta_2\right)}$$

$$\{\text{multiplying by the conjugate of the denominator}\}$$

$$= \frac{r_1\left[\left(\cos\theta_1\,\cos\theta_2 + \sin\theta_1\,\sin\theta_2\right) + i\left(\sin\theta_1\,\cos\theta_2 - \sin\theta_2\,\cos\theta_1\right)\right]}{r_2\left[\left(\cos^2\theta_2 + \sin^2\theta_2\right) + i\left(\sin\theta_2\,\cos\theta_2 - \cos\theta_2\,\sin\theta_2\right)\right]}$$

$$= \frac{r_1\left[\cos\left(\theta_1 - \theta_2\right) + i\sin\left(\theta_1 - \theta_2\right)\right]}{r_2\left(1 + 0i\right)} = \frac{r_1}{r_2}\operatorname{cis}\left(\theta_1 - \theta_2\right).$$

[67] The unknown quantity is V: $I = V/Z \Rightarrow$

$$V = IZ = (10 \operatorname{cis} 35°)(3 \operatorname{cis} 20°) = (10 \times 3) \operatorname{cis} (35° + 20°) = 30 \operatorname{cis} 55° \approx 17.21 + 24.57i.$$

[69] The unknown quantity is Z:

$$I = \frac{V}{Z} \Rightarrow Z = \frac{V}{I} = \frac{115 \operatorname{cis} 45°}{8 \operatorname{cis} 5°} = (115 \div 8) \operatorname{cis} (45° - 5°) = 14.375 \operatorname{cis} 40° \approx 11.01 + 9.24i$$

[71] $Z = 14 - 13i \Rightarrow |Z| = \sqrt{14^2 + (-13)^2} = \sqrt{365} \approx 19.1$ ohms

[73] $I = \frac{V}{Z} \Rightarrow V = IZ = (4 \operatorname{cis} 90°)[18 \operatorname{cis} (-78°)] = 72 \operatorname{cis} 12° \approx 70.43 + 14.97i$

4.4 Exercises

Note: In this section, it is assumed that the reader can transform complex numbers to their trigonometric from. If this is not true, see Exercises 4.3.

[1] $(3 + 3i)^5 = (3\sqrt{2} \operatorname{cis} \frac{\pi}{4})^5 = (3\sqrt{2})^5 \operatorname{cis} (5 \cdot \frac{\pi}{4}) = (3\sqrt{2})^5 \operatorname{cis} \frac{5\pi}{4} =$

$$972\sqrt{2}\left(-\frac{\sqrt{2}}{2} - \frac{\sqrt{2}}{2}i\right) = -972 - 972i$$

[3] $(1 - i)^{10} = (\sqrt{2} \operatorname{cis} \frac{7\pi}{4})^{10} = (\sqrt{2})^{10} \operatorname{cis} (10 \cdot \frac{7\pi}{4}) = (\sqrt{2})^{10} \operatorname{cis} \frac{35\pi}{2} =$

$$2^5 \operatorname{cis} (16\pi + \frac{3\pi}{2}) = 32 \operatorname{cis} \frac{3\pi}{2} = 32(0 - i) = -32i$$

[7] $\left(-\frac{\sqrt{2}}{2} + \frac{\sqrt{2}}{2}i\right)^{15} = (1 \operatorname{cis} \frac{3\pi}{4})^{15} = 1^{15} \operatorname{cis} \frac{45\pi}{4} = \operatorname{cis} \frac{5\pi}{4} = -\frac{\sqrt{2}}{2} - \frac{\sqrt{2}}{2}i$

[9] $\left(-\frac{\sqrt{3}}{2} - \frac{1}{2}i\right)^{20} = (1 \operatorname{cis} \frac{7\pi}{6})^{20} = 1^{20} \operatorname{cis} \frac{70\pi}{3} = \operatorname{cis} \frac{4\pi}{3} = -\frac{1}{2} - \frac{\sqrt{3}}{2}i$

[13] $1 + \sqrt{3}i = 2 \operatorname{cis} 60°$. $w_k = \sqrt{2} \operatorname{cis} \left(\frac{60° + 360°k}{2}\right)$ for $k = 0, 1$.

$$w_0 = \sqrt{2} \operatorname{cis} 30° = \sqrt{2}\left(\frac{\sqrt{3}}{2} + \frac{1}{2}i\right) = \frac{\sqrt{6}}{2} + \frac{\sqrt{2}}{2}i.$$

$$w_1 = \sqrt{2} \operatorname{cis} 210° = \sqrt{2}\left(-\frac{\sqrt{3}}{2} - \frac{1}{2}i\right) = -\frac{\sqrt{6}}{2} - \frac{\sqrt{2}}{2}i.$$

[15] $-1 - \sqrt{3}i = 2 \operatorname{cis} 240°$. $w_k = \sqrt[4]{2} \operatorname{cis} \left(\frac{240° + 360°k}{4}\right)$ for $k = 0, 1, 2, 3$.

$$w_0 = \sqrt[4]{2} \operatorname{cis} 60° = \sqrt[4]{2}\left(\frac{1}{2} + \frac{\sqrt{3}}{2}i\right) = \frac{\sqrt[4]{2}}{2} + \frac{\sqrt[4]{18}}{2}i.$$

$$\{ \text{since } \sqrt[4]{2} \cdot \sqrt{3} = \sqrt[4]{2} \cdot \sqrt[4]{9} = \sqrt[4]{18} \}$$

$$w_1 = \sqrt[4]{2} \operatorname{cis} 150° = \sqrt[4]{2}\left(-\frac{\sqrt{3}}{2} + \frac{1}{2}i\right) = -\frac{\sqrt[4]{18}}{2} + \frac{\sqrt[4]{2}}{2}i.$$

$$w_2 = \sqrt[4]{2} \operatorname{cis} 240° = \sqrt[4]{2}\left(-\frac{1}{2} - \frac{\sqrt{3}}{2}i\right) = -\frac{\sqrt[4]{2}}{2} - \frac{\sqrt[4]{18}}{2}i.$$

$$w_3 = \sqrt[4]{2} \operatorname{cis} 330° = \sqrt[4]{2}\left(\frac{\sqrt{3}}{2} - \frac{1}{2}i\right) = \frac{\sqrt[4]{18}}{2} - \frac{\sqrt[4]{2}}{2}i.$$

$\boxed{17}$ $-27i = 27\operatorname{cis}270°$. $w_k = \sqrt[3]{27}\operatorname{cis}\left(\dfrac{270° + 360°k}{3}\right)$ for $k = 0, 1, 2$.

$w_0 = 3\operatorname{cis}90° = 3(0 + i) = 3i$.

$w_1 = 3\operatorname{cis}210° = 3\left(-\dfrac{\sqrt{3}}{2} - \dfrac{1}{2}i\right) = -\dfrac{3\sqrt{3}}{2} - \dfrac{3}{2}i$.

$w_2 = 3\operatorname{cis}330° = 3\left(\dfrac{\sqrt{3}}{2} - \dfrac{1}{2}i\right) = \dfrac{3\sqrt{3}}{2} - \dfrac{3}{2}i$.

$\boxed{19}$ $1 = 1\operatorname{cis}0°$. $w_k = \sqrt[6]{1}\operatorname{cis}\left(\dfrac{0° + 360°k}{6}\right)$ for $k = 0, 1, 2, 3, 4, 5$.

$w_0 = 1\operatorname{cis}0° = 1 + 0i$. $w_1 = 1\operatorname{cis}60° = \dfrac{1}{2} + \dfrac{\sqrt{3}}{2}i$.

$w_2 = 1\operatorname{cis}120° = -\dfrac{1}{2} + \dfrac{\sqrt{3}}{2}i$. $w_3 = 1\operatorname{cis}180° = -1 + 0i$.

$w_4 = 1\operatorname{cis}240° = -\dfrac{1}{2} - \dfrac{\sqrt{3}}{2}i$. $w_5 = 1\operatorname{cis}300° = \dfrac{1}{2} - \dfrac{\sqrt{3}}{2}i$.

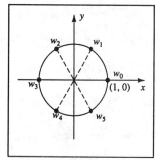

Figure 19

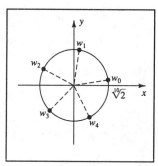

Figure 21

$\boxed{21}$ $1 + i = \sqrt{2}\operatorname{cis}45°$. $w_k = \sqrt{\sqrt[5]{2}}\operatorname{cis}\left(\dfrac{45° + 360°k}{5}\right)$ for $k = 0, 1, 2, 3, 4$.

$$w_k = \sqrt[10]{2}\operatorname{cis}\theta \text{ with } \theta = 9°, 81°, 153°, 225°, 297°.$$

$\boxed{23}$ $x^4 - 16 = 0 \Rightarrow x^4 = 16$. The problem is now to find the 4 fourth roots of 16.

$16 = 16 + 0i = 16\operatorname{cis}0°$. $w_k = \sqrt[4]{16}\operatorname{cis}\left(\dfrac{0° + 360°k}{4}\right)$ for $k = 0, 1, 2, 3$.

$w_0 = 2\operatorname{cis}0° = 2(1 + 0i) = 2$. $w_1 = 2\operatorname{cis}90° = 2(0 + i) = 2i$.

$w_2 = 2\operatorname{cis}180° = 2(-1 + 0i) = -2$. $w_3 = 2\operatorname{cis}270° = 2(0 - i) = -2i$.

$\boxed{25}$ $x^6 + 64 = 0 \Rightarrow x^6 = -64$. The problem is now to find the 6 sixth roots of -64.

$-64 = -64 + 0i = 64\operatorname{cis}180°$. $w_k = \sqrt[6]{64}\operatorname{cis}\left(\dfrac{180° + 360°k}{6}\right)$ for $k = 0, 1, \ldots, 5$.

$w_0 = 2\operatorname{cis}30° = 2\left(\dfrac{\sqrt{3}}{2} + \dfrac{1}{2}i\right) = \sqrt{3} + i$.

$w_1 = 2\operatorname{cis}90° = 2(0 + i) = 2i$.

$w_2 = 2\operatorname{cis}150° = 2\left(-\dfrac{\sqrt{3}}{2} + \dfrac{1}{2}i\right) = -\sqrt{3} + i$.

(continued)

$$w_3 = 2\operatorname{cis}210° = 2\left(-\frac{\sqrt{3}}{2} - \frac{1}{2}i\right) = -\sqrt{3} - i.$$

$$w_4 = 2\operatorname{cis}270° = 2(0 - i) = -2i.$$

$$w_5 = 2\operatorname{cis}330° = 2\left(\frac{\sqrt{3}}{2} - \frac{1}{2}i\right) = \sqrt{3} - i.$$

27 $x^3 + 8i = 0 \Rightarrow x^3 = -8i$. The problem is now to find the 3 cube roots of $-8i$.

$$-8i = 0 - 8i = 8\operatorname{cis}270°. \quad w_k = \sqrt[3]{8}\operatorname{cis}\left(\frac{270° + 360°k}{3}\right) \text{ for } k = 0, 1, 2.$$

$$w_0 = 2\operatorname{cis}90° = 2(0 + i) = 2i.$$

$$w_1 = 2\operatorname{cis}210° = 2\left(-\frac{\sqrt{3}}{2} - \frac{1}{2}i\right) = -\sqrt{3} - i.$$

$$w_2 = 2\operatorname{cis}330° = 2\left(\frac{\sqrt{3}}{2} - \frac{1}{2}i\right) = \sqrt{3} - i.$$

29 $x^5 - 243 = 0 \Rightarrow x^5 = 243$. The problem is now to find the 5 fifth roots of 243.

$$243 = 243 + 0i = 243\operatorname{cis}0°. \quad w_k = \sqrt[5]{243}\operatorname{cis}\left(\frac{0° + 360°k}{5}\right) \text{ for } k = 0, 1, 2, 3, 4.$$

$$w_k = 3\operatorname{cis}\theta \text{ with } \theta = 0°, 72°, 144°, 216°, 288°.$$

4.5 Exercises

1 $\mathbf{a} + \mathbf{b} = \langle 2, -3\rangle + \langle 1, 4\rangle = \langle 2 + 1, -3 + 4\rangle = \langle 3, 1\rangle.$

$\mathbf{a} - \mathbf{b} = \langle 2, -3\rangle - \langle 1, 4\rangle = \langle 2 - 1, -3 - 4\rangle = \langle 1, -7\rangle.$

$4\mathbf{a} + 5\mathbf{b} = 4\langle 2, -3\rangle + 5\langle 1, 4\rangle = \langle 4(2), 4(-3)\rangle + \langle 5(1), 5(4)\rangle$

$$= \langle 8, -12\rangle + \langle 5, 20\rangle = \langle 8 + 5, -12 + 20\rangle = \langle 13, 8\rangle.$$

From above, $4\mathbf{a} - 5\mathbf{b} = \langle 8, -12\rangle - \langle 5, 20\rangle = \langle 8 - 5, -12 - 20\rangle = \langle 3, -32\rangle.$

3 Simplifying $\mathbf{a}$ and $\mathbf{b}$ first, we have $\mathbf{a} = -\langle 7, -2\rangle = \langle -(7), -(-2)\rangle = \langle -7, 2\rangle$ and

$$\mathbf{b} = 4\langle -2, 1\rangle = \langle 4(-2), 4(1)\rangle = \langle -8, 4\rangle.$$

$\mathbf{a} + \mathbf{b} = \langle -7, 2\rangle + \langle -8, 4\rangle = \langle -7 + (-8), 2 + 4\rangle = \langle -15, 6\rangle.$

$\mathbf{a} - \mathbf{b} = \langle -7, 2\rangle - \langle -8, 4\rangle = \langle -7 - (-8), 2 - 4\rangle = \langle 1, -2\rangle.$

$4\mathbf{a} + 5\mathbf{b} = 4\langle -7, 2\rangle + 5\langle -8, 4\rangle = \langle -28, 8\rangle + \langle -40, 20\rangle = \langle -68, 28\rangle.$

From above, $4\mathbf{a} - 5\mathbf{b} = \langle -28, 8\rangle - \langle -40, 20\rangle = \langle 12, -12\rangle.$

5 $\mathbf{a} + \mathbf{b} = (\mathbf{i} + 2\mathbf{j}) + (3\mathbf{i} - 5\mathbf{j}) = (1 + 3)\mathbf{i} + (2 - 5)\mathbf{j} = 4\mathbf{i} - 3\mathbf{j}.$

$\mathbf{a} - \mathbf{b} = (\mathbf{i} + 2\mathbf{j}) - (3\mathbf{i} - 5\mathbf{j}) = (1 - 3)\mathbf{i} + (2 - (-5))\mathbf{j} = -2\mathbf{i} + 7\mathbf{j}.$

$4\mathbf{a} + 5\mathbf{b} = 4(\mathbf{i} + 2\mathbf{j}) + 5(3\mathbf{i} - 5\mathbf{j}) = (4\mathbf{i} + 8\mathbf{j}) + (15\mathbf{i} - 25\mathbf{j}) = 19\mathbf{i} - 17\mathbf{j}.$

From above, $4\mathbf{a} - 5\mathbf{b} = (4\mathbf{i} + 8\mathbf{j}) - (15\mathbf{i} - 25\mathbf{j}) = -11\mathbf{i} + 33\mathbf{j}.$

$\boxed{7}$ $\mathbf{a} = 3\mathbf{i} + 2\mathbf{j}$ and $\mathbf{b} = -\mathbf{i} + 5\mathbf{j} \Rightarrow \mathbf{a} + \mathbf{b} = 2\mathbf{i} + 7\mathbf{j}$, $2\mathbf{a} = 6\mathbf{i} + 4\mathbf{j}$, and $-3\mathbf{b} = 3\mathbf{i} - 15\mathbf{j}$.

Terminal points of the vectors are $(3, 2)$, $(-1, 5)$, $(2, 7)$, $(6, 4)$, and $(3, -15)$.

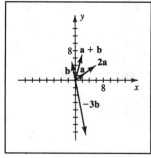

Figure 7

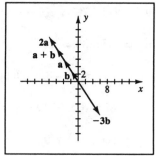

Figure 9

$\boxed{9}$ $\mathbf{a} = \langle -4, 6 \rangle$ and $\mathbf{b} = \langle -2, 3 \rangle \Rightarrow \mathbf{a} + \mathbf{b} = \langle -6, 9 \rangle$, $2\mathbf{a} = \langle -8, 12 \rangle$, and $-3\mathbf{b} = \langle 6, -9 \rangle$.

Terminal points of the vectors are $(-4, 6)$, $(-2, 3)$, $(-6, 9)$, $(-8, 12)$, and $(6, -9)$.

$\boxed{11}$ $\mathbf{a} + \mathbf{b} = \langle 2, 0 \rangle + \langle -1, 0 \rangle = \langle 1, 0 \rangle = -\langle -1, 0 \rangle = -\mathbf{b}$

$\boxed{13}$ $\mathbf{b} + \mathbf{e} = \langle -1, 0 \rangle + \langle 2, 2 \rangle = \langle 1, 2 \rangle = \mathbf{f}$

$\boxed{15}$ $\mathbf{b} + \mathbf{d} = \langle -1, 0 \rangle + \langle 0, -1 \rangle = \langle -1, -1 \rangle = -\frac{1}{2}\langle 2, 2 \rangle = -\frac{1}{2}\mathbf{e}$

$\boxed{17}$ $\mathbf{a} + (\mathbf{b} + \mathbf{c}) = \langle a_1, a_2 \rangle + (\langle b_1, b_2 \rangle + \langle c_1, c_2 \rangle)$

$= \langle a_1, a_2 \rangle + \langle b_1 + c_1, b_2 + c_2 \rangle$

$= \langle a_1 + b_1 + c_1, a_2 + b_2 + c_2 \rangle$

$= \langle a_1 + b_1, a_2 + b_2 \rangle + \langle c_1, c_2 \rangle$

$= (\langle a_1, a_2 \rangle + \langle b_1, b_2 \rangle) + \langle c_1, c_2 \rangle = (\mathbf{a} + \mathbf{b}) + \mathbf{c}$

$\boxed{21}$ $(mn)\mathbf{a} = (mn)\langle a_1, a_2 \rangle$

$= \langle (mn)a_1, (mn)a_2 \rangle$

$= \langle mna_1, mna_2 \rangle$

$= m\langle na_1, na_2 \rangle$ or $n\langle ma_1, ma_2 \rangle$

$= m(n\langle a_1, a_2 \rangle)$ or $n(m\langle a_1, a_2 \rangle)$

$= m(n\mathbf{a})$ or $n(m\mathbf{a})$

$\boxed{25}$ $-(\mathbf{a} + \mathbf{b}) = -(\langle a_1, a_2 \rangle + \langle b_1, b_2 \rangle)$

$= -(\langle a_1 + b_1, a_2 + b_2 \rangle)$

$= \langle -(a_1 + b_1), -(a_2 + b_2) \rangle$

$= \langle -a_1 - b_1, -a_2 - b_2 \rangle$

$= \langle -a_1, -a_2 \rangle + \langle -b_1, -b_2 \rangle$

$= -\mathbf{a} + (-\mathbf{b}) = -\mathbf{a} - \mathbf{b}$

$\boxed{27}$ $\|2\mathbf{v}\| = \|2\langle a, b \rangle\| = \|\langle 2a, 2b \rangle\| = \sqrt{(2a)^2 + (2b)^2} = \sqrt{4a^2 + 4b^2} =$

$2\sqrt{a^2 + b^2} = 2\|\langle a, b \rangle\| = 2\|\mathbf{v}\|$

$\boxed{29}$ $\|\mathbf{a}\| = \sqrt{3^2 + (-3)^2} = \sqrt{18} = 3\sqrt{2}$. $\tan\theta = \frac{-3}{3} = -1$ and θ in QIV $\Rightarrow \theta = \frac{7\pi}{4}$.

31 $\|\mathbf{a}\| = 5$. The terminal side of θ is on the negative x-axis $\Rightarrow \theta = \pi$.

33 $\|\mathbf{a}\| = \sqrt{41}$. $\tan\theta = \frac{5}{-4}$ and θ in QII $\Rightarrow \theta = \tan^{-1}\left(-\frac{5}{4}\right) + \pi$.

35 $\|\mathbf{a}\| = 18$. The terminal side of θ is on the negative y-axis $\Rightarrow \theta = \frac{3\pi}{2}$.

Note: Exercises 37–42: Each resultant force is found by completing the parallelogram and then applying the law of cosines.

37 $\|\mathbf{r}\| = \sqrt{40^2 + 70^2 - 2(40)(70)\cos 135°} = \sqrt{6500 + 2800\sqrt{2}} \approx 102.3$, or 102 lb.

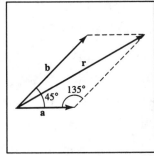

Figure 37

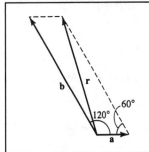

Figure 39

39 $\|\mathbf{r}\| = \sqrt{2^2 + 8^2 - 2(2)(8)\cos 60°} = \sqrt{68 - 16} = \sqrt{52} \approx 7.2$ kg.

41 $\|\mathbf{r}\| = \sqrt{90^2 + 60^2 - 2(90)(60)\cos 70°} \approx 89.48$, or 89 kg.

Using the law of cosines, $\alpha = \cos^{-1}\left(\frac{90^2 + \|\mathbf{r}\|^2 - 60^2}{2(90)(\|\mathbf{r}\|)}\right) \approx$ $\cos^{-1}(0.7765) \approx 39°$, which is 24° under the negative x-axis.

This angle is 204°, or S66°W.

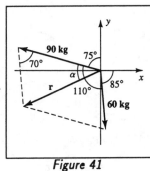

Figure 41

43 We will use a component approach for this exercise.

(a) $= \langle 6\cos 110°, 6\sin 110° \rangle \approx \langle -2.05, 5.64 \rangle$.

(b) $= \langle 2\cos 215°, 2\sin 215° \rangle \approx \langle -1.64, -1.15 \rangle$.

$\mathbf{a} + \mathbf{b} \approx \langle -3.69, 4.49 \rangle$ and $\|\mathbf{a} + \mathbf{b}\| \approx 5.8$ lb. $\tan\theta \approx \frac{4.49}{-3.69} \Rightarrow \theta \approx 129°$ since θ is in QII.

45 Horizontal $= 50\cos 35° \approx 40.96$. Vertical $= 50\sin 35° \approx 28.68$.

47 Horizontal $= 20\cos 108° \approx -6.18$. Vertical $= 20\sin 108° \approx 19.02$.

49 (a) $\mathbf{F} = \mathbf{F_1} + \mathbf{F_2} + \mathbf{F_3} = \langle 4, 3 \rangle + \langle -2, -3 \rangle + \langle 5, 2 \rangle = \langle 7, 2 \rangle$.

(b) $\mathbf{F} + \mathbf{G} = \mathbf{0} \Rightarrow \mathbf{G} = -\mathbf{F} = \langle -7, -2 \rangle$.

51 (a) $\mathbf{F} = \mathbf{F_1} + \mathbf{F_2} = \langle 6\cos 130°, 6\sin 130° \rangle + \langle 4\cos(-120°), 4\sin(-120°) \rangle \approx \langle -5.86, 1.13 \rangle$.

(b) $\mathbf{F} + \mathbf{G} = \mathbf{0} \Rightarrow \mathbf{G} = -\mathbf{F} \approx \langle 5.86, -1.13 \rangle$.

53 The vertical components of the forces must add up to zero for the large ship to move along the line ℓ. The vertical component of the smaller tug is $3200 \sin(-30°) = -1600$. The vertical component of the larger tug is $4000 \sin\theta$.

$$4000 \sin\theta = 1600 \Rightarrow \theta = \sin^{-1}\left(\frac{1600}{4000}\right) = \sin^{-1}(0.4) \approx 23.6°.$$

Note: Exercises 55–60: Measure angles from the positive x-axis.

55 $\mathbf{p} = \langle 200\cos 40°, 200\sin 40° \rangle \approx \langle 153.21, 128.56 \rangle$. $\mathbf{w} = \langle 40\cos 0°, 40\sin 0° \rangle = \langle 40, 0 \rangle$.

$\mathbf{p} + \mathbf{w} \approx \langle 193.21, 128.56 \rangle$ and $\|\mathbf{p} + \mathbf{w}\| \approx 232.07$, or 232 mi/hr.

$\tan\theta \approx \frac{128.56}{193.21} \Rightarrow \theta \approx 34°$. The true course is then N(90° − 34°)E, or N56°E.

57 $\mathbf{w} = \langle 50\cos 90°, 50\sin 90° \rangle = \langle 0, 50 \rangle$.

$\mathbf{r} = \langle 400\cos 200°, 400\sin 200° \rangle \approx \langle -375.88, -136.81 \rangle$, where $\mathbf{r}$ is the desired resultant of $\mathbf{p} + \mathbf{w}$. Since $\mathbf{r} = \mathbf{p} + \mathbf{w}$, $\mathbf{p} = \mathbf{r} - \mathbf{w} \approx \langle -375.88, -186.81 \rangle$. $\|\mathbf{p}\| \approx 419.74$, or 420 mi/hr.

$\tan\theta \approx \frac{-186.81}{-375.88}$ and θ is in QIII $\Rightarrow \theta \approx 206°$ from the positive x-axis,

or 244° using the directional form.

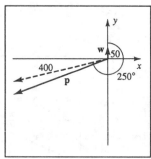

Figure 57

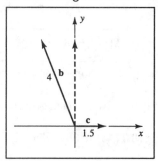

Figure 59

59 Let the vectors $\mathbf{c}$, $\mathbf{b}$, and $\mathbf{r}$ denote the current, the boat, and the resultant, respectively. $\mathbf{c} = \langle 1.5\cos 0°, 1.5\sin 0° \rangle = \langle 1.5, 0 \rangle$. $\mathbf{r} = \langle s\cos 90°, s\sin 90° \rangle = \langle 0, s \rangle$, where s is the resulting speed. $\mathbf{b} = \langle 4\cos\theta, 4\sin\theta \rangle$. Also, $\mathbf{b} = \mathbf{r} - \mathbf{c} = \langle -1.5, s \rangle$.

$$4\cos\theta = -1.5 \Rightarrow \theta \approx 112°, \text{ or N22°W.}$$

61 In the figure in the text, suppose $\mathbf{v}_1$ was pointing upwards instead of downwards. If we then write $\mathbf{v}_1$ in terms of its vertical and horizontal components, we have $\mathbf{v}_1 = \|\mathbf{v}_1\|\cos\theta_1\,\mathbf{j} - \|\mathbf{v}_1\|\sin\theta_1\,\mathbf{i}$. We want the negative of this vector since $\mathbf{v}_1$ is actually pointing downward, so using $\|\mathbf{v}_1\| = 8.2$ and $\theta_1 = 30°$, we obtain $\mathbf{v}_1 = \|\mathbf{v}_1\|\sin\theta_1\,\mathbf{i} - \|\mathbf{v}_1\|\cos\theta_1\,\mathbf{j} = 8.2(\frac{1}{2})\mathbf{i} - 8.2(\sqrt{3}/2)\mathbf{j} = 4.1\mathbf{i} - 4.1\sqrt{3}\,\mathbf{j} \approx 4.1\mathbf{i} - 7.10\mathbf{j}$. The angle θ_2 can now be computed using the given relationship in the text.

$\dfrac{\|\mathbf{v}_1\|}{\|\mathbf{v}_2\|} = \dfrac{\tan\theta_1}{\tan\theta_2} \Rightarrow \tan\theta_2 = \dfrac{\|\mathbf{v}_2\|}{\|\mathbf{v}_1\|}\tan\theta_1 = \dfrac{3.8}{8.2} \times \dfrac{1}{\sqrt{3}} \Rightarrow \tan\theta_2 \approx 0.2676 \Rightarrow \theta_2 \approx 14.98°.$

We now write $\mathbf{v}_2$ in terms of its horizontal and vertical components, and using $\|\mathbf{v}_2\| = 3.8$, it follows that $\mathbf{v}_2 = \|\mathbf{v}_2\|\sin\theta_2\,\mathbf{i} - \|\mathbf{v}_2\|\cos\theta_2\,\mathbf{j} \approx 0.98\mathbf{i} - 3.67\mathbf{j}$.

[63] (a) $\mathbf{a} = 15\cos 40°\mathbf{i} + 15\sin 40°\mathbf{j} \approx 11.49\,\mathbf{i} + 9.64\,\mathbf{j}$.

$\mathbf{b} = 17\cos 40°\mathbf{i} + 17\sin 40°\mathbf{j} \approx 13.02\,\mathbf{i} + 10.93\,\mathbf{j}$.

$$\overrightarrow{PR} = \mathbf{a} + \mathbf{b} \approx 24.51\,\mathbf{i} + 20.57\,\mathbf{j} \Rightarrow R \approx (24.51, 20.57).$$

(b) $\mathbf{c} = 15\cos(40° + 85°)\mathbf{i} + 15\sin 125°\mathbf{j} \approx -8.60\,\mathbf{i} + 12.29\,\mathbf{j}$.

$\mathbf{d} = 17\cos(40° + 85° + 35°)\mathbf{i} + 17\sin 160°\mathbf{j} \approx -15.97\,\mathbf{i} + 5.81\,\mathbf{j}$.

$$\overrightarrow{PR} = \mathbf{c} + \mathbf{d} \approx -24.57\,\mathbf{i} + 18.10\,\mathbf{j} \Rightarrow R \approx (-24.57, 18.10).$$

[65] Break the force into a horizontal and a vertical component. The group of 550 people had to contribute a force equal to the vertical component up the ramp. The vertical component is $99{,}000\sin 9° \approx 15{,}487$ lb. $\dfrac{15{,}487}{550} \approx 28.2$ lb/person. (The actual force would have been more since friction was ignored.)

4.6 Exercises

[1] (a) The dot product of the two vectors is

$$\langle -2,\, 5\rangle \cdot \langle 3,\, 6\rangle = (-2)(3) + (5)(6) = -6 + 30 = 24.$$

(b) The angle between the two vectors is

$$\theta = \cos^{-1}\left(\frac{\langle -2,\, 5\rangle \cdot \langle 3,\, 6\rangle}{\|\langle -2,\, 5\rangle\|\,\|\langle 3,\, 6\rangle\|}\right) = \cos^{-1}\left(\frac{24}{\sqrt{29}\,\sqrt{45}}\right) \approx 48°22'.$$

[3] (a) $(4\mathbf{i} - \mathbf{j}) \cdot (-3\mathbf{i} + 2\mathbf{j}) = (4)(-3) + (-1)(2) = -12 - 2 = -14$

(b) $\theta = \cos^{-1}\left(\dfrac{(4\mathbf{i} - \mathbf{j}) \cdot (-3\mathbf{i} + 2\mathbf{j})}{\|4\mathbf{i} - \mathbf{j}\|\,\|-3\mathbf{i} + 2\mathbf{j}\|}\right) = \cos^{-1}\left(\dfrac{-14}{\sqrt{17}\,\sqrt{13}}\right) \approx 160°21'$

[7] (a) $\langle 10,\, 7\rangle \cdot \langle -2,\, -\frac{7}{5}\rangle = (10)(-2) + (7)(-\frac{7}{5}) = -\frac{149}{5}$

(b) $\theta = \cos^{-1}\left(\dfrac{\langle 10,\, 7\rangle \cdot \langle -2,\, -\frac{7}{5}\rangle}{\|\langle 10,\, 7\rangle\|\,\|\langle -2,\, -\frac{7}{5}\rangle\|}\right) = \cos^{-1}\left(\dfrac{-149/5}{\sqrt{149}\,\sqrt{149/25}}\right) = \cos^{-1}(-1) = 180°$

This result $\{\theta = 180°\}$ indicates that the vectors have the opposite direction.

[9] $\langle 4,\, -1\rangle \cdot \langle 2,\, 8\rangle = 8 - 8 = 0 \Rightarrow$ vectors are orthogonal since their dot product is zero.

[11] $(-4\mathbf{j}) \cdot (-7\mathbf{i}) = 0 + 0 = 0 \Rightarrow$ vectors are orthogonal.

[13] We first find the angle between the two vectors.

If this angle is 0 or π, then the vectors are parallel.

$$\cos\theta = \frac{\mathbf{a} \cdot \mathbf{b}}{\|\mathbf{a}\|\,\|\mathbf{b}\|} = \frac{(3)(-\frac{12}{7}) + (-5)(\frac{20}{7})}{\sqrt{9 + 25}\,\sqrt{\frac{144}{49} + \frac{400}{49}}} = \frac{-\frac{136}{7}}{\sqrt{\frac{18{,}496}{49}}} = \frac{-\frac{136}{7}}{\frac{136}{7}} = -1 \Rightarrow$$

$\theta = \cos^{-1}(-1) = \pi$. $\mathbf{b} = m\mathbf{a} \Rightarrow -\frac{12}{7}\mathbf{i} + \frac{20}{7}\mathbf{j} = 3m\mathbf{i} - 5m\mathbf{j} \Rightarrow$

$3m = -\frac{12}{7}$ and $-5m = \frac{20}{7} \Rightarrow m = -\frac{4}{7} < 0 \Rightarrow \mathbf{a}$ and $\mathbf{b}$ have the opposite direction.

15 $\cos\theta = \dfrac{\mathbf{a}\cdot\mathbf{b}}{\|\mathbf{a}\|\|\mathbf{b}\|} = \dfrac{(\frac{2}{3})(8)+(\frac{1}{2})(6)}{\sqrt{\frac{4}{9}+\frac{1}{4}}\sqrt{64+36}} = \dfrac{\frac{25}{3}}{\sqrt{\frac{25}{36}}\cdot 10} = 1 \Rightarrow \theta = \cos^{-1}1 = 0.$

$\mathbf{b} = m\mathbf{a} \Rightarrow 8\mathbf{i}+6\mathbf{j} = \frac{2}{3}m\mathbf{i}+\frac{1}{2}m\mathbf{j} \Rightarrow 8 = \frac{2}{3}m$ and $6 = \frac{1}{2}m \Rightarrow m = 12 > 0 \Rightarrow$

$\mathbf{a}$ and $\mathbf{b}$ have the same direction.

17 We need to have the dot product of the two vectors equal 0.

$$(3\mathbf{i}-2\mathbf{j})\cdot(4\mathbf{i}+5m\mathbf{j}) = 0 \Rightarrow 12-10m = 0 \Rightarrow m = \tfrac{6}{5}.$$

19 $(9\mathbf{i}-16m\mathbf{j})\cdot(\mathbf{i}+4m\mathbf{j}) = 0 \Rightarrow 9-64m^2 = 0 \Rightarrow m^2 = \frac{9}{64} \Rightarrow m = \pm\frac{3}{8}.$

21 (a) $\mathbf{a}\cdot(\mathbf{b}+\mathbf{c}) = \langle 2,-3\rangle\cdot(\langle 3,4\rangle+\langle -1,5\rangle) = \langle 2,-3\rangle\cdot\langle 2,9\rangle = 4-27 = -23$

(b) $\mathbf{a}\cdot\mathbf{b}+\mathbf{a}\cdot\mathbf{c} = \langle 2,-3\rangle\cdot\langle 3,4\rangle+\langle 2,-3\rangle\cdot\langle -1,5\rangle = (6-12)+(-2-15) = -23$

23 $(2\mathbf{a}+\mathbf{b})\cdot(3\mathbf{c}) = (2\langle 2,-3\rangle+\langle 3,4\rangle)\cdot(3\langle -1,5\rangle)$

$= (\langle 4,-6\rangle+\langle 3,4\rangle)\cdot\langle -3,15\rangle$

$= \langle 7,-2\rangle\cdot\langle -3,15\rangle = -21-30 = -51$

25 $\text{comp}_{\mathbf{c}}\mathbf{b} = \dfrac{\mathbf{b}\cdot\mathbf{c}}{\|\mathbf{c}\|} = \dfrac{\langle 3,4\rangle\cdot\langle -1,5\rangle}{\|\langle -1,5\rangle\|} = \dfrac{17}{\sqrt{26}} \approx 3.33$

27 $\text{comp}_{\mathbf{b}}(\mathbf{a}+\mathbf{c}) = \dfrac{(\mathbf{a}+\mathbf{c})\cdot\mathbf{b}}{\|\mathbf{b}\|} = \dfrac{(\langle 2,-3\rangle+\langle -1,5\rangle)\cdot\langle 3,4\rangle}{\|\langle 3,4\rangle\|} = \dfrac{\langle 1,2\rangle\cdot\langle 3,4\rangle}{5} = \dfrac{11}{5} = 2.2$

29 $\mathbf{c}\cdot\overrightarrow{PQ} = \langle 3,4\rangle\cdot\langle 5,-2\rangle = 15-8 = 7.$

31 We want a vector with initial point at the origin and terminal point located so that this vector has the same magnitude and direction as $\overrightarrow{PQ}$. Following the hint in the text, $\mathbf{b} = \overrightarrow{PQ} \Rightarrow \langle b_1,b_2\rangle = \langle 4-2, 3-(-1)\rangle \Rightarrow \langle b_1,b_2\rangle = \langle 2,4\rangle.$

$$\mathbf{c}\cdot\mathbf{b} = \langle 6,4\rangle\cdot\langle 2,4\rangle = 12+16 = 28.$$

33 The force is described by the vector $\langle 0,4\rangle$.

The work done is $\langle 0,4\rangle\cdot\langle 8,3\rangle = 0+12 = 12.$

35 $\mathbf{a}\cdot\mathbf{a} = \langle a_1,a_2\rangle\cdot\langle a_1,a_2\rangle = a_1^2+a_2^2 = (\sqrt{a_1^2+a_2^2})^2 = \|\mathbf{a}\|^2$

37 $(m\mathbf{a})\cdot\mathbf{b} = (m\langle a_1,a_2\rangle)\cdot\langle b_1,b_2\rangle$

$= \langle ma_1,ma_2\rangle\cdot\langle b_1,b_2\rangle$

$= ma_1b_1+ma_2b_2$

$= m(a_1b_1+a_2b_2) = m(\mathbf{a}\cdot\mathbf{b})$

41 Using the horizontal and vertical components of a vector from Section 4.5, we have the force vector as $\langle 20\cos 30°, 20\sin 30°\rangle = \langle 10\sqrt{3}, 10\rangle.$

The distance (direction vector) can be described by the vector $\langle 100,0\rangle$.

The work done is $\langle 10\sqrt{3},10\rangle\cdot\langle 100,0\rangle = 1000\sqrt{3} \approx 1732$ ft-lb.

43 (a) The horizontal component has magnitude 93×10^6 and the vertical component has magnitude 0.432×10^6.

Thus, $\mathbf{v} = (93 \times 10^6)\mathbf{i} + (0.432 \times 10^6)\mathbf{j}$ and $\mathbf{w} = (93 \times 10^6)\mathbf{i} - (0.432 \times 10^6)\mathbf{j}$.

(b) $\cos\theta = \dfrac{\mathbf{v} \cdot \mathbf{w}}{\|\mathbf{v}\|\|\mathbf{w}\|} = \dfrac{(93 \times 10^6)^2 - (0.432 \times 10^6)^2}{\sqrt{(93 \times 10^6)^2 + (0.432 \times 10^6)^2}\sqrt{(93 \times 10^6)^2 + (0.432 \times 10^6)^2}} \approx$

$0.99995684 \Rightarrow \theta \approx 0.53°.$

45 $\mathbf{R} = 2(\mathbf{N} \cdot \mathbf{L})\mathbf{N} - \mathbf{L} = 2(\langle 0, 1\rangle \cdot \langle -\frac{4}{5}, \frac{3}{5}\rangle)\langle 0, 1\rangle - \langle -\frac{4}{5}, \frac{3}{5}\rangle = 2(\frac{3}{5})\langle 0, 1\rangle - \langle -\frac{4}{5}, \frac{3}{5}\rangle =$

$\langle 0, \frac{6}{5}\rangle - \langle -\frac{4}{5}, \frac{3}{5}\rangle = \langle \frac{4}{5}, \frac{3}{5}\rangle$

47 Let horizontal ground be represented by $\mathbf{b} = \langle 1, 0\rangle$ (it could be any $\langle a, 0\rangle$).

$\text{comp}_{\mathbf{b}}\,\mathbf{a} = \dfrac{\mathbf{a} \cdot \mathbf{b}}{\|\mathbf{b}\|} = \dfrac{\langle 2.6, 4.5\rangle \cdot \langle 1, 0\rangle}{\|\langle 1, 0\rangle\|} = \dfrac{2.6}{1} = 2.6 \Rightarrow |\text{comp}_{\mathbf{b}}\,\mathbf{a}| = 2.6$

49 Let the direction of the ground be represented by $\mathbf{b} = \langle \cos\theta, \sin\theta\rangle = \langle \cos 12°, \sin 12°\rangle$.

$\text{comp}_{\mathbf{b}}\,\mathbf{a} = \dfrac{\mathbf{a} \cdot \mathbf{b}}{\|\mathbf{b}\|} = \dfrac{\langle 25.7, -3.9\rangle \cdot \langle \cos 12°, \sin 12°\rangle}{\|\langle \cos 12°, \sin 12°\rangle\|} \approx \dfrac{24.33}{1} = 24.33 \Rightarrow |\text{comp}_{\mathbf{b}}\,\mathbf{a}| = 24.33$

51 From the "Theorem on the Dot Product," we know that $\mathbf{F} \cdot \mathbf{v} = \|\mathbf{F}\|\|\mathbf{v}\|\cos\theta$, so

$P = \frac{1}{550}(\mathbf{F} \cdot \mathbf{v}) = \frac{1}{550}\|\mathbf{F}\|\|\mathbf{v}\|\cos\theta = \frac{1}{550}(2200)(8)\cos 30° = 16\sqrt{3} \approx 27.7$ horsepower.

Chapter 4 Review Exercises

1 We are given 2 sides of a triangle and the angle between them—so we use the law of cosines to find the third side.

$a = \sqrt{b^2 + c^2 - 2bc\cos\alpha} = \sqrt{6^2 + 7^2 - 2(6)(7)\cos 60°} = \sqrt{43}.$

$\beta = \cos^{-1}\left(\dfrac{a^2 + c^2 - b^2}{2ac}\right) = \cos^{-1}\left(\dfrac{43 + 49 - 36}{2\sqrt{43}(7)}\right) = \cos^{-1}\left(\dfrac{4}{\sqrt{43}}\right).$

$\gamma = \cos^{-1}\left(\dfrac{a^2 + b^2 - c^2}{2ab}\right) = \cos^{-1}\left(\dfrac{43 + 36 - 49}{2\sqrt{43}(6)}\right) = \cos^{-1}\left(\dfrac{5}{2\sqrt{43}}\right).$

2 $\dfrac{\sin\alpha}{a} = \dfrac{\sin\gamma}{c} \Rightarrow \alpha = \sin^{-1}\left(\dfrac{a\sin\gamma}{c}\right) = \sin^{-1}\left(\dfrac{2\sqrt{3} \cdot \frac{1}{2}}{2}\right) = \sin^{-1}\left(\dfrac{\sqrt{3}}{2}\right) = 60°$ or $120°$.

There are two triangles possible since in either case $\alpha + \gamma < 180°$.

$\beta = (180° - \gamma) - \alpha = (180° - 30°) - (60°\text{ or }120°) = 90°$ or $30°$.

$\dfrac{b}{\sin\beta} = \dfrac{c}{\sin\gamma} \Rightarrow b = \dfrac{c\sin\beta}{\sin\gamma} = \dfrac{2\sin(90°\text{ or }30°)}{\sin 30°} = 4$ or 2.

[3] $\gamma = 180° - \alpha - \beta = 180° - 60° - 45° = 75°.$

$$\frac{a}{\sin \alpha} = \frac{b}{\sin \beta} \Rightarrow a = \frac{b \sin \alpha}{\sin \beta} = \frac{100 \sin 60°}{\sin 45°} = \frac{100 \cdot (\sqrt{3}/2)}{\sqrt{2}/2} \cdot \frac{\sqrt{2}}{\sqrt{2}} = 50\sqrt{6}.$$

$$\frac{c}{\sin \gamma} = \frac{b}{\sin \beta} \Rightarrow c = \frac{b \sin \gamma}{\sin \beta} = \frac{100 \sin (45° + 30°)}{\sqrt{2}/2} =$$

$$100\sqrt{2} \,(\sin 45° \cos 30° + \cos 45° \sin 30°) = 100\sqrt{2}\left(\frac{\sqrt{2}}{2} \cdot \frac{\sqrt{3}}{2} + \frac{\sqrt{2}}{2} \cdot \frac{1}{2}\right) =$$

$$\tfrac{100}{4}\sqrt{2}\,(\sqrt{6} + \sqrt{2}) = 25(2\sqrt{3} + 2) = 50(1 + \sqrt{3}).$$

[4] $\alpha = \cos^{-1}\left(\dfrac{b^2 + c^2 - a^2}{2bc}\right) = \cos^{-1}\left(\dfrac{9 + 16 - 4}{2(3)(4)}\right) = \cos^{-1}\left(\dfrac{7}{8}\right).$

$\beta = \cos^{-1}\left(\dfrac{a^2 + c^2 - b^2}{2ac}\right) = \cos^{-1}\left(\dfrac{4 + 16 - 9}{2(2)(4)}\right) = \cos^{-1}\left(\dfrac{11}{16}\right).$

$\gamma = \cos^{-1}\left(\dfrac{a^2 + b^2 - c^2}{2ab}\right) = \cos^{-1}\left(\dfrac{4 + 9 - 16}{2(2)(3)}\right) = \cos^{-1}\left(-\dfrac{1}{4}\right).$

[6] $\dfrac{\sin \gamma}{c} = \dfrac{\sin \alpha}{a} \Rightarrow \gamma = \sin^{-1}\left(\dfrac{c \sin \alpha}{a}\right) = \sin^{-1}\left(\dfrac{125 \sin 23°30'}{152}\right) \approx \sin^{-1}(0.3279) \approx$

19°10' or 160°50' { rounded to the nearest 10 minutes }. Reject 160°50' because then

$\alpha + \gamma \geq 180°.$ $\beta = 180° - \alpha - \gamma \approx 180° - 23°30' - 19°10' = 137°20'.$

$$\frac{b}{\sin \beta} = \frac{a}{\sin \alpha} \Rightarrow b = \frac{a \sin \beta}{\sin \alpha} = \frac{152 \sin 137°20'}{\sin 23°30'} \approx 258.3, \text{ or } 258.$$

[9] Since we are given two sides and the included angle of a triangle,

we use the formula for the area of a triangle from Section 4.2.

$$\mathcal{A} = \tfrac{1}{2}bc \sin \alpha = \tfrac{1}{2}(20)(30) \sin 75° \approx 289.8, \text{ or } 290 \text{ square units.}$$

[10] Given the three sides of a triangle, we apply Heron's formula to find the area.

$s = \tfrac{1}{2}(a + b + c) = \tfrac{1}{2}(4 + 7 + 10) = 10.5.$

$$\mathcal{A} = \sqrt{s(s - a)(s - b)(s - c)} = \sqrt{(10.5)(6.5)(3.5)(0.5)} \approx 10.9 \text{ square units.}$$

[11] $z = -10 + 10i \Rightarrow r = \sqrt{(-10)^2 + 10^2} = \sqrt{200} = 10\sqrt{2}.$

$$\tan \theta = \frac{10}{-10} = -1 \text{ and } \theta \text{ in QII} \Rightarrow \theta = \frac{3\pi}{4}. \quad z = 10\sqrt{2}\, \text{cis} \frac{3\pi}{4}.$$

[13] $z = -17 \Rightarrow r = 17.$ θ on the negative x-axis $\Rightarrow \theta = \pi.$ $z = 17 \,\text{cis}\, \pi.$

[16] $z = 4 + 5i \Rightarrow r = \sqrt{4^2 + 5^2} = \sqrt{41}.$ $\tan \theta = \frac{5}{4}$ and θ in QI $\Rightarrow \theta = \tan^{-1} \frac{5}{4}.$

$$z = \sqrt{41}\, \text{cis}\left(\tan^{-1} \tfrac{5}{4}\right).$$

[17] $20\left(\cos \frac{11\pi}{6} + i \sin \frac{11\pi}{6}\right) = 20\left(\dfrac{\sqrt{3}}{2} - \dfrac{1}{2}i\right) = 10\sqrt{3} - 10i$

19 $z_1 = -3\sqrt{3} - 3i = 6\operatorname{cis}\frac{7\pi}{6}$ and $z_2 = 2\sqrt{3} + 2i = 4\operatorname{cis}\frac{\pi}{6}$.

$$z_1 z_2 = 6 \cdot 4 \operatorname{cis}\left(\frac{7\pi}{6} + \frac{\pi}{6}\right) = 24\operatorname{cis}\frac{4\pi}{3} = 24\left(-\frac{1}{2} - \frac{\sqrt{3}}{2}i\right) = -12 - 12\sqrt{3}\,i.$$

$$\frac{z_1}{z_2} = \frac{6}{4}\operatorname{cis}\left(\frac{7\pi}{6} - \frac{\pi}{6}\right) = \frac{3}{2}\operatorname{cis}\pi = \frac{3}{2}(-1 + 0i) = -\frac{3}{2}.$$

21 $(-\sqrt{3} + i)^9 = (2\operatorname{cis}\frac{5\pi}{6})^9 = 2^9 \operatorname{cis}\left(9 \cdot \frac{5\pi}{6}\right) = 2^9\operatorname{cis}\frac{15\pi}{2} = 512\operatorname{cis}\frac{3\pi}{2} = 512(0 - i) = -512i$

23 $(3 - 3i)^5 = (3\sqrt{2}\operatorname{cis}\frac{7\pi}{4})^5 = (3\sqrt{2})^5\operatorname{cis}\frac{35\pi}{4} = 972\sqrt{2}\operatorname{cis}\frac{3\pi}{4} = 972\sqrt{2}\left(-\frac{\sqrt{2}}{2} + \frac{\sqrt{2}}{2}i\right) =$
$$-972 + 972i$$

25 $-27 + 0i = 27\operatorname{cis}180°$. $w_k = \sqrt[3]{27}\operatorname{cis}\left(\frac{180° + 360°k}{3}\right)$ for $k = 0, 1, 2$.

$$w_0 = 3\operatorname{cis}60° = 3\left(\frac{1}{2} + \frac{\sqrt{3}}{2}i\right) = \frac{3}{2} + \frac{3\sqrt{3}}{2}i.$$

$$w_1 = 3\operatorname{cis}180° = 3(-1 + 0i) = -3.$$

$$w_2 = 3\operatorname{cis}300° = 3\left(\frac{1}{2} - \frac{\sqrt{3}}{2}i\right) = \frac{3}{2} - \frac{3\sqrt{3}}{2}i.$$

27 $x^5 - 32 = 0 \Rightarrow x^5 = 32$. The problem is now to find the 5 fifth roots of 32.

$$32 = 32 + 0i = 32\operatorname{cis}0°. \quad w_k = \sqrt[5]{32}\operatorname{cis}\left(\frac{0° + 360°k}{5}\right) \text{ for } k = 0, 1, 2, 3, 4.$$

$$w_k = 2\operatorname{cis}\theta \text{ with } \theta = 0°, 72°, 144°, 216°, 288°.$$

29 (a) $4\mathbf{a} + \mathbf{b} = 4(2\mathbf{i} + 5\mathbf{j}) + (4\mathbf{i} - \mathbf{j}) = 8\mathbf{i} + 20\mathbf{j} + 4\mathbf{i} - \mathbf{j} = 12\mathbf{i} + 19\mathbf{j}$.

(b) $2\mathbf{a} - 3\mathbf{b} = 2(2\mathbf{i} + 5\mathbf{j}) - 3(4\mathbf{i} - \mathbf{j}) = 4\mathbf{i} + 10\mathbf{j} - 12\mathbf{i} + 3\mathbf{j} = -8\mathbf{i} + 13\mathbf{j}$.

(c) $\|\mathbf{a} - \mathbf{b}\| = \|(2\mathbf{i} + 5\mathbf{j}) - (4\mathbf{i} - \mathbf{j})\| = \|-2\mathbf{i} + 6\mathbf{j}\| = \sqrt{(-2)^2 + 6^2} = \sqrt{40} = 2\sqrt{10} \approx$
$$6.32.$$

(d) $\|\mathbf{a}\| - \|\mathbf{b}\| = \|2\mathbf{i} + 5\mathbf{j}\| - \|4\mathbf{i} - \mathbf{j}\| = \sqrt{2^2 + 5^2} - \sqrt{4^2 + (-1)^2} = \sqrt{29} - \sqrt{17} \approx 1.26$.

30 $\|\mathbf{r} - \mathbf{a}\| = c \Rightarrow \|\langle x, y\rangle - \langle a_1, a_2\rangle\| = c \Rightarrow \|\langle x - a_1, y - a_2\rangle\| = c \Rightarrow$

$$\sqrt{(x - a_1)^2 + (y - a_2)^2} = c \Rightarrow (x - a_1)^2 + (y - a_2)^2 = c^2.$$

This is a circle with center (a_1, a_2) and radius c.

33 S60°E is equivalent to 330° and N74°E is equivalent to 16°.

$$\langle 72\cos 330°, 72\sin 330°\rangle + \langle 46\cos 16°, 46\sin 16°\rangle = \mathbf{r} \approx \langle 106.57, -23.32\rangle.$$

$$\|\mathbf{r}\| \approx 109 \text{ kg.} \quad \tan\theta \approx \frac{-23.32}{106.57} \Rightarrow \theta \approx -12°, \text{ or equivalently, S78°E.}$$

[34] $\mathbf{p} = \langle 400\cos 10°, 400\sin 10° \rangle \approx \langle 393.92, 69.46 \rangle$.

$\mathbf{r} = \langle 390\cos 0°, 390\sin 0° \rangle = \langle 390, 0 \rangle$.

$\mathbf{w} = \mathbf{r} - \mathbf{p} \approx \langle -3.92, -69.46 \rangle$ and $\|\mathbf{w}\| \approx 69.57$, or 70 mi/hr.

$\tan \theta \approx \dfrac{-69.46}{-3.92} \Rightarrow \theta \approx 267°$, or in the direction of 183°.

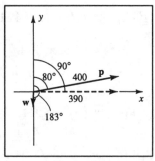

Figure 34

[36] (a) $(2\mathbf{a} - 3\mathbf{b}) \cdot \mathbf{a} = \big[2(6\mathbf{i} - 2\mathbf{j}) - 3(\mathbf{i} + 3\mathbf{j}) \big] \cdot (6\mathbf{i} - 2\mathbf{j})$

$\qquad = (9\mathbf{i} - 13\mathbf{j}) \cdot (6\mathbf{i} - 2\mathbf{j}) = 54 + 26 = 80.$

(b) $\mathbf{c} = \mathbf{a} + \mathbf{b} = (6\mathbf{i} - 2\mathbf{j}) + (\mathbf{i} + 3\mathbf{j}) = 7\mathbf{i} + \mathbf{j}$. The angle between $\mathbf{a}$ and $\mathbf{c}$ is

$$\theta = \cos^{-1}\left(\frac{\mathbf{a} \cdot \mathbf{c}}{\|\mathbf{a}\|\|\mathbf{c}\|} \right) = \cos^{-1}\left(\frac{(6\mathbf{i} - 2\mathbf{j}) \cdot (7\mathbf{i} + \mathbf{j})}{\|6\mathbf{i} - 2\mathbf{j}\|\,\|7\mathbf{i} + \mathbf{j}\|} \right) = \cos^{-1}\left(\frac{40}{\sqrt{40}\,\sqrt{50}} \right) \approx 26°34'.$$

(c) $\text{comp}_{\mathbf{a}}(\mathbf{a} + \mathbf{b}) = \text{comp}_{\mathbf{a}}\mathbf{c} = \dfrac{\mathbf{c} \cdot \mathbf{a}}{\|\mathbf{a}\|} = \dfrac{40}{\sqrt{40}} = \sqrt{40} = 2\sqrt{10} \approx 6.32.$

[38] $\dfrac{\sin \gamma}{150} = \dfrac{\sin 27.4°}{200} \Rightarrow$

$\gamma = \sin^{-1}\left(\dfrac{150\sin 27.4°}{200} \right) \approx \sin^{-1}(0.3451) \approx 20.2°.$

$\beta = 180° - \alpha - \beta \approx 180° - 27.4° - 20.2° = 132.4°.$

The angle between the hill and the horizontal is then

$180° - 132.4° = 47.6°.$

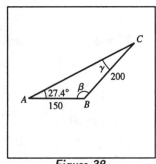

Figure 38

[39] Let a be the Earth–Venus distance, b be the Earth–sun distance, and c be the Venus–sun distance. Then, by the law of cosines (with a, b, and c in millions), $a^2 = b^2 + c^2 - 2bc\cos \alpha = 93^2 + 67^2 - 2(93)(67)\cos 34° \approx 2807 \Rightarrow$ $a \approx 53$—that is, 53,000,000 miles.

[40] Let P denote the point at the base of the shorter building, S the point at the top of the shorter building, T the point at the top of the skyscraper, Q the point 50 feet up the side of the skyscraper, and h the height of the skyscraper.

(a) $\angle SPT = 90° - 62° = 28°$. $\angle PST = 90° + 59° = 149°$.

$\qquad$ Thus, $\angle STP = 180° - 28° - 149° = 3°$. $\dfrac{\overline{ST}}{\sin 28°} = \dfrac{50}{\sin 3°} \Rightarrow \overline{ST} \approx 448.52$, or 449 ft.

(b) $h = \overline{QT} + 50 = \overline{ST}\sin 59° + 50 \approx 434.45$, or 434 ft.

42 Let E denote the middle point. $\angle CDA = \angle BDC - \angle BDA = 125° - 100° = 25°$.

In $\triangle CAD$, $\angle CAD = 180° - \angle ACD - \angle CDA = 180° - 115° - 25° = 40°$.

$\dfrac{\overline{AD}}{\sin 115°} = \dfrac{120}{\sin 40°} \Rightarrow \overline{AD} \approx 169.20$. $\angle DCB = \angle ACD - \angle ACB = 115° - 92° = 23°$.

In $\triangle DBC$, $\angle DBC = 180° - \angle BDC - \angle DCB = 180° - 125° - 23° = 32°$.

$\dfrac{\overline{BD}}{\sin 23°} = \dfrac{120}{\sin 32°} \Rightarrow \overline{BD} \approx 88.48$.

In $\triangle ADB$, $\overline{AB}^2 = \overline{AD}^2 + \overline{BD}^2 - 2(\overline{AD})(\overline{BD})\cos \angle BDA \Rightarrow$

$$\overline{AB} \approx \sqrt{(169.20)^2 + (88.48)^2 - 2(169.20)(88.48)\cos 100°} \approx 204.1, \text{ or } 204 \text{ ft.}$$

43 If d denotes the distance each girl walks before losing contact with each other,

then $d = 5t$, where t is in hours. Using the law of cosines,

$10^2 = d^2 + d^2 - 2(d)(d)\cos 105° \Rightarrow 100 = 2d^2(1 - \cos 105°) \Rightarrow d \approx 6.30 \Rightarrow$

$$t = d/5 \approx 1.26 \text{ hours, or 1 hour and 16 minutes.}$$

44 (a) Draw a vertical line l through C and label its x-intercept D. Since we have

alternate interior angles, $\angle ACD = \theta_1$. $\angle DCP = 180° - \theta_2$.

$$\text{Thus } \angle ACP = \angle ACD + \angle DCP = \theta_1 + (180° - \theta_2) = 180° - (\theta_2 - \theta_1).$$

(b) Let $k = d(A, P)$. $k^2 = 17^2 + 17^2 - 2(17)(17)\cos\left[180° - (\theta_2 - \theta_1)\right]$.

Since $\cos(180° - \alpha) = \cos 180° \cos \alpha + \sin 180° \sin \alpha = -\cos \alpha$, we have

$k^2 = 578 + 578\cos(\theta_2 - \theta_1) = 578\left[1 + \cos(\theta_2 - \theta_1)\right]$. Using the distance

formula with the points $A(0, 26)$ and $P(x, y)$, we also have $k^2 = x^2 + (y - 26)^2$.

$$\text{Hence, } 578\left[1 + \cos(\theta_2 - \theta_1)\right] = x^2 + (y - 26)^2 \Rightarrow 1 + \cos(\theta_2 - \theta_1) = \frac{x^2 + (y - 26)^2}{578}.$$

(c) If $x = 25$, $y = 4$, and $\theta_1 = 135°$, then $1 + \cos(\theta_2 - 135°) = \dfrac{25^2 + (-22)^2}{578} = \dfrac{1109}{578} \Rightarrow$

$$\cos(\theta_2 - 135°) = \tfrac{531}{578} \Rightarrow \theta_2 - 135° \approx 23.3° \Rightarrow \theta_2 \approx 158.3°, \text{ or } 158°.$$

45 (a) Let d denote the length of the rescue tunnel. Using the law of cosines,

$d^2 = 45^2 + 50^2 - 2(45)(50)\cos 78° \Rightarrow d \approx 59.91$ ft. Now using the law of sines,

$$\frac{\sin \theta}{45} = \frac{\sin 78°}{d} \Rightarrow \theta = \sin^{-1}\left(\frac{45\sin 78°}{d}\right) \approx 47.28°, \text{ or } 47°.$$

(b) If x denotes the number of hours needed, then

$$d \text{ ft} = (3 \text{ ft/hr})(x \text{ hr}) \Rightarrow x = \tfrac{1}{3}d = \tfrac{1}{3}(59.91) \approx 20 \text{ hr.}$$

46 (a) $\angle CBA = 180° - 136° = 44°$ and

$$d = \overline{AC} = \sqrt{22.9^2 + 17.2^2 - 2(22.9)(17.2)\cos 44°} \approx 15.9. \text{ Let } \alpha = \angle BAC.$$

Using the law of sines, $\dfrac{\sin \alpha}{22.9} = \dfrac{\sin 44°}{d} \Rightarrow \alpha = \sin^{-1}\left(\dfrac{22.9 \sin 44°}{d}\right) \approx 87.4°.$

Let $\beta = \angle CAD$. Using the law of cosines, $5.7^2 = d^2 + 16^2 - 2(d)(16)\cos \beta \Rightarrow$

$$\beta = \cos^{-1}\left(\dfrac{d^2 + 16^2 - 5.7^2}{2(d)(16)}\right) \approx 20.6°. \quad \phi \approx 180° - 87.4° - 20.6° = 72°.$$

(b) The area of $ABCD$ is the sum of the areas of $\triangle CBA$ and $\triangle ADC$.

Area $= \frac{1}{2}(\text{base } \overline{BC})(\text{height to } A) + \frac{1}{2}(\text{base } \overline{AC})(\text{height to } D)$

$= \frac{1}{2}(\overline{BC})(\overline{BA})\sin \angle CBA + \frac{1}{2}(\overline{AC})(\overline{AD})\sin \angle CAD$

$= \frac{1}{2}(22.9)(17.2)\sin 44° + \frac{1}{2}(15.9)(16)\sin 20.6° \approx 136.8 + 44.8 = 181.6 \text{ ft}^2.$

(c) Let h denote the perpendicular distance from $\overline{BA}$ to C.

$$\sin 44° = \dfrac{h}{22.9} \Rightarrow h \approx 15.9. \text{ The wing span } \overline{CC'} \text{ is } 2h + 5.8 \approx 37.6 \text{ ft.}$$

Chapter 4 Discussion Exercises

1 (a) $\dfrac{a}{\sin \alpha} = \dfrac{c}{\sin \gamma} \Rightarrow \dfrac{a}{c} = \dfrac{\sin \alpha}{\sin \gamma}$ and $\dfrac{b}{\sin \beta} = \dfrac{c}{\sin \gamma} \Rightarrow \dfrac{b}{c} = \dfrac{\sin \beta}{\sin \gamma}.$

Adding the equations yields $\dfrac{a}{c} + \dfrac{b}{c} = \dfrac{\sin \alpha}{\sin \gamma} + \dfrac{\sin \beta}{\sin \gamma} \Rightarrow \dfrac{a+b}{c} = \dfrac{\sin \alpha + \sin \beta}{\sin \gamma}.$

(b) $\dfrac{a+b}{c} = \dfrac{\sin \alpha + \sin \beta}{\sin \gamma} \Rightarrow \dfrac{a+b}{c} = \dfrac{\overset{[\text{S1}]}{2 \sin \frac{1}{2}(\alpha + \beta) \cos \frac{1}{2}(\alpha - \beta)}}{2 \sin \frac{1}{2}\gamma \cos \frac{1}{2}\gamma}.$

Now $\gamma = 180° - (\alpha + \beta) \Rightarrow \frac{1}{2}\gamma = \left[90° - \frac{1}{2}(\alpha + \beta)\right]$ and

$$\sin \tfrac{1}{2}(\alpha + \beta) = \cos\left[90° - \tfrac{1}{2}(\alpha + \beta)\right] = \cos \tfrac{1}{2}\gamma. \text{ Thus, } \dfrac{a+b}{c} = \dfrac{\cos \frac{1}{2}(\alpha - \beta)}{\sin \frac{1}{2}\gamma}.$$

Note: This is an interesting result and gives an answer to the question "How can I check these triangle problems?" Some of my students have written programs for their graphing calculators to utilize this check.

3 Example 4 in Section 4.4 illustrates the case $a = 1$.

Algebraic: $\sqrt[3]{a}, \ \sqrt[3]{a} \operatorname{cis} \frac{2\pi}{3}, \ \sqrt[3]{a} \operatorname{cis} \frac{4\pi}{3}$

Geometric: All roots lie on a circle of radius $\sqrt[3]{a}$, they are all 120° apart,

one is on the real axis, one is on $\theta = \frac{2\pi}{3}$, and one is on $\theta = \frac{4\pi}{3}$

$\boxed{5}$ (a) $\mathbf{c} = \mathbf{b} + \mathbf{a} = (\|\mathbf{b}\|\cos\alpha\,\mathbf{i} + \|\mathbf{b}\|\sin\alpha\,\mathbf{j}) + (\|\mathbf{a}\|\cos(-\beta)\,\mathbf{i} + \|\mathbf{a}\|\sin(-\beta)\,\mathbf{j})$

$\qquad = \|\mathbf{b}\|\cos\alpha\,\mathbf{i} + \|\mathbf{b}\|\sin\alpha\,\mathbf{j} + \|\mathbf{a}\|\cos\beta\,\mathbf{i} - \|\mathbf{a}\|\sin\beta\,\mathbf{j}$

$\qquad = (\|\mathbf{b}\|\cos\alpha + \|\mathbf{a}\|\cos\beta)\mathbf{i} + (\|\mathbf{b}\|\sin\alpha - \|\mathbf{a}\|\sin\beta)\mathbf{j}$

(b) $\|\mathbf{c}\|^2 = (\|\mathbf{b}\|\cos\alpha + \|\mathbf{a}\|\cos\beta)^2 + (\|\mathbf{b}\|\sin\alpha - \|\mathbf{a}\|\sin\beta)^2$

$\qquad = \|\mathbf{b}\|^2\cos^2\alpha + 2\|\mathbf{a}\|\|\mathbf{b}\|\cos\alpha\cos\beta + \|\mathbf{a}\|^2\cos^2\beta +$

$\qquad\qquad\qquad\qquad \|\mathbf{b}\|^2\sin^2\alpha - 2\|\mathbf{a}\|\|\mathbf{b}\|\sin\alpha\sin\beta + \|\mathbf{a}\|^2\sin^2\beta$

$\qquad = (\|\mathbf{b}\|^2\cos^2\alpha + \|\mathbf{b}\|^2\sin^2\alpha) + (\|\mathbf{a}\|^2\cos^2\beta + \|\mathbf{a}\|^2\sin^2\beta) +$

$\qquad\qquad\qquad\qquad 2\|\mathbf{a}\|\|\mathbf{b}\|\cos\alpha\cos\beta - 2\|\mathbf{a}\|\|\mathbf{b}\|\sin\alpha\sin\beta$

$\qquad = \|\mathbf{b}\|^2 + \|\mathbf{a}\|^2 + 2\|\mathbf{a}\|\|\mathbf{b}\|(\cos\alpha\cos\beta - \sin\alpha\sin\beta)$

$\qquad = \|\mathbf{a}\|^2 + \|\mathbf{b}\|^2 + 2\|\mathbf{a}\|\|\mathbf{b}\|\cos(\alpha + \beta)$

$\qquad = \|\mathbf{a}\|^2 + \|\mathbf{b}\|^2 + 2\|\mathbf{a}\|\|\mathbf{b}\|\cos(\pi - \gamma) \ \{\alpha + \beta + \gamma = \pi\}$

$\qquad = \|\mathbf{a}\|^2 + \|\mathbf{b}\|^2 - 2\|\mathbf{a}\|\|\mathbf{b}\|\cos\gamma \ \{\cos(\pi - \gamma) = -\cos\gamma\}$

(c) From part (a), we let $(\|\mathbf{b}\|\sin\alpha - \|\mathbf{a}\|\sin\beta) = 0$.

$\qquad\qquad$ Thus, $\|\mathbf{b}\|\sin\alpha = \|\mathbf{a}\|\sin\beta$, and $\dfrac{\sin\alpha}{\|\mathbf{a}\|} = \dfrac{\sin\beta}{\|\mathbf{b}\|}$.

Chapter 5: Exponential and Logarithmic Functions

$\boxed{3}$ $3^{2x+3} = 3^{(x^2)} \Rightarrow 2x + 3 = x^2 \Rightarrow x^2 - 2x - 3 = 0 \Rightarrow (x-3)(x+1) = 0 \Rightarrow x = -1, 3$

$\boxed{5}$ We need to obtain the same base on each side of the equals sign, then we can apply part (2) of the theorem about exponential functions being one-to-one.

$2^{-100x} = (0.5)^{x-4} \Rightarrow (2^{-1})^{100x} = \left(\frac{1}{2}\right)^{x-4} \Rightarrow \left(\frac{1}{2}\right)^{100x} = \left(\frac{1}{2}\right)^{x-4} \Rightarrow$

$$100x = x - 4 \Rightarrow 99x = -4 \Rightarrow x = -\frac{4}{99}$$

$\boxed{7}$ $4^{x-3} = 8^{4-x} \Rightarrow (2^2)^{x-3} = (2^3)^{4-x} \Rightarrow 2^{2x-6} = 2^{12-3x} \Rightarrow$

$$2x - 6 = 12 - 3x \Rightarrow 5x = 18 \Rightarrow x = \frac{18}{5}$$

$\boxed{9}$ (a) Let $g = f(x) = 2^x$ for reference purposes.

The graph of g goes through the points $(-1, \frac{1}{2})$, $(0, 1)$, and $(1, 2)$.

(b) $f(x) = -2^x$ •

Reflect g through the x-axis since f is just $-1 \cdot 2^x$. Do not confuse this function

with $(-2)^x$ — remember, the base is positive for exponential functions.

(c) $f(x) = 3 \cdot 2^x$ • vertically stretch g by a factor of 3

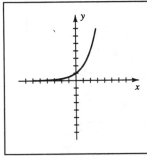

Figure 9(a)

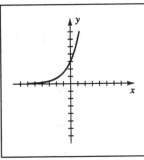

Figure 9(b)

Figure 9(c)

(d) $f(x) = 2^{x+3}$ • shift g left 3 units since f is $g(x+3)$

(e) $f(x) = 2^x + 3$ • vertically shift g up 3 units

(f) $f(x) = 2^{x-3}$ • shift g right 3 units since f is $g(x-3)$

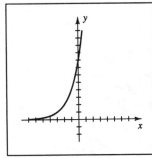

Figure 9(d)

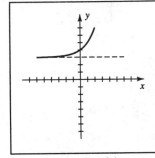

Figure 9(e)

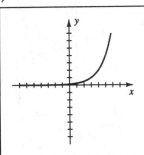

Figure 9(f)

(g) $f(x) = 2^x - 3$ •   vertically shift g down 3 units

(h) $f(x) = 2^{-x}$ •   reflect g through the y-axis since f is $g(-x)$

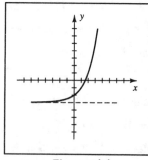

Figure 9(g)

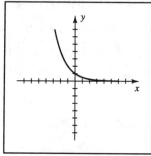

Figure 9(h)

(i) $f(x) = \left(\frac{1}{2}\right)^x$ • $\left(\frac{1}{2}\right)^x = (2^{-1})^x = 2^{-x}$, same graph as in part (h)

(j) $f(x) = 2^{3-x}$ • $2^{3-x} = 2^{-(x-3)}$, shift g right 3 units and reflect through the

line $x = 3$. Alternatively, $2^{3-x} = 2^3 2^{-x} = 8\left(\frac{1}{2}\right)^x$, vertically stretch $y = \left(\frac{1}{2}\right)^x$ (the

graph in part (i))by a factor of 8.

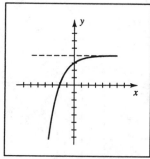

Figure 9(i)

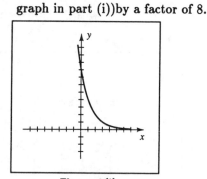

Figure 9(j)

$\boxed{13}$ $f(x) = -\left(\frac{1}{2}\right)^x + 4$ •   reflect $y = \left(\frac{1}{2}\right)^x$ through the x-axis and shift up 4 units

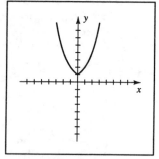

Figure 13

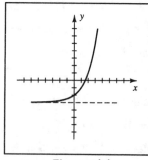

Figure 15

$\boxed{15}$ $f(x) = 2^{|x|} = \begin{cases} 2^x & \text{if } x \geq 0 \\ 2^{-x} & \text{if } x < 0 \end{cases} = \begin{cases} 2^x & \text{if } x \geq 0 \\ \left(\frac{1}{2}\right)^x & \text{if } x < 0 \end{cases}$

Use the portion of $y = 2^x$ with $x \geq 0$ and reflect it through the y-axis since f is even.

Note: For Exercises 17, 18, and 5 of the review exercises, refer to Example 5 in the text

for the basic graph of $y = a^{-x^2} = (a^{-1})^{(x^2)} = \left(\frac{1}{a}\right)^{(x^2)}$, where $a > 1$.

[17] $f(x) = 3^{1-x^2} = 3^1 3^{-x^2} = 3\left(\frac{1}{3}\right)^{(x^2)}$ • stretch $y = \left(\frac{1}{3}\right)^{(x^2)}$ by a factor of 3

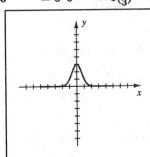

Figure 17

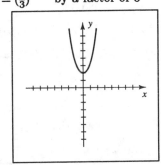

Figure 19

[19] $f(x) = 3^x + 3^{-x}$ • Adding the functions $g(x) = 3^x$ and $h(x) = 3^{-x} = \left(\frac{1}{3}\right)^x$ together, we see that the y-intercept will be (0, 2). If $x > 0$, f looks like $y = 3^x$ since 3^x dominates 3^{-x} (3^{-x} gets close to 0 and 3^x grows very large). If $x < 0$, f looks like $y = 3^{-x}$ since 3^{-x} dominates 3^x.

[23] (a) 8:00 A.M. corresponds to $t = 1$ and $f(1) = 600\sqrt{3} \approx 1039$.

10:00 A.M. corresponds to $t = 3$ and $f(3) = 600(3\sqrt{3}) = 1800\sqrt{3} \approx 3118$.

11:00 A.M. corresponds to $t = 4$ and $f(4) = 600(9) = 5400$.

(b) The graph of f is an increasing exponential that passes through (0, 600) and the

points in part (a).

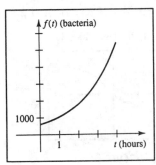

Figure 23

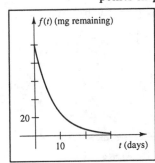

Figure 25

[25] (a) $f(5) = 100(2)^{-1} = 50$ mg; $f(10) = 100(2)^{-2} = 25$ mg;

$$f(12.5) = 100(2)^{-2.5} = \frac{100}{4\sqrt{2}} = \frac{25}{2}\sqrt{2} \approx 17.7 \text{ mg}$$

(b) The endpoints are (0, 100) and (30, 1.5625).

[27] A half-life of 1600 years means that when $t = 1600$, the amount remaining, $q(t)$, will be one-half the original amount—that is, $\frac{1}{2}q_0$. $q(t) = \frac{1}{2}q_0$ when $t = 1600 \Rightarrow$ $\frac{1}{2}q_0 = q_0 2^{k(1600)} \Rightarrow 2^{-1} = 2^{1600k} \Rightarrow 1600k = -1 \Rightarrow k = -\frac{1}{1600}$.

29 Using $A = P\left(1 + \frac{r}{n}\right)^{nt}$, we have $P = 1000$, $r = 0.12$, and $n = 12$.

Consider A to be a function of t, that is, $A(t) = 1000\left(1 + \frac{0.12}{12}\right)^{12t} = 1000(1.01)^{12t}$.

(a) $A(\frac{1}{12}) = \$1010.00$ (d) $A(20) \approx \$10{,}892.55$

31 $C = 10{,}000 \Rightarrow V(t) = 7800(0.85)^{t-1}$

(a) $V(1) = \$7800$ (b) $V(4) \approx \$4790.18$, or $\$4790$ (c) $V(7) \approx \$2941.77$, or $\$2942$

33 $t = 2006 - 1626 = 380$; $A = \$24(1 + 0.06/4)^{4 \cdot 380} = \$161{,}657{,}351{,}965.80$.

That's right—$161 billion!

35 (a) Examine the pattern formed by the value y in the year n.

year (n)	value (y)
0	y_0
1	$(1-a)y_0 = y_1$
2	$(1-a)y_1 = (1-a)\left[(1-a)y_0\right] = (1-a)^2 y_0 = y_2$
3	$(1-a)y_2 = (1-a)\left[(1-a)^2 y_0\right] = (1-a)^3 y_0 = y_3$

(b) $s = (1-a)^T y_0 \Rightarrow (1-a)^T = s/y_0 \Rightarrow 1 - a = \sqrt[T]{s/y_0} \Rightarrow a = 1 - \sqrt[T]{s/y_0}$

37 (a) $r = 0.12$, $t = 30$, $L = 90{,}000 \Rightarrow k \approx 35.95$, $M \approx 925.75$

(b) (360 payments) $\times \$925.75 - \$90{,}000 = \$243{,}270$

43 Part (b) may be interpreted as doubling an investment at 8.5%.

(a) If $y = (1.085)^x$ and $x = 40$, then $y \approx 26.13$.   (b) If $y = 2$, then $x \approx 8.50$.

[0, 60] by [0, 40]   [−3, 3] by [−2, 2]

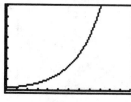   Xscl $= 5$ 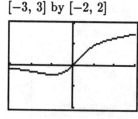   Xscl $= 1$
               Yscl $= 5$ Yscl $= 1$

Figure 43   Figure 47

47 (a) f is not one-to-one since

the horizontal line $y = -0.1$ intersects the graph of f more than once.

(b) The only zero of f is $x = 0$.

51 *Figure 51* is a graph of $N(t) = 1000(0.9)^t$.

By tracing and zooming, we can determine that $N = 500$ when $t \approx 6.58$ yr.

[0, 10] by [0, 1000]   [0, 7.5] by [0, 5]

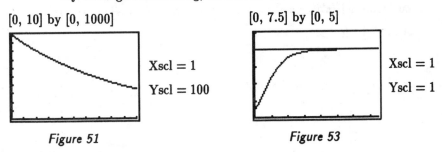

Xscl = 1   Xscl = 1

Yscl = 100   Yscl = 1

Figure 51   *Figure 53*

53 Graph $y = 4(0.125)^{(0.25^x)}$. The line $y = k = 4$ is a horizontal asymptote for the

Gompertz function. The maximum number of sales of the product approaches k.

55 From the graph, we determine that $A = 100,000$ when $n \approx 32.8$.

[0, 40] by [0, 200,000]

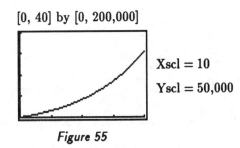

Xscl = 10

Yscl = 50,000

Figure 55

57 (a) Let $x = 0$ correspond to 1910, $x = 20$ to 1930, ... , and $x = 85$ to 1995.

Graph the data together with the functions

$f(x) = 0.809(1.094)^x$ and $g(x) = 0.375x^2 - 18.4x + 88.1$.

[−10, 90] by [−200, 1500]   [−10, 90] by [−200, 1500]

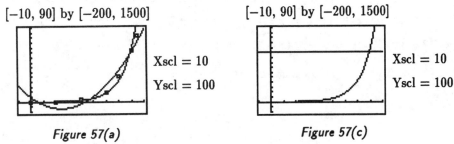

Xscl = 10   Xscl = 10

Yscl = 100   Yscl = 100

Figure 57(a)   *Figure 57(c)*

(b) The exponential function f best models the data.

(c) Graph $Y_1 = f(x)$ and $Y_2 = 1000$. The graphs intersect at $x \approx 79$, or in 1989.

5.2 Exercises

Note: Examine Figure 13 in this section to reinforce the idea that $y = e^x$ is just a special case of $y = a^x$ with $a > 1$.

$\boxed{3}$ (a) $f(x) = e^{x+4}$ • shift $y = e^x$ left 4 units

    (b) $f(x) = e^x + 4$ • shift $y = e^x$ up 4 units

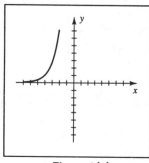

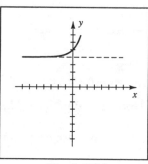

                  Figure 3(a)   *Figure 3(b)*

$\boxed{5}$ $A = Pe^{rt} = 1000e^{(0.0825)(5)} \approx \$1510.59$

$\boxed{7}$ $100{,}000 = Pe^{(0.11)(18)} \Rightarrow P = \dfrac{100{,}000}{e^{1.98}} \approx \$13{,}806.92$

$\boxed{9}$ $13{,}464 = 1000e^{(r)(20)} \Rightarrow e^{20r} = 13.464.$ Using a trial and error approach on a scientific calculator or tracing and zooming or an intersect feature on a graphing calculator, we determine that $e^x \approx 13.464$ if $x \approx 2.6$. Thus, $20r = 2.6$ and $r = 0.13$ or 13%.

$\boxed{11}$ $e^{(x^2)} = e^{7x-12} \Rightarrow x^2 = 7x - 12 \Rightarrow x^2 - 7x + 12 = 0 \Rightarrow (x-3)(x-4) = 0 \Rightarrow x = 3,\, 4$

$\boxed{13}$ $xe^x + e^x = 0 \Rightarrow e^x(x+1) = 0 \Rightarrow x = -1$ { Note that $e^x \neq 0$. }

$\boxed{15}$ $x^3(4e^{4x}) + 3x^2e^{4x} = 0 \Rightarrow x^2e^{4x}(4x+3) = 0 \Rightarrow x = -\frac{3}{4},\, 0.$ Note that $e^{4x} \neq 0$.

$\boxed{17}$ $\dfrac{(e^x + e^{-x})(e^x + e^{-x}) - (e^x - e^{-x})(e^x - e^{-x})}{(e^x + e^{-x})^2} =$

$$\dfrac{(e^{2x} + 2 + e^{-2x}) - (e^{2x} - 2 + e^{-2x})}{(e^x + e^{-x})^2} = \dfrac{4}{(e^x + e^{-x})^2}$$

$\boxed{19}$ $W(t) = W_0 e^{kt};\ t = 30 \Rightarrow W(30) = 68e^{(0.2)(30)} \approx 27{,}433$ mg, or 27.43 grams

$\boxed{21}$ The year 2010 corresponds to $t = 2010 - 1980 = 30$. Using the law of growth formula on page 329 with $q_0 = 227$ and $r = 0.007$, we have $N(t) = 227e^{0.007t}$.

                                Thus, $N(30) = 227e^{(0.007)(30)} \approx 280.0$ million.

$\boxed{23}$ $N(10) = N_0 e^{-2}$. The percentage of the original number still alive after 10 years is

$$100 \times \left(\dfrac{N(10)}{N_0} \right) = 100e^{-2} \approx 13.5\%.$$

$\boxed{25}$ The year 2010 corresponds to $t = 2010 - 1978 = 32$.

$$N(32) = 5000e^{(0.0036)(32)} = 5000e^{(0.1152)} \approx 5610.$$

$\boxed{27}$ $h = 40,000 \Rightarrow p = 29e^{(-0.000034)(40,000)} = 29e^{(-1.36)} \approx 7.44$ in.

$\boxed{29}$ $x = 1 \Rightarrow y = 79.041 + 6.39 - e^{2.268} \approx 75.77$ cm.

$$x = 1 \Rightarrow R = 6.39 + 0.993e^{2.268} \approx 15.98 \text{ cm/yr.}$$

$\boxed{31}$ $2010 - 1971 = 39 \Rightarrow t = 39$ years. Using the continuously compounded interest

formula with $P = 1.60$ and $r = 0.05$, we have $A = 1.60e^{(0.05)(39)} \approx \$11.25$ per hour.

$\boxed{33}$ (a) Note here that the amount of money invested is not of interest and that we are

only concerned with the percent of growth.

$$\left(1 + \tfrac{0.07}{4}\right)^{4 \cdot 1} \approx 1.0719. \quad (1.0719 - 1) \times 100\% = 7.19\%$$

(b) $e^{(0.07)(1)} \approx 1.0725.$ $(1.0725 - 1) \times 100\% = 7.25\%.$ The results indicate that we

would receive an extra 0.06% in interest by investing our money in an account

that is compounded continuously rather than quarterly. This is only an extra 6

cents on a \$100 investment, but \$600 extra on a \$1,000,000 investment (actually

\$649.15 if the computations are carried beyond 0.01%).

$\boxed{35}$ It may be of interest to compare this graph with the graph of $y = (1.085)^x$ in

Exercise 43 of §5.1. Both are compounding functions with $r = 8.5\%$.

Note that $e^{0.085x} = (e^{0.085})^x \approx (1.0887)^x > (1.085)^x$ for $x > 0$.

(a) If $y = e^{0.085x}$ and $x = 40$, then $y \approx 29.96.$ (b) If $y = 2$, then $x \approx 8.15.$

[0, 60] by [0, 40]

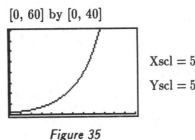

Xscl = 5
Yscl = 5

Figure 35

$\boxed{37}$ (a) As $x \to \infty$, $e^{-x} \to 0$ and f will resemble $\tfrac{1}{2}e^x$.

As $x \to -\infty$, $e^x \to 0$ and f will resemble $-\tfrac{1}{2}e^x$.

See *Figure 37(a)* on the next page.

(b) At $x = 0$, $f(x) = 0$, and g will have a vertical asymptote since g is undefined (division by 0). As $x \to \infty$, $f(x) \to \infty$, and since the reciprocal of a large positive number is a small positive number, we have $g(x) \to 0$. As $x \to -\infty$, $f(x) \to -\infty$, and since the reciprocal of a large negative number is a small negative number, we have $g(x) \to 0$.

$[-7.5, 7.5]$ by $[-5, 5]$

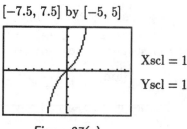

Xscl $= 1$

Yscl $= 1$

Figure 37(a)

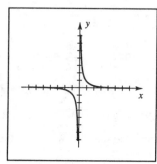

Figure 37(b)

39 (a) $f(x) = \dfrac{e^x - e^{-x}}{e^x + e^{-x}} = \dfrac{e^x - 1/e^x}{e^x + 1/e^x} \cdot \dfrac{e^x}{e^x} = \dfrac{e^{2x} - 1}{e^{2x} + 1}$. At $x = 0$, $f(x) = 0$. As $x \to \infty$,

$f(x) \to 1$ since the numerator and denominator are nearly the same number. As $x \to -\infty$, $f(x) \to -1$.

(b) At $x = 0$, we will have a vertical asymptote. As $x \to \infty$, $f(x) \to 1$, and since g is the reciprocal of f, $g(x) \to 1$. Similarly, as $x \to -\infty$, $g(x) \to -1$.

$[-4.5, 4.5]$ by $[-3, 3]$

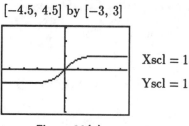

Xscl $= 1$

Yscl $= 1$

Figure 39(a)

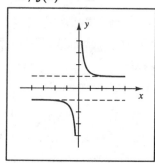

Figure 39(b)

41 The approximate coordinates of the points where the graphs of f and g intersect are $(-1.04, -0.92)$, $(2.11, 2.44)$, and $(8.51, 70.42)$. The region near the origin in *Figure 41(a)* is enhanced in *Figure 41(b)*. Thus, the solutions are $x \approx -1.04$, 2.11, and 8.51.

$[-3, 11]$ by $[-10, 80]$ $[-2.26, 3.34]$ by $[-7.14, 8.57]$

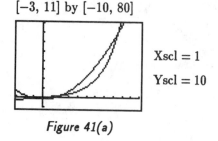

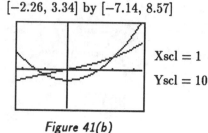

Xscl $= 1$ Xscl $= 1$

Yscl $= 10$ Yscl $= 10$

Figure 41(a)   Figure 41(b)

[45] From the graph, we see that f has zeros at $x \approx 0.11$, 0.79, and 1.13.

[-2, 2.5] by [-1, 2]

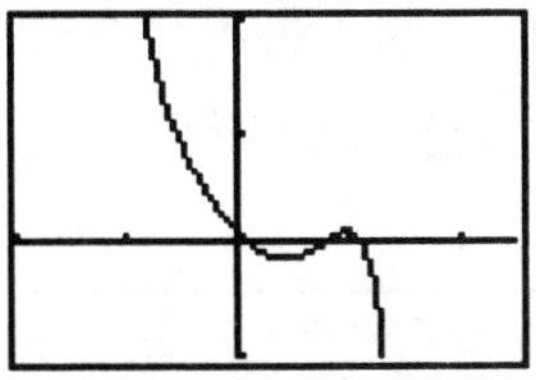

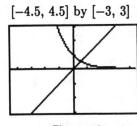

Xscl = 1
Yscl = 1

[0, 200] by [0, 8]

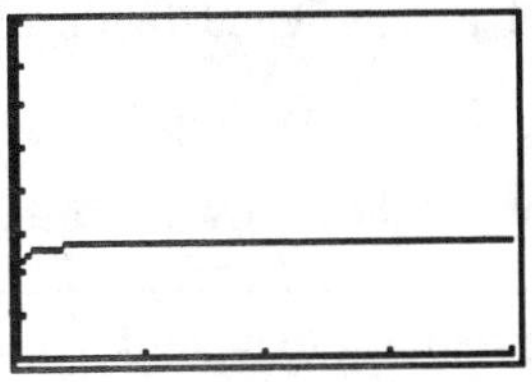

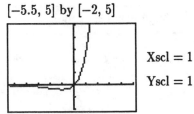

Xscl = 50
Yscl = 1

Figure 45

Figure 47

[47] From the graph, there is a horizontal asymptote of $y \approx 2.71$.

f is approaching the value of e asymptotically.

[49] $e^{-x} = x$ when $x \approx 0.567$.

[-4.5, 4.5] by [-3, 3]

Xscl = 1
Yscl = 1

[-5.5, 5] by [-2, 5]

Xscl = 1
Yscl = 1

Figure 49

Figure 51

[51] f is increasing on $[-1, \infty)$ and f is decreasing on $(-\infty, -1]$.

[53] (a) When $y = 0$ and $z = 0$, the equation becomes $C = \dfrac{2Q}{\pi vab} e^{-h^2/(2b^2)}$. As h increases, the exponent becomes a larger *negative* value, and hence the concentration C decreases.

(b) When $z = 0$, the equation becomes $C = \dfrac{2Q}{\pi vab} e^{-y^2/(2a^2)} e^{-h^2/(2b^2)}$.

As y increases, the concentration C decreases.

[55] (a) Chose two arbitrary points that appear to lie on the curve such as $(0, 1.225)$ and $(10{,}000, 0.414)$. $f(0) = Ce^0 = C = 1.225$ and $f(10{,}000) = 1.225e^{10,000k} = 0.414$. To solve the last equation, graph $Y_1 = 1.225e^{10,000x}$ and $Y_2 = 0.414$. The graphs intersect at $x \approx -0.0001085$. Thus, $f(x) = 1.225e^{-0.0001085x}$.

(b) $f(3000) \approx 0.885$ and $f(9000) \approx 0.461$.

[-1000, 10,100] by [0, 1.5]

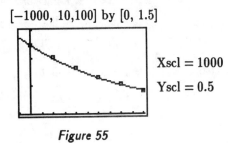

Xscl = 1000
Yscl = 0.5

Figure 55

Note: Exercises 1–4 are designed to familiarize the reader with the definition of $\log_a$ in this section. It is very important that you can generalize your understanding of this definition to the following case:

$$\log_{\text{base}}(\text{argument}) = \text{exponent} \quad \textit{is equivalent to} \quad (\text{base})^{\text{exponent}} = \text{argument}$$

Later in this section, you will also want to be able to use the following two special cases with ease:

$$\log(\text{argument}) = \text{exponent} \quad \textit{is equivalent to} \quad (10)^{\text{exponent}} = \text{argument}$$

$$\ln(\text{argument}) = \text{exponent} \quad \textit{is equivalent to} \quad (e)^{\text{exponent}} = \text{argument}$$

$\boxed{1}$ (e) In this case, the *base* is 5, the *exponent* is $7t$, and the *argument* is $\frac{a+b}{a}$. Thus,

$$5^{7t} = \frac{a+b}{a} \quad \text{is equivalent to} \quad \log_5 \frac{a+b}{a} = 7t.$$

$\boxed{3}$ (e) In this case, the *base* is 2, the *argument* is m, and the *exponent* is $3x+4$. Thus,

$$\log_2 m = 3x+4 \quad \text{is equivalent to} \quad 2^{3x+4} = m.$$

$\boxed{7}$ In order to solve for t, we must isolate the expression containing t—in this case, that expression is the exponential a^{Ct}. $A = Ba^{Ct} + D \Rightarrow A - D = Ba^{Ct} \Rightarrow$

$$\frac{A-D}{B} = a^{Ct} \Rightarrow Ct = \log_a\left(\frac{A-D}{B}\right) \Rightarrow t = \frac{1}{C}\log_a\left(\frac{A-D}{B}\right).$$

The confusing step to most students in the above solution is $\frac{A-D}{B} = a^{Ct} \Rightarrow Ct = \log_a\left(\frac{A-D}{B}\right)$. This is similar to $y = a^x \Rightarrow \log_a y = x$, except x and y are more complicated expressions.

$\boxed{9}$ (a) Changing $10^5 = 100{,}000$ to logarithmic form gives us $\log_{10} 100{,}000 = 5$.

Since this is a common logarithm, we denote it as $\log 100{,}000 = 5$.

(e) Changing $e^{2t} = 3 - x$ to logarithmic form gives us $\log_e(3 - x) = 2t$.

Since this is a natural logarithm, we denote it as $\ln(3 - x) = 2t$.

$\boxed{11}$ (b) Remember that $\log x = 20t$ is the same as $\log_{10} x = 20t$.

Changing to exponential form, we have $10^{20t} = x$.

(d) Remember that $\ln w = 4 + 3x$ is the same as $\log_e w = 4 + 3x$.

Changing to exponential form, we have $e^{4+3x} = w$.

[13] (c) Remember that you cannot take the logarithm, any base, of a negative number.

Hence, $\log_4(-2)$ is undefined.

(g) We will change the form of $\frac{1}{16}$ so that it can be written as an exponential expression with the same base as the logarithm—in this case, that base is 4.

$$\log_4 \frac{1}{16} = \log_4 4^{-2} = -2$$

[15] Parts (a)–(f) are direct applications of the properties in the chart on page 342.

For part (g), we use a property of exponents that will enable us to use the property $e^{\ln x} = x$. (g) $e^{2+\ln 3} = e^2 e^{\ln 3} = e^2(3) = 3e^2$

[17] $\log_4 x = \log_4(8-x) \Rightarrow x = 8-x$ {since the logarithm function is one-to-one} $\Rightarrow 2x = 8 \Rightarrow x = 4$. We must check to make sure that all proposed solutions do not make any of the original expressions undefined. Checking $x = 4$ in the original equation gives us $\log_4 4 = \log_4(8-4)$, which is a true statement, so $x = 4$ is a valid solution.

[19] $\log_5(x-2) = \log_5(3x+7) \Rightarrow x-2 = 3x+7 \Rightarrow 2x = -9 \Rightarrow x = -\frac{9}{2}$. The value $x = -\frac{9}{2}$ is extraneous since it makes either of the original logarithm expressions undefined. Hence, there is no solution.

[21] $\log x^2 = \log(-3x-2) \Rightarrow x^2 = -3x-2 \Rightarrow x^2+3x+2 = 0 \Rightarrow (x+1)(x+2) = 0 \Rightarrow x = -1, -2$. Checking -1 and -2, we find that both are valid solutions.

[23] $\log_3(x-4) = 2 \Rightarrow x-4 = 3^2 \Rightarrow x = 13$

[25] $\log_9 x = \frac{3}{2} \Rightarrow x = 9^{3/2} = (9^{1/2})^3$ {remember, root first, power second} $= 3^3 = 27$

[27] $\ln x^2 = -2 \Rightarrow x^2 = e^{-2} = \frac{1}{e^2} \Rightarrow x = \pm\frac{1}{e}$

[29] $e^{2\ln x} = 9 \Rightarrow (e^{\ln x})^2 = 9 \Rightarrow x^2 = 9 \Rightarrow x = \pm 3$; -3 is extraneous

[31] (a) $f(x) = \log_4 x$ • This graph has a vertical asymptote of $x = 0$ and goes through $(\frac{1}{4}, -1)$, $(1, 0)$, and $(4, 1)$. For reference purposes, call this $F(x)$.

(b) $f(x) = -\log_4 x$ • reflect F through the x-axis

(c) $f(x) = 2\log_4 x$ • vertically stretch F by a factor of 2

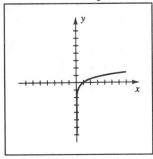

Figure 31(a)

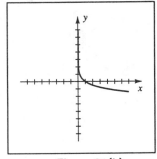

Figure 31(b)

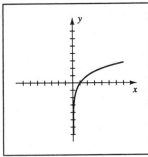

Figure 31(c)

(d) $f(x) = \log_4(x+2)$ •   shift F left 2 units

(e) $f(x) = (\log_4 x) + 2$ •   shift F up 2 units

(f) $f(x) = \log_4(x-2)$ •   shift F right 2 units

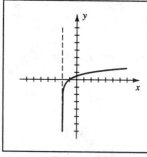

Figure 31(d)

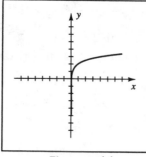

Figure 31(e)

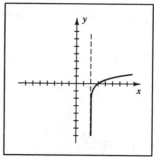

Figure 31(f)

(g) $f(x) = (\log_4 x) - 2$ •   shift F down 2 units

(h) $f(x) = \log_4 |x|$ •   include the reflection of F through the y-axis since x may
be positive or negative, but $\log_4 |x|$ will give the same result

(i) $f(x) = \log_4(-x)$ • x must be negative so that $-x$ is positive,

reflect F through the y-axis

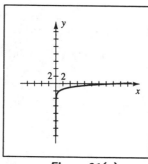

Figure 31(g)

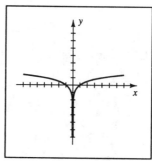

Figure 31(h)

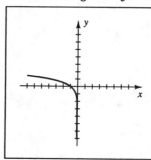

Figure 31(i)

(j) $f(x) = \log_4(3-x) = \log_4[-(x-3)]$ ●   Shift F right 3 units and reflect through the line $x = 3$. It may be helpful to determine the domain of this function. We know that $3 - x$ must be positive for the function to be defined. Thus, $3 - x > 0 \Rightarrow 3 > x$, or, equivalently, $x < 3$.

(k) $f(x) = |\log_4 x|$ ●

reflect points with negative y-coordinates through the x-axis

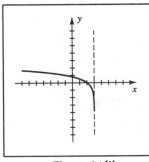
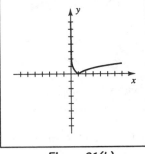
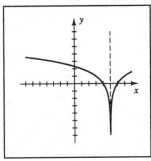

Figure 31(j)   *Figure 31(k)*   *Figure 35*

35 $f(x) = \log_2|x-5|$ ●   shift $y = \log_2|x|$ right 5 units

37 This is the basic "logarithm with base 2" graph, call it $F(x) = \log_2 x$.   ★ $f(x) = \log_2 x$

39 shift F down 1 unit   ★ $f(x) = \log_2 x - 1$

41 the reflection of F through the y-axis   ★ $f(x) = \log_2(-x)$

43 F appears to be stretched. Since the graph goes through $(2, 2)$ instead of $(2, 1)$ and $(4, 4)$ instead of $(4, 2)$, we might guess that the y-coordinates of F are doubled, that is, $f(x) = 2\log_2 x$. $f(x) \neq \log_2 x^2$ since the domain of f is $(0, \infty)$ and the domain of $g(x) = \log_2 x^2$ is $\mathbf{R} - \{0\}$.   ★ $f(x) = 2\log_2 x$

45 (a) $\log x = 3.6274 \Rightarrow x = 10^{3.6274} \approx 4240.333$, or 4240 to three significant figures

   (f) $\ln x = -1.6 \Rightarrow x = e^{-1.6} \approx 0.2019$, or 0.202

47 $q = q_0(2)^{-t/1600} \Rightarrow \frac{q}{q_0} = 2^{-t/1600}$ { change to logarithm form } $\Rightarrow$

$$-\frac{t}{1600} = \log_2\left(\frac{q}{q_0}\right) \text{ \{ multiply by } -1600 \text{ \}} \Rightarrow t = -1600\log_2\left(\frac{q}{q_0}\right)$$

49 $I = 20e^{-Rt/L} \Rightarrow \frac{I}{20} = e^{-Rt/L} \Rightarrow \ln\left(\frac{I}{20}\right) = -\frac{Rt}{L} \Rightarrow t = -\frac{L}{R}\ln\left(\frac{I}{20}\right)$

51 $I = 10^a I_0 \Rightarrow R = \log\left(\frac{I}{I_0}\right) = \log\left(\frac{10^a I_0}{I_0}\right) = \log 10^a = a.$

Hence, for $10^2 I_0$, $10^4 I_0$, and $10^5 I_0$, the answers are: (a) 2   (b) 4   (c) 5

53 We will find a general formula for α first.

$$I = 10^a I_0 \Rightarrow \alpha = 10\log\left(\frac{I}{I_0}\right) = 10\log\left(\frac{10^a I_0}{I_0}\right) = 10\,(\log 10^a) = 10(a) = 10a.$$

Hence, for $10^1 I_0$, $10^3 I_0$, and $10^4 I_0$, the answers are: (a) 10   (b) 30   (c) 40

[55] 1980 corresponds to $t = 0$ and $N(0) = 227$ million. $2 \cdot 227 = 227e^{0.007t} \Rightarrow$

$2 = e^{0.007t} \Rightarrow \ln 2 = 0.007t \Rightarrow t \approx 99$, which corresponds to the year 2079.

Alternatively, using the doubling time formula, $t = (\ln 2)/r = (\ln 2)/0.007 \approx 99$.

[57] (a) $\ln W = \ln 2.4 + (1.84)h$ { change to exponential form } $\Rightarrow W = e^{[\ln 2.4 + (1.84)h]} \Rightarrow$

$W = e^{\ln 2.4}e^{1.84h}$ { since $e^x e^y = e^{x+y}$ } $\Rightarrow W = 2.4e^{1.84h}$

(b) $h = 1.5 \Rightarrow W = 2.4e^{(1.84)(1.5)} = 2.4e^{2.76} \approx 37.92$ kg

[59] (a) $10 = 14.7e^{-0.0000385h} \Rightarrow \frac{10}{14.7} = e^{-0.0000385h} \Rightarrow \ln\left(\frac{10}{14.7}\right) = -0.0000385h \Rightarrow$

$h = -\frac{1}{0.0000385}\ln\left(\frac{10}{14.7}\right) \approx 10,007$ ft.

(b) At sea level, $h = 0$, and $p(0) = 14.7$. Setting $p(h)$ equal to $\frac{1}{2}(14.7)$,

and solving as in part (a), we have $h = -\frac{1}{0.0000385}\ln\left(\frac{1}{2}\right) \approx 18,004$ ft.

[61] (a) $t = 0 \Rightarrow W = 2600(1 - 0.51)^3 = 2600(0.49)^3 \approx 305.9$ kg

(b) (1) From the graph, if $W = 1800$, t appears to be about 20.

(2) Solving the equation for t, we have $1800 = 2600(1 - 0.51e^{-0.075t})^3 \Rightarrow$

$\frac{1800}{2600} = (1 - 0.51e^{-0.075t})^3 \Rightarrow \sqrt[3]{\frac{9}{13}} = 1 - 0.51e^{-0.075t} \Rightarrow$

$e^{-0.075t} = (1 - \sqrt[3]{\frac{9}{13}})(\frac{100}{51})$ { call this A } $\Rightarrow (-0.075)t = \ln A \Rightarrow t \approx 19.8$ yr.

[63] $D = 2 \Rightarrow 5.5e^{-0.1x} = 2 \Rightarrow e^{-0.1x} = \frac{4}{11} \Rightarrow -0.1x = \ln\frac{4}{11} \Rightarrow x = -10\ln\frac{4}{11}$ mi ≈ 10.1 mi

[65] Since the half-life is eight days, $A(t) = \frac{1}{2}A_0$ when $t = 8$.

Thus, $\frac{1}{2}A_0 = A_0 a^{-8} \Rightarrow a^{-8} = \frac{1}{2} \Rightarrow \frac{1}{a^8} = \frac{1}{2} \Rightarrow$

$a^8 = 2 \Rightarrow a = 2^{1/8}$ { take the eighth root } ≈ 1.09.

[67] (a) Since $\log P$ is an increasing function, increasing the population increases the

walking speed. Pedestrians have faster average walking speeds in large cities.

(b) $S = 5 \Rightarrow 5 = 0.05 + 0.86\log P \Rightarrow 4.95 = 0.86\log P \Rightarrow$

$\frac{4.95}{0.86} = \log P \Rightarrow P = 10^{4.95/0.86} \approx 570,000$

[71] *Figure 71* shows a graph of $Y_1 = x \ln x$ and $Y_2 = 1$. By tracing and zooming or using

an intersect feature, we determine that $x \ln x = 1$ when $x \approx 1.763$. Alternatively, we

could graph $Y_1 = x \ln x - 1$, and find the zero of that graph.

$[0, 4]$ by $[-1, 1.67]$

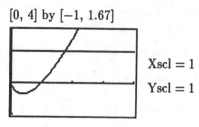

Xscl $= 1$

Yscl $= 1$

Figure 71

73̄ The domain of $f(x) = 2.2\log(x+2)$ is $x > -2$. The domain of $g(x) = \ln x$ is $x > 0$. From *Figure 73*, we determine that f intersects g at about 14.90. Thus, $f(x) \geq g(x)$ on approximately $(0, 14.90)$, not $(-2, 14.90)$ since g is not defined if $x \leq 0$. In general, the larger the base of the logarithm, the slower its graph will rise. In this case, we have base 10 and base $e \approx 2.72$, so we know that g will eventually intersect f even if we don't see this intersection in our first window.

[−2, 16] by [−4, 8]

[3, 5] by [0, 1]

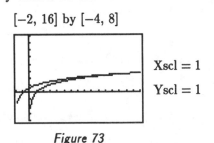

Xscl = 1
Yscl = 1

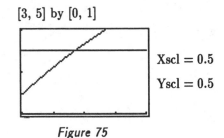

Xscl = 0.5
Yscl = 0.5

Figure 73

Figure 75

75̄ (a) $R = 2.07\ln\frac{242}{78} - 2.04 \approx 0.3037 \approx 30\%$.

(b) Graph $Y_1 = 2.07\ln x - 2.04$ and $Y_2 = 0.75$.

From the graph, $R \approx 0.75$ when $x \approx 3.85$.

5.4 Exercises

3̄ $\log_a\frac{x^3 w}{y^2 z^4} = \log_a x^3 w - \log_a y^2 z^4 = \log_a x^3 + \log_a w - (\log_a y^2 + \log_a z^4) =$
$$3\log_a x + \log_a w - 2\log_a y - 4\log_a z$$

The most common mistake is to not have the minus sign in front of $4\log_a z$.

This error results from not having the parentheses in the correct place.

5̄ $\log\frac{\sqrt[3]{z}}{x\sqrt{y}} = \log\sqrt[3]{z} - \log x\sqrt{y} = \log z^{1/3} - \log x - \log y^{1/2} = \frac{1}{3}\log z - \log x - \frac{1}{2}\log y$

7̄ $\ln\sqrt[4]{\frac{x^7}{y^5 z}} = \ln x^{7/4} - \ln y^{5/4}z^{1/4} = \ln x^{7/4} - \ln y^{5/4} - \ln z^{1/4} = \frac{7}{4}\ln x - \frac{5}{4}\ln y - \frac{1}{4}\ln z$

As a generalization for exercises similar to those in 1–8, if the exponents on the variables are positive, then the sign in front of the individual logarithms will be positive if the variable was originally in the numerator and negative if the variable was originally in the denominator.

11̄ $2\log_a x + \frac{1}{3}\log_a(x-2) - 5\log_a(2x+3)$

$\quad = \log_a x^2 + \log_a(x-2)^{1/3} - \log_a(2x+3)^5 \qquad$ { logarithm law (3) }

$\quad = \log_a x^2\sqrt[3]{x-2} - \log_a(2x+3)^5 \qquad$ { logarithm law (1) }

$\quad = \log_a\frac{x^2\sqrt[3]{x-2}}{(2x+3)^5} \qquad$ { logarithm law (2) }

13 $\log(x^3 y^2) - 2\log x \sqrt[3]{y} - 3\log\left(\frac{x}{y}\right) = \log(x^3 y^2) - \left[\log(x\sqrt[3]{y})^2 + \log\left(\frac{x}{y}\right)^3\right]$

$$= \log(x^3 y^2) - \left[\log(x^2 y^{2/3} \cdot (x^3/y^3))\right]$$

$$= \log\frac{x^3 y^2}{x^5 y^{-7/3}} = \log\frac{y^{13/3}}{x^2}$$

15 $\ln y^3 + \frac{1}{3}\ln(x^3 y^6) - 5\ln y = \ln y^3 + \ln(xy^2) - \ln y^5 = \ln\left[(xy^5)/y^5\right] = \ln x$

17 $\log_6(2x-3) = \log_6 12 - \log_6 3 \Rightarrow \log_6(2x-3) = \log_6\frac{12}{3} \Rightarrow 2x - 3 = 4 \Rightarrow x = \frac{7}{2}$

19 $2\log_3 x = 3\log_3 5 \Rightarrow \log_3 x^2 = \log_3 5^3 \Rightarrow x^2 = 125 \Rightarrow x = \pm 5\sqrt{5}$;

$-5\sqrt{5}$ is extraneous since it would make $\log_3 x$ undefined

21 $\log x - \log(x+1) = 3\log 4 \Rightarrow \log\frac{x}{x+1} = \log 64 \Rightarrow$

$\frac{x}{x+1} = 64 \Rightarrow x = 64x + 64 \Rightarrow x = -\frac{64}{63}$; $-\frac{64}{63}$ is extraneous, no solution

23 $\ln(-4-x) + \ln 3 = \ln(2-x) \Rightarrow \ln(-12-3x) = \ln(2-x) \Rightarrow -12 - 3x = 2 - x \Rightarrow$

$2x = -14 \Rightarrow x = -7$. Remember, the solution of a logarithmic equation may be negative—you must examine what happens to the original logarithm expressions. In this case, we have $\ln 3 + \ln 3 = \ln 9$, which is true.

25 $\log_2(x+7) + \log_2 x = 3 \Rightarrow \log_2(x^2 + 7x) = 3 \Rightarrow x^2 + 7x = 2^3 \Rightarrow x^2 + 7x - 8 = 0 \Rightarrow$

$(x+8)(x-1) = 0 \Rightarrow x = -8,\ 1$; -8 is extraneous

27 $\log_3(x+3) + \log_3(x+5) = 1 \Rightarrow \log_3\left[(x+3)(x+5)\right] = 1 \Rightarrow \log_3(x^2 + 8x + 15) = 1 \Rightarrow$

$x^2 + 8x + 15 = 3 \Rightarrow x^2 + 8x + 12 = 0 \Rightarrow (x+2)(x+6) = 0 \Rightarrow x = -6,\ -2$;

-6 is extraneous

29 $\log(x+3) = 1 - \log(x-2) \Rightarrow \log(x+3) + \log(x-2) = 1 \Rightarrow$

$\log\left[(x+3)(x-2)\right] = 1 \Rightarrow x^2 + x - 6 = 10^1 \Rightarrow$

$x^2 + x - 16 = 0 \Rightarrow x = \frac{-1 + \sqrt{65}}{2} \approx 3.53$; $\frac{-1-\sqrt{65}}{2} \approx -4.53$ is extraneous

31 $\ln x = 1 - \ln(x+2) \Rightarrow \ln x + \ln(x+2) = 1 \Rightarrow \ln\left[x(x+2)\right] = 1 \Rightarrow x^2 + 2x = e^1 \Rightarrow$

$x^2 + 2x - e = 0 \Rightarrow x = \frac{-2 \pm \sqrt{4 + 4e}}{2} = \frac{-2 \pm 2\sqrt{1+e}}{2} = -1 \pm \sqrt{1+e}$.

$x = -1 + \sqrt{1+e} \approx 0.93$ is a valid solution,

but $x = -1 - \sqrt{1+e} \approx -2.93$ is extraneous.

33 $f(x) = \log_3(3x) = \log_3 3 + \log_3 x = \log_3 x + 1$ • shift $y = \log_3 x$ up 1 unit

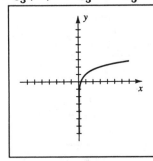

Figure 33

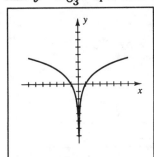

Figure 37

37 $f(x) = \log_3(x^2) = 2\log_3 x$ • Vertically stretch $y = \log_3 x$ by a factor of 2 and include its reflection through the y-axis since the domain of the original function, $f(x) = \log_3(x^2)$, is $\mathbf{R} - \{0\}$. Keep in mind that the laws of logarithms are established for positive real numbers, so that when we make the step $\log_3(x^2) = 2\log_3 x$, it is only for positive x.

41 $f(x) = \log_2 \sqrt{x} = \log_2 x^{1/2} = \frac{1}{2}\log_2 x$ •

vertically compress $y = \log_2 x$ by a factor of $1/(1/2) = 2$

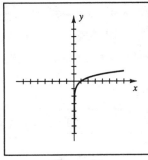

Figure 41

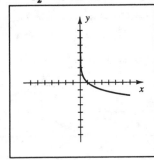

Figure 43

43 $f(x) = \log_3\left(\frac{1}{x}\right) = \log_3 x^{-1} = -\log_3 x$ • reflect $y = \log_3 x$ through the x-axis

45 The values of $F(x) = \log_2 x$ are doubled and the reflection of F through the y-axis is included. The domain of f is $\mathbf{R} - \{0\}$ and $f(x) = \log_2 x^2$.

47 $F(x) = \log_2 x$ is shifted up 3 units since $(1, 0)$ on F is $(1, 3)$ on the graph.

Hence, $f(x) = 3 + \log_2 x = \log_2 2^3 + \log_2 x = \log_2(8x)$.

49 $\log y = \log b - k\log x \Rightarrow \log y = \log b - \log x^k \Rightarrow \log y = \log\frac{b}{x^k} \Rightarrow y = \frac{b}{x^k}$

53 (a) $R(x_0) = a\log\left(\frac{x_0}{x_0}\right) = a\log 1 = a \cdot 0 = 0$

(b) $R(2x) = a\log\left(\frac{2x}{x_0}\right) = a\left[\log 2 + \log\left(\frac{x}{x_0}\right)\right] = a\log 2 + a\log\left(\frac{x}{x_0}\right) = R(x) + a\log 2$

55 $\ln I_0 - \ln I = kx \Rightarrow \ln\frac{I_0}{I} = kx \Rightarrow x = \frac{1}{k}\ln\frac{I_0}{I} = \frac{1}{0.39}\ln 1.12 \approx 0.29$ cm.

57 From the graph, the coordinates of the points of intersection are approximately

(1.01, 0.48) and (2.4, 0.86). $f(x) \geq g(x)$ on the intervals $(0, 1.01]$ and $[2.4, \infty)$.

[0, 6] by [−1, 3]   [0, 8] by [−1.67, 3.67]

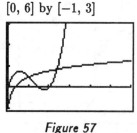
Xscl = 1
Yscl = 1

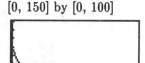
Xscl = 1
Yscl = 1

Figure 57   *Figure 59*

59 Graph $y = e^{-x} - 2\log(1 + x^2) + 0.5x$ and estimate any x-intercepts. From the graph,

we see that the roots of the equation are approximately $x \approx 1.41, 6.59$.

63 Graph $Y_1 = x\log x - \log x$ and $Y_2 = 5$. The graphs intersect at $x \approx 6.94$.

[−5, 10] by [−2, 8]   [0, 150] by [0, 100]

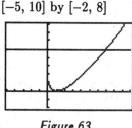
Xscl = 1
Yscl = 1

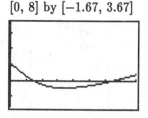
Xscl = 10
Yscl = 10

Figure 63   *Figure 65*

65 Let $d = x$. Graph $Y_1 = I_0 - 20\log x - kx = 70 - 20\log x - 0.076x$ and $Y_2 = 20$.

At the point of intersection, $x \approx 115.3$. The distance is approximately 115 meters.

5.5 Exercises

1 (a) $5^x = 8 \Rightarrow \log 5^x = \log 8 \Rightarrow x\log 5 = \log 8 \Rightarrow x = \dfrac{\log 8}{\log 5} \approx 1.29$

(b) $5^x = 8 \Rightarrow x = \log_5 8 = \dfrac{\log 8}{\log 5} \approx 1.29$

3 (a) $3^{4-x} = 5 \Rightarrow \log(3^{4-x}) = \log 5 \Rightarrow (4-x)\log 3 = \log 5 \Rightarrow$

$4 - x = \dfrac{\log 5}{\log 3} \Rightarrow x = 4 - \dfrac{\log 5}{\log 3} \approx 2.54$. *Note:* The answer could also

be written as $4 - \dfrac{\log 5}{\log 3} = \dfrac{4\log 3 - \log 5}{\log 3} = \dfrac{\log 81 - \log 5}{\log 3} = \dfrac{\log \frac{81}{5}}{\log 3}$.

(b) $3^{4-x} = 5 \Rightarrow 4 - x = \log_3 5 \Rightarrow x = 4 - \dfrac{\log 5}{\log 3} \approx 2.54$.

7 $\log_9 0.2 = \dfrac{\log 0.2}{\log 9} \left\{ \text{or, equivalently, } \dfrac{\ln 0.2}{\ln 9} \right\} \approx -0.7325$.

Note that either log or ln can be used here. You should get comfortable using both.

9 $\dfrac{\log_5 16}{\log_5 4} = \log_4 16 = \log_4 4^2 = 2$

13 The steps are similar to those in Example 4. $2^{2x-3} = 5^{x-2} \Rightarrow$

$\log(2^{2x-3}) = \log(5^{x-2}) \Rightarrow (2x-3)\log 2 = (x-2)\log 5 \Rightarrow$

$2x\log 2 - 3\log 2 = x\log 5 - 2\log 5 \Rightarrow 2x\log 2 - x\log 5 = 3\log 2 - 2\log 5 \Rightarrow$

$$x(2\log 2 - \log 5) = \log 2^3 - \log 5^2 \Rightarrow x = \frac{\log 8 - \log 25}{\log 4 - \log 5} \Rightarrow x = \frac{\log \frac{8}{25}}{\log \frac{4}{5}} \approx 5.11$$

17 $\log x = 1 - \log(x-3) \Rightarrow \log x + \log(x-3) = 1 \Rightarrow \log(x^2 - 3x) = 1 \Rightarrow$

$$x^2 - 3x = 10^1 \Rightarrow (x-5)(x+2) = 0 \Rightarrow x = 5, -2; \; -2 \text{ is extraneous}$$

19 $\log(x^2+4) - \log(x+2) = 2 + \log(x-2) \Rightarrow \log\left(\frac{x^2+4}{x+2}\right) - \log(x-2) = 2 \Rightarrow$

$\log\left(\frac{x^2+4}{x^2-4}\right) = 2 \Rightarrow \frac{x^2+4}{x^2-4} = 10^2 \Rightarrow x^2 + 4 = 100x^2 - 400 \Rightarrow 404 = 99x^2 \Rightarrow$

$$x = \pm\sqrt{\frac{404}{99}} = \pm\frac{2}{3}\sqrt{\frac{101}{11}} \approx \pm 2.02; \; -\frac{2}{3}\sqrt{\frac{101}{11}} \text{ is extraneous}$$

21 See Example 5 for more detail concerning this type of exercise.

$5^x + 125(5^{-x}) = 30 \; \{\text{multiply by } 5^x\} \Rightarrow$

$(5^x)^2 - 30(5^x) + 125 = 0 \; \{\text{recognize as a quadratic in } 5^x \text{ and factor}\} \Rightarrow$

$$(5^x - 5)(5^x - 25) = 0 \Rightarrow 5^x = 5, 25 \Rightarrow 5^x = 5^1, 5^2 \Rightarrow x = 1, 2$$

23 $4^x - 3(4^{-x}) = 8 \; \{\text{multiply by } 4^x\} \Rightarrow$

$(4^x)^2 - 8(4^x) - 3 = 0 \; \{\text{recognize as a quadratic in } 4^x\} \Rightarrow$

$4^x = \frac{8 \pm \sqrt{76}}{2} = \frac{8 \pm 2\sqrt{19}}{2} = 4 \pm \sqrt{19}.$

Since since 4^x is positive and $4 - \sqrt{19}$ is negative, $4 - \sqrt{19}$ is discarded.

Continuing, $4^x = 4 + \sqrt{19} \Rightarrow x = \log_4(4 + \sqrt{19}) = \frac{\log(4 + \sqrt{19})}{\log 4} \; \{\text{use the change of}$

base formula to approximate$\} \approx 1.53$

25 $\log(x^2) = (\log x)^2 \Rightarrow 2\log x = (\log x)^2 \Rightarrow (\log x)^2 - 2\log x = 0 \Rightarrow$

$$(\log x)(\log x - 2) = 0 \Rightarrow \log x = 0, 2 \Rightarrow x = 10^0, 10^2 \Rightarrow x = 1 \text{ or } 100$$

27 Don't confuse $\log(\log x)$ with $(\log x)(\log x)$. The first expression is

the log of the log of x, whereas the second expression is the log of x times itself.

$$\log(\log x) = 2 \Rightarrow \log x = 10^2 = 100 \Rightarrow x = 10^{100}$$

29 $x^{\sqrt{\log x}} = 10^8 \; \{\text{take the log of both sides}\} \Rightarrow \log\left(x^{\sqrt{\log x}}\right) = \log 10^8 \Rightarrow$

$\sqrt{\log x}\,(\log x) = 8 \Rightarrow (\log x)^{1/2}(\log x)^1 = 8 \Rightarrow (\log x)^{3/2} = 8 \Rightarrow$

$$\left[(\log x)^{3/2}\right]^{2/3} = (8)^{2/3} \Rightarrow \log x = (\sqrt[3]{8})^2 = 4 \Rightarrow x = 10{,}000$$

Note: For Exercises 31–38 and 39–40 of the Chapter Review Exercises, let D denote the domain of the function determined by the original equation, and R its range. These are then the range and domain, respectively, of the equation listed in the answer.

31 $y = \dfrac{10^x + 10^{-x}}{2} \left\{ D = \mathbf{R},\, R = [1, \infty) \right\} \Rightarrow$

$2y = 10^x + 10^{-x} \left\{ \text{since } 10^{-x} = \frac{1}{10^x}, \text{ multiply by } 10^x \text{ to eliminate denominator} \right\} \Rightarrow$

$10^{2x} - 2y\, 10^x + 1 = 0 \left\{ \text{treat as a quadratic in } 10^x \right\} \Rightarrow$

$$10^x = \frac{2y \pm \sqrt{4y^2 - 4}}{2} = y \pm \sqrt{y^2 - 1} \Rightarrow x = \log\left(y \pm \sqrt{y^2 - 1}\,\right)$$

33 $y = \dfrac{10^x - 10^{-x}}{10^x + 10^{-x}} \left\{ D = \mathbf{R},\, R = (-1, 1) \right\} \Rightarrow y\, 10^x + y\, 10^{-x} = 10^x - 10^{-x} \Rightarrow$

$y\, 10^{2x} + y = 10^{2x} - 1 \Rightarrow (y - 1)\, 10^{2x} = -1 - y \Rightarrow$

$$10^{2x} = \frac{-1 - y}{y - 1} \Rightarrow 2x = \log\left(\frac{1 + y}{1 - y} \right) \Rightarrow x = \tfrac{1}{2} \log\left(\frac{1 + y}{1 - y} \right)$$

35 $y = \dfrac{e^x - e^{-x}}{2} \left\{ D = R = \mathbf{R} \right\} \Rightarrow 2y = e^x - e^{-x} \Rightarrow$

$e^{2x} - 2y\, e^x - 1 = 0 \Rightarrow e^x = \dfrac{2y \pm \sqrt{4y^2 + 4}}{2} = y \pm \sqrt{y^2 + 1};$

$$\sqrt{y^2 + 1} > y, \text{ so } y - \sqrt{y^2 + 1} < 0, \text{ but } e^x > 0 \text{ and thus, } x = \ln\left(y + \sqrt{y^2 + 1}\,\right)$$

37 $y = \dfrac{e^x + e^{-x}}{e^x - e^{-x}} \left\{ D = \mathbf{R} - \{0\},\, R = (-\infty, -1) \cup (1, \infty) \right\} \Rightarrow$

$ye^x - ye^{-x} = e^x + e^{-x} \Rightarrow ye^{2x} - y = e^{2x} + 1 \Rightarrow$

$$(y - 1)e^{2x} = y + 1 \Rightarrow e^{2x} = \frac{y + 1}{y - 1} \Rightarrow 2x = \ln\left(\frac{y + 1}{y - 1} \right) \Rightarrow x = \tfrac{1}{2} \ln\left(\frac{y + 1}{y - 1} \right)$$

39 $f(x) = \log_2(x + 3) \quad \bullet \quad x = 0 \Rightarrow y\text{-intercept} = \log_2 3 = \dfrac{\log 3}{\log 2} \approx 1.5850$

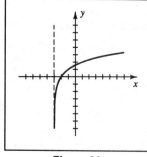

Figure 39

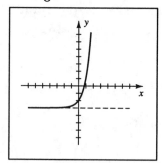

Figure 41

41 $f(x) = 4^x - 3 \quad \bullet \quad y = 0 \Rightarrow 4^x = 3 \Rightarrow x\text{-intercept} = \log_4 3 = \dfrac{\log 3}{\log 4} \approx 0.7925$

43 (a) vinegar: pH $\approx -\log(6.3 \times 10^{-3}) = -(\log 6.3 + \log 10^{-3}) = -(\log 6.3 - 3) =$

$$3 - \log 6.3 \approx 2.2$$

(b) carrots: pH $\approx -\log(1.0 \times 10^{-5}) = 5 - \log 1.0 = 5$

(c) sea water: pH $\approx -\log(5.0 \times 10^{-9}) = 9 - \log 5.0 \approx 8.3$

45 $[H^+] < 10^{-7} \Rightarrow \log[H^+] < \log 10^{-7}$ {since log is an increasing function} $\Rightarrow$

$\log[H^+] < -7 \Rightarrow -\log[H^+] > -(-7) \Rightarrow$ pH > 7 for basic solutions;

similarly, pH < 7 for acidic solutions.

47 Solving $A = P(1 + \frac{r}{n})^{nt}$ for t with $A = 2P$, $r = 0.06$, and $n = 12$ yields

$2P = P(1 + \frac{0.06}{12})^{12t}$ {divide by P} $\Rightarrow 2 = (1.005)^{12t}$ {take the ln of both sides} $\Rightarrow$

$\ln 2 = \ln(1.005)^{12t} \Rightarrow \ln 2 = 12t \ln(1.005) \Rightarrow$

$$t = \frac{\ln 2}{12 \ln(1.005)} \approx 11.58 \text{ yr, or about 11 years and 7 months.}$$

49 50% of the light reaching a depth of 13 meters corresponds to the equation

$\frac{1}{2}I_0 = I_0 c^{13}$. Solving for c, we have $c^{13} = \frac{1}{2} \Rightarrow c = \sqrt[13]{\frac{1}{2}} = 2^{-1/13}$.

Now letting $I = 0.01 I_0$, $c = 2^{-1/13}$, and using the formula from Example 6,

$$x = \frac{\log(I/I_0)}{\log c} = \frac{\log[(0.01 I_0)/I_0]}{\log 2^{-1/13}} = \frac{\log 10^{-2}}{-\frac{1}{13}\log 2} = \frac{26}{\log 2} \approx 86.4 \text{ m.}$$

51 (a) A is an *increasing* exponential that contains $(0, 0)$, $(5, \approx 41)$, and $(10, \approx 65)$.

(b) $A = 50 \Rightarrow 50 = 100[1 - (0.9)^t] \Rightarrow$

$1 - (0.9)^t = 0.5 \Rightarrow (0.9)^t = 0.5 \Rightarrow$

$t = \log_{0.9}(0.5) = \frac{\log 0.5}{\log 0.9} \approx 6.58$ min.

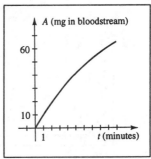

Figure 51

53 (a) $F = F_0(1-m)^t \Rightarrow (1-m)^t = \frac{F}{F_0} \Rightarrow \log(1-m)^t = \log\left(\frac{F}{F_0}\right) \Rightarrow$

$$t \log(1-m) = \log\left(\frac{F}{F_0}\right) \Rightarrow t = \frac{\log(F/F_0)}{\log(1-m)}$$

(b) Using part (a) with $F = \frac{1}{2}F_0$ and $m = 0.00005$,

$$t = \frac{\log(F/F_0)}{\log(1-m)} = \frac{\log(\frac{1}{2}F_0/F_0)}{\log(1-0.00005)} = \frac{\log\frac{1}{2}}{\log 0.99995} \approx 13{,}863 \text{ generations.}$$

$\boxed{55}$ (a) $t = 10 \Rightarrow h = \dfrac{120}{1 + 200e^{-2}} \approx 4.28$ ft

(b) $h = 50 \Rightarrow 50 = \dfrac{120}{1 + 200e^{-0.2t}} \Rightarrow 1 + 200e^{-0.2t} = \dfrac{120}{50} \Rightarrow 200e^{-0.2t} = \dfrac{12}{5} - 1 \Rightarrow$

$e^{-0.2t} = \dfrac{7}{5} \cdot \dfrac{1}{200} \Rightarrow e^{-0.2t} = 0.007 \Rightarrow -0.2t = \ln 0.007 \Rightarrow t = \dfrac{\ln 0.007}{-0.2} \approx 24.8$ yr

$\boxed{57}$ $\dfrac{v_0}{v_1} = \left(\dfrac{h_0}{h_1}\right)^P \Rightarrow \ln \dfrac{v_0}{v_1} = P \ln \dfrac{h_0}{h_1} \Rightarrow P = \dfrac{\ln (v_0/v_1)}{\ln (h_0/h_1)} = \dfrac{\ln (25/6)}{\ln (200/35)} \approx 0.82$

$\boxed{59}$ When $x = 0$, $y = c2^0 = c = 4$. Thus, $y = 4(2)^{kx}$. Similarly, $x = 1 \Rightarrow$ $y = 4(2)^k = 3.249 \Rightarrow k = \log_2\left(\dfrac{3.249}{4}\right) \approx -0.300$. Thus, $y = 4(2)^{-0.3x}$. Checking the remaining two points, we see that $x = 2 \Rightarrow y \approx 2.639$ and $x = 3 \Rightarrow y \approx 2.144$. The points all lie on the graph of $y = 4(2)^{-0.3x}$ to within three-decimal-place accuracy.

$\boxed{61}$ When $x = 0$, $y = c \log 10 = c = 1.5$. Thus, $y = 1.5 \log (kx + 10)$. Similarly, $x = 1 \Rightarrow$ $y = 1.5 \log (k + 10) = 1.619 \Rightarrow k + 10 = 10^{1.619/1.5} \Rightarrow k = 10^{1.619/1.5} - 10 \approx 2.004$.

Thus, $y = 1.5 \log (2.004x + 10)$. Checking the remaining two points, we see that $x = 2 \Rightarrow y \approx 1.720$, and $x = 3 \Rightarrow y \approx 1.807$. The points do not all lie on the graph of $y = c2^{kx}$ to within three-decimal-place accuracy.

$\boxed{63}$ $h(5.3) = \log_4 5.3 - 2 \log_8 (1.2 \times 5.3) = \dfrac{\ln 5.3}{\ln 4} - \dfrac{2 \ln 6.36}{\ln 8} \approx -0.5764$

$\boxed{65}$ From the graph of $y = x - \ln (0.3x) - 3 \log_3 x$,

we see that there are *no* x-intercepts, and hence, no roots of the equation.

$[-1, 17]$ by $[-1, 11]$ $[-1, 17]$ by $[-1, 11]$

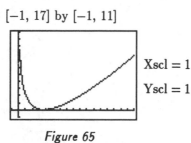   Xscl = 1   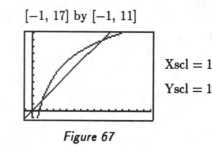   Xscl = 1

                  Yscl = 1   Yscl = 1

Figure 65   *Figure 67*

$\boxed{67}$ The graphs of $f(x) = x$ and $g(x) = 3 \log_2 x$ intersect at approximately $(1.37, 1.37)$ and $(9.94, 9.94)$. Hence, the solutions of the equation $f(x) = g(x)$ are 1.37 and 9.94.

69 From the graph, we see that the graphs of f and g intersect at three points. Their coordinates are approximately $(-0.32, 0.50)$, $(1.52, -1.33)$, and $(6.84, -6.65)$. The region near the origin in *Figure 69(a)* is enhanced in *Figure 69(b)*. Thus, $f(x) > g(x)$ on $(-\infty, -0.32)$ and $(1.52, 6.84)$.

$[-5, 10]$ by $[-8, 2]$ $[-1.53, 2.26]$ by $[-2.92, 1.05]$

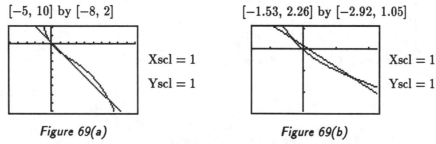

Xscl $= 1$ Xscl $= 1$
Yscl $= 1$ Yscl $= 1$

Figure 69(a)   *Figure 69(b)*

71 (1) The graph of $n(t) = 85e^{t/3}$ is increasing rapidly and soon becomes greater than 100. It is doubtful that the average score would improve dramatically without any review.

$[0, 5]$ by $[0, 200]$ $[0, 5]$ by $[0, 100]$

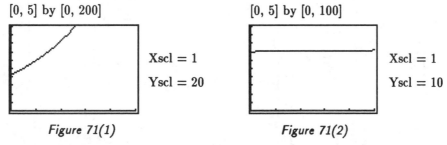

Xscl $= 1$ Xscl $= 1$
Yscl $= 20$ Yscl $= 10$

Figure 71(1)   *Figure 71(2)*

(2) The graph of $n(t) = 70 + \ln{(t+1)}$ is also increasing. It is incorrect because $n(0) = 70 \neq 85$. Also, one would not expect the average score to improve without any review.

(3) The graph of $n(t) = 86 - e^{t}$ decreases rapidly to zero. The *average* score probably would not be zero after 5 weeks.

$[0, 5]$ by $[0, 100]$ $[0, 5]$ by $[0, 100]$

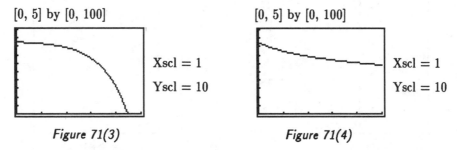

Xscl $= 1$ Xscl $= 1$
Yscl $= 10$ Yscl $= 10$

Figure 71(3)   *Figure 71(4)*

(4) The graph of $n(t) = 85 - 15\ln{(t+1)}$ is decreasing. During the first weeks it decreases most rapidly and then starts to level off. Of the four functions, this function seems to best model the situation.

[73] (a) The ozone level is decreasing by 11% per year. The fraction of ozone remaining x years after April 1992 is given by the function $f(x) = (0.89)^x$. We must approximate x when $f(x) = 0.5$. Using a table, $f(x) = 0.5$ when $x \approx 6$. Thus, in 1998 the ozone level would be 50% of its normal amount.

x	1	2	3	4	5	6	7
$f(x)$	0.89	0.79	0.70	0.63	0.56	0.50	0.44

(b) $(0.89)^t = 0.5 \Rightarrow \ln(0.89)^t = \ln 0.5 \Rightarrow t \ln 0.89 = \ln 0.5 \Rightarrow$

$$t = \frac{\ln 0.5}{\ln 0.89} \approx 5.948, \text{ or in 1998.}$$

Chapter 5 Review Exercises

[3] $f(x) = \left(\frac{3}{2}\right)^{-x} = \left(\frac{2}{3}\right)^x$ • goes through $\left(-1, \frac{3}{2}\right)$, $(0, 1)$, and $\left(1, \frac{2}{3}\right)$

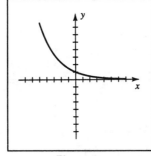

Figure 3

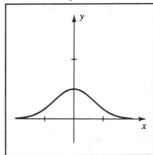

Figure 5

[5] $f(x) = 3^{-x^2} = (3^{-1})^{(x^2)} = \left(\frac{1}{3}\right)^{(x^2)}$ • see the note in §5.1 before Exercise 17

[7] $f(x) = e^{x/2} = (e^{1/2})^x \approx (1.65)^x$ • goes through $(-1, 1/\sqrt{e})$, $(0, 1)$, and $(1, \sqrt{e})$; or approximately $(-1, 0.61)$, $(0, 1)$, and $(1, 1.65)$

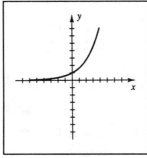

Figure 7

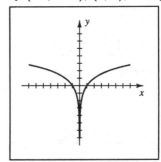

Figure 13

[13] $f(x) = \log_4(x^2) = 2\log_4 x$ • stretch $y = \log_4 x$ by a factor of 2 and include its reflection through the y-axis since the domain of the original function is $\mathbf{R} - \{0\}$

$\boxed{17}$ (a) $\log_2 \frac{1}{16} = \log_2 2^{-4} = -4$ (b) $\log_\pi 1 = 0$ (c) $\ln e = 1$

     (d) $6^{\log_6 4} = 4$ (e) $\log 1{,}000{,}000 = \log 10^6 = 6$

     (f) $10^{3\log 2} = 10^{\log 2^3} = 2^3 = 8$ (g) $\log_4 2 = \log_4 4^{1/2} = \frac{1}{2}$

$\boxed{19}$ $2^{3x-1} = \frac{1}{2} \Rightarrow 2^{3x-1} = 2^{-1} \Rightarrow 3x - 1 = -1 \Rightarrow 3x = 0 \Rightarrow x = 0$

$\boxed{22}$ $\log_4(x+1) = 2 + \log_4(3x-2) \Rightarrow \log_4(x+1) - \log_4(3x-2) = 2 \Rightarrow$

$$\log_4\left(\frac{x+1}{3x-2}\right) = 2 \Rightarrow \frac{x+1}{3x-2} = 16 \Rightarrow x + 1 = 48x - 32 \Rightarrow x = \frac{33}{47}$$

$\boxed{23}$ $2\ln(x+3) - \ln(x+1) = 3\ln 2 \Rightarrow \ln\frac{(x+3)^2}{x+1} = \ln 2^3 \Rightarrow (x+3)^2 = 8(x+1) \Rightarrow$

$$x^2 + 6x + 9 = 8x + 8 \Rightarrow x^2 - 2x + 1 = 0 \Rightarrow (x-1)^2 = 0 \Rightarrow x = 1$$

$\boxed{26}$ $3^{(x^2)} = 7 \Rightarrow x^2 = \log_3 7 \Rightarrow x^2 = \frac{\log 7}{\log 3} \Rightarrow x = \pm\sqrt{\frac{\log 7}{\log 3}}$

$\boxed{27}$ $2^{5x+3} = 3^{2x+1} \Rightarrow \log(2^{5x+3}) = \log(3^{2x+1}) \Rightarrow (5x+3)\log 2 = (2x+1)\log 3 \Rightarrow$

$5x\log 2 + 3\log 2 = 2x\log 3 + \log 3 \Rightarrow 5x\log 2 - 2x\log 3 = \log 3 - 3\log 2 \Rightarrow$

$$x(5\log 2 - 2\log 3) = \log 3 - \log 2^3 \Rightarrow x = \frac{\log 3 - \log 8}{\log 32 - \log 9} = \frac{\log\frac{3}{8}}{\log\frac{32}{9}}$$

$\boxed{29}$ $\log_4 x = \sqrt[3]{\log_4 x} \Rightarrow \log_4 x = (\log_4 x)^{1/3} \Rightarrow (\log_4 x)^3 = \log_4 x \Rightarrow$

$(\log_4 x)^3 - \log_4 x = 0 \Rightarrow \log_4 x\left[(\log_4 x)^2 - 1\right] = 0 \Rightarrow$

$$\log_4 x = 0 \text{ or } \log_4 x = \pm 1 \Rightarrow x = 1 \text{ or } x = 4, \frac{1}{4}$$

$\boxed{31}$ $10^{2\log x} = 5 \Rightarrow 10^{\log x^2} = 5 \Rightarrow x^2 = 5 \Rightarrow x = \pm\sqrt{5}; \ -\sqrt{5}$ is extraneous

$\boxed{34}$ $e^x + 2 = 8e^{-x} \Rightarrow e^{2x} + 2e^x - 8 = 0 \Rightarrow (e^x + 4)(e^x - 2) = 0 \Rightarrow e^x = -4, 2 \Rightarrow$

$$x = \ln 2 \text{ since } e^x \neq -4$$

$\boxed{35}$ (a) $\log x^2 = \log(6-x) \Rightarrow x^2 = 6 - x \Rightarrow x^2 + x - 6 = 0 \Rightarrow (x+3)(x-2) = 0 \Rightarrow$

$$x = -3, 2$$

     (b) $2\log x = \log(6-x) \Rightarrow \log x^2 = \log(6-x)$, which is the equation in part (a).

   This equation has the same solutions provided they are in the domain.

                            But -3 is extraneous, so 2 is the only solution.

$\boxed{36}$ (a) $\ln(e^x)^2 = 16 \Rightarrow 2\ln e^x = 16 \Rightarrow 2x = 16 \Rightarrow x = 8$

     (b) $\ln e^{(x^2)} = 16 \Rightarrow x^2 = 16 \Rightarrow x = \pm 4$

$\boxed{38}$ $\log(x^2/y^3) + 4\log y - 6\log\sqrt{xy} = \log\left(\frac{x^2}{y^3}\right) + \log y^4 - \log(x^3 y^3)$

$$= \log\left(\frac{x^2 y}{x^3 y^3}\right) = \log\left(\frac{1}{xy^2}\right) = -\log(xy^2)$$

39 $y = \dfrac{1}{10^x + 10^{-x}} \left\{ D = \mathbf{R}, \; R = (0, \tfrac{1}{2}] \right\} \Rightarrow y\,10^x + y\,10^{-x} = 1 \Rightarrow$

$y\,10^{2x} + y = 10^x \Rightarrow y\,10^{2x} - 10^x + y = 0 \Rightarrow 10^x = \dfrac{1 \pm \sqrt{1 - 4y^2}}{2y} \Rightarrow$

$$x = \log\!\left(\dfrac{1 \pm \sqrt{1 - 4y^2}}{2y} \right)$$

43 (a) For $y = \log_2(x+1)$, $D = (-1, \infty)$ and $R = \mathbf{R}$.

(b) $y = \log_2(x+1) \Rightarrow x = \log_2(y+1) \Rightarrow 2^x = y + 1 \Rightarrow y = 2^x - 1$,

$$D = \mathbf{R}, \; R = (-1, \infty)$$

45 (a) $Q(0) = 2(3^0) = 2$ { in thousands }, or 2000

(b) $Q(\tfrac{10}{60}) = 2000(3^{1/6}) \approx 2401$; $Q(\tfrac{30}{60}) = 2000(3^{1/2}) \approx 3464$; $Q(1) = 2(3) = 6000$

47 (a) $N = 64(0.5)^{t/8} = 64\left[(0.5)^{1/8}\right]^t \approx 64(0.917)^t$

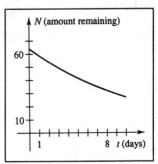

(b) $N = \tfrac{1}{2}N_0 \Rightarrow \tfrac{1}{2}N_0 = N_0(0.5)^{t/8} \Rightarrow$

$(\tfrac{1}{2})^1 = (\tfrac{1}{2})^{t/8} \Rightarrow 1 = t/8 \Rightarrow t = 8$ days

Figure 47

49 (a) Using $A = Pe^{rt}$ with $A = \$35{,}000$, $P = \$10{,}000$, and $r = 11\%$, we have

$$35{,}000 = 10{,}000e^{0.11t} \Rightarrow e^{0.11t} = 3.5 \Rightarrow 0.11t = \ln 3.5 \Rightarrow t = \tfrac{1}{0.11}\ln 3.5 \approx 11.39 \text{ yr.}$$

(b) $2 \cdot 10{,}000 = 10{,}000e^{0.11t} \Rightarrow e^{0.11t} = 2 \Rightarrow 0.11t = \ln 2 \Rightarrow t \approx 6.30$ yr

Alternatively, using the doubling time formula, $t = (\ln 2)/r = (\ln 2)/0.11 \approx 6.30$.

51 (a) $\alpha = 10\log\!\left(\dfrac{I}{I_0}\right) \Rightarrow \dfrac{\alpha}{10} = \log\!\left(\dfrac{I}{I_0}\right) \Rightarrow 10^{\alpha/10} = \dfrac{I}{I_0} \Rightarrow I = I_0 10^{\alpha/10}$

(b) Let $I(\alpha)$ be the intensity corresponding to α decibels.

$$I(\alpha + 1) = I_0 10^{(\alpha + 1)/10} = I_0 10^{\alpha/10} 10^{1/10} = I(\alpha)\, 10^{1/10} \approx 1.26\, I(\alpha),$$

which represents a 26% increase in $I(\alpha)$.

53 $R = 2.3\log(A + 3000) - 5.1 \Rightarrow R + 5.1 = 2.3\log(A + 3000) \Rightarrow$

$\dfrac{R + 5.1}{2.3} = \log(A + 3000) \Rightarrow$

$A + 3000 = 10^{(R + 5.1)/2.3}$ { change to exponential form } $\Rightarrow A = 10^{(R + 5.1)/2.3} - 3000$

55 $R = 4 \Rightarrow 2.3\log(A + 14{,}000) - 6.6 = 4 \Rightarrow \log(A + 14{,}000) = \dfrac{10.6}{2.3} \Rightarrow$

$$A = 10^{106/23} - 14{,}000 \approx 26{,}615.9 \text{ mi}^2.$$

[57] Substituting $v = 0$ and $m = m_1 + m_2$ in $v = -a \ln m + b$ yields

$0 = -a \ln (m_1 + m_2) + b$. Thus, $b = a \ln (m_1 + m_2)$. At burnout, $m = m_1$,

and hence, $v = -a \ln m_1 + b = -a \ln m_1 + a \ln (m_1 + m_2)$

$$= a[\ln (m_1 + m_2) - \ln m_1] \Rightarrow v = a \ln\left(\frac{m_1 + m_2}{m_1}\right)$$

[59] (a) $\log E = 11.4 + (1.5)R \Rightarrow E = 10^{11.4 + 1.5R}$ { merely change the form }

(b) $R = 8.4 \Rightarrow E = 10^{11.4 + 1.5(8.4)} = 10^{24}$ ergs

[62] $I = \frac{V}{R}(1 - e^{-Rt/L}) \Rightarrow \frac{RI}{V} = 1 - e^{-Rt/L} \Rightarrow$

$$e^{-Rt/L} = \left(\frac{V - RI}{V}\right) \Rightarrow -\frac{Rt}{L} = \ln\left(\frac{V - RI}{V}\right) \Rightarrow t = -\frac{L}{R}\ln\left(\frac{V - RI}{V}\right)$$

[63] (a) $x = 4\% \Rightarrow T = -8310 \ln(0.04) \approx 26{,}749$ yr.

(b) $T = 10{,}000 \Rightarrow 10{,}000 = -8310 \ln x \Rightarrow -\frac{10{,}000}{8310} = \ln x \Rightarrow$

$$x = e^{-1000/831} \approx 0.30, \text{ or } 30\%.$$

[65] $N(t) = \frac{1}{2}N_0 \Rightarrow \frac{1}{2}N_0 = N_0(0.805)^t \Rightarrow 2^{-1} = (0.805)^t \Rightarrow \ln(2^{-1}) = \ln(0.805)^t \Rightarrow$

$$-\ln 2 = t \ln(0.805) \Rightarrow t = -\frac{\ln 2}{\ln(0.805)} \approx 3.196 \text{ millennia, or } 3196 \text{ yr}$$

Chapter 5 Discussion Exercises

[1] (a) The y-intercept is a, so it increases as a does. The graph flattens out as a increases.

(b) Graph $Y_1 = \frac{a}{2}(e^{x/a} + e^{-x/a}) + (30 - a)$ on $[-20, 20]$ by $[30, 32]$ and check for $Y_1 < 32$ at $x = 20$. In this case it turns out that $a = 101$ is the smallest integer value that satisfies the conditions, so an equation is

$$y = \frac{101}{2}(e^{x/101} + e^{-x/101}) - 71.$$

[3] (a) $x^y = y^x \Rightarrow \ln(x^y) = \ln(y^x) \Rightarrow y \ln x = x \ln y \Rightarrow \frac{\ln x}{x} = \frac{\ln y}{y}$

(b) Once you find two values of $(\ln x)/x$ that are the same (such as 0.36652), you know that the corresponding x-values, x_1 and x_2, satisfy the relationship $(x_1)^{x_2} \approx (x_2)^{x_1}$. In particular, when $(\ln x)/x \approx 0.36652$, we find that $x_1 \approx 2.50$ and $x_2 \approx 2.97$. Note that $2.50^{2.97} \approx 2.97^{2.50} \approx 15.20$.

(c) Note that $f(e) = \frac{1}{e}$. Any horizontal line $y = k$, with $0 < k < \frac{1}{e}$, will intersect the graph at the two points $\left(x_1, \frac{\ln x_1}{x_1}\right)$ and $\left(x_2, \frac{\ln x_2}{x_2}\right)$, where $1 < x_1 < e$ and $x_2 > e$.

5 Logarithm law 3 states that it is valid only for positive real numbers, so $y = \log_3(x^2)$ is equivalent to $y = 2\log_3 x$ only for $x > 0$. The domain of $y = \log_3(x^2)$ is $\mathbf{R} - \{0\}$, whereas the domain of $y = 2\log_3 x$ is $x > 0$.

7 There are 4 points of intersection. Listed in order of difficulty to find we have: $(-0.9999011,\ 0.00999001)$, $(-0.0001,\ 0.01)$, $(100,\ 0.01105111)$, and $(36{,}102.844,\ 4.6928 \times 10^{13})$. Exponential function values (with base > 1) are greater than polynomial function values (with leading term positive) for very large values of x.

9 $60{,}000 = 40{,}000 b^5 \Rightarrow b = \sqrt[5]{1.5} \approx 1.0844717712$, or 8.44717712%. Mentally—it would take $70/8.5 \approx 8^+$ years to double and there would be about $40/8 = 5$ doubling periods, $2^5 = 32$ and $32 \cdot \$40{,}000 = \$1{,}280{,}000$.

$$\text{Actual} = \$40{,}000(1 + 0.0844717712)^{40} = \$1{,}025{,}156.25.$$

11 On the TI-82/83, enter the sum of the days, $\{0,\ 5168,\ 6728,\ 8136,\ 8407,\ 8735,\ 8857,\ 9010,\ 9274\}$, and the averages, $\{1003.16,\ 2002.25,\ 3004.46,\ 4003.33,\ 5023.55,\ 6010.00,\ 7022.44,\ 8038.88,\ 9033.23\}$, in L_1 and L_2, respectively. Use ExpReg under STAT CALC to obtain $y = ab^x$, where $a = 809.1200949$ and $b = 1.000229099$. Plot the data along with the exponential function and the line $y = 15{,}000$. The functions intersect at approximately $12{,}746$. This value corresponds to October 8, 2007. *Note:* The TI-83 has a convenient "day between dates" function for problems of this nature.

Using the first and last milestone figures and the continuously compounded interest formula gives us $A = Pe^{rt} \Leftrightarrow 9033.23 = 1003.16e^{r(9274)} \Rightarrow$
$r \approx 0.0002369802655$ (daily) or about 8.65% annually. *Note:* The Dow was 100.25 on 1/12/1906. You may want to include this information and examine the differences it makes in any type of prediction.

A discussion of practical considerations should lead to mention of crashes, corrections, and the validity of any model over too long of a period of time.

Chapter 6: Topics from Analytic Geometry

Note: For Exercises 1–12, we will put each parabola equation in one of the forms listed on page 378—either

$$(x - h)^2 = 4p(y - k) \quad \text{or} \quad (y - k)^2 = 4p(x - h).$$

Once in one of those forms, the information concerning the vertex, focus, and directrix is easily obtainable and illustrated in the chart in the text. Let V, F, and l denote the vertex, focus, and directrix, respectively.

$\boxed{3}$ $2y^2 = -3x \Rightarrow (y - 0)^2 = -\frac{3}{2}(x - 0) \Rightarrow 4p = -\frac{3}{2} \Rightarrow p = -\frac{3}{8}.$ We know that the parabola opens either right or left since the variable "y" is squared. Since p is negative, we know that the parabola opens left and that the focus is $\frac{3}{8}$ unit to the left of the vertex. The directrix is $\frac{3}{8}$ unit to the right of the vertex.

$$V(0, 0); \ F(-\tfrac{3}{8}, 0); \ l: x = \tfrac{3}{8}$$

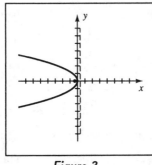

Figure 3

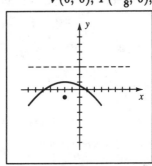

Figure 5

$\boxed{5}$ $(x + 2)^2 = -8(y - 1) \Rightarrow 4p = -8 \Rightarrow p = -2.$ The $(x + 2)$ and $(y - 1)$ factors indicate that we need to shift the vertex, of the parabola having equation $x^2 = -8y$, 2 units left and 1 unit up—that is, move it from $(0, 0)$ to $(-2, 1)$.

$$V(-2, 1); \ F(-2, -1); \ l: y = 3$$

$\boxed{7}$ $(y-2)^2 = \frac{1}{4}(x-3) \Rightarrow 4p = \frac{1}{4} \Rightarrow p = \frac{1}{16}$. "$y$" squared and p positive imply that the parabola opens to the right and the focus is to the right of the vertex. The $(x-3)$ and $(y-2)$ factors indicate that we need to shift the vertex 3 units right and 2 units up from $(0,0)$ to $(3,2)$. $V(3,2)$; $F(\frac{49}{16}, 2)$; l: $x = \frac{47}{16}$

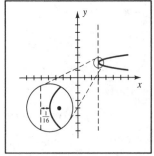

Figure 7

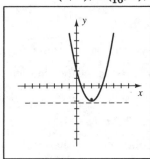

Figure 9

$\boxed{9}$ For this exercise, we need to "complete the square" in order to get the equation in proper form. The term we need to add is $\left[\frac{1}{2}(\text{coefficient of } x)\right]^2$. In this case, that value is $\left[\frac{1}{2}(-4)\right]^2 = 4$. Notice that we add and subtract the value 4 from the same side of the equation as opposed to adding 4 to both sides of the equation. $y = x^2 - 4x + 2 = (x^2 - 4x + \underline{4}) + 2 - \underline{4} = (x-2)^2 - 2 \Rightarrow (y+2) = 1(x-2)^2 \Rightarrow$ $4p = 1 \Rightarrow p = \frac{1}{4}$. $V(2,-2)$; $F(2, -\frac{7}{4})$; l: $y = -\frac{9}{4}$

$\boxed{13}$ Since the vertex is at $(1,0)$ and the parabola has a horizontal axis, the standard equation has the form $y^2 = 4p(x-1)$. The distance from the focus $F(6,0)$ to the vertex $V(1,0)$ is $6-1 = 5$, which is the value of p. Hence, an equation of the parabola is $y^2 = 20(x-1)$.

$\boxed{15}$ $V(-2,3) \Rightarrow (x+2)^2 = 4p(y-3)$.

$$x = 2, \ y = 2 \Rightarrow 16 = 4p(-1) \Rightarrow p = -4. \ \ (x+2)^2 = -16(y-3).$$

$\boxed{17}$ The distance from the directrix to the focus is $2 - (-2) = 4$ units. The vertex $V(0,0)$ is halfway between the directrix and the focus—that is, 2 units from either one. Since the focus is 2 units to the right of the vertex, p is 2. Using one of the forms of an equation of a parabola with vertex at (h, k), we have $(y-0)^2 = 4p(x-0)$, or, equivalently, $y^2 = 8x$.

$\boxed{19}$ $F(6,4)$ and l: $y = -2 \Rightarrow p = 3$ and $V(6,1)$.

$$(x-6)^2 = 4p(y-1) \Rightarrow (x-6)^2 = 12(y-1).$$

$\boxed{21}$ $V(3,-5)$ and l: $x = 2 \Rightarrow p = 1$. $(y+5)^2 = 4p(x-3) \Rightarrow (y+5)^2 = 4(x-3)$.

$\boxed{23}$ $V(-1,0)$ and $F(-4,0) \Rightarrow p = -3$. $(y-0)^2 = 4p(x+1) \Rightarrow y^2 = -12(x+1)$.

[25] The vertex at the origin and symmetric to the y-axis imply that the equation is of the form $y = ax^2$. Substituting $x = 2$ and $y = -3$ into that equation yields
$$-3 = a \cdot 4 \Rightarrow a = -\tfrac{3}{4}. \text{ Thus, an equation is } y = -\tfrac{3}{4}x^2, \text{ or } 3x^2 = -4y.$$

[27] The vertex at $(-3, 5)$ and axis parallel to the x-axis imply that the equation is of the form $(y - 5)^2 = 4p(x + 3)$. Substituting $x = 5$ and $y = 9$ into that equation yields $16 = 4p \cdot 8 \Rightarrow p = \tfrac{1}{2}$. Thus, an equation is $(y - 5)^2 = 2(x + 3)$.

[31] Refer to the definition of a parabola. The point $P(-6, 3)$ is the fixed point (focus) and the line $l\!: x = -2$ is the fixed line (directrix). The vertex is halfway between the focus and the directrix—that is, at $V(-4, 3)$. An equation is of the form $(y - 3)^2 = 4p(x + 4)$. The distance from the vertex to the focus is $p = -6 - (-4) = -2$. Thus, an equation is $(y - 3)^2 = -8(x + 4)$.

Note: To find an equation for a lower or upper half, we need to solve for y (use $-$ or $+$ respectively). For the left or right half, solve for x (use $-$ or $+$ respectively).

[33] $(y + 1)^2 = x + 3 \Rightarrow y + 1 = \pm\sqrt{x + 3} \Rightarrow y = -\sqrt{x + 3} - 1$

[35] $(x + 1)^2 = y - 4 \Rightarrow x + 1 = \pm\sqrt{y - 4} \Rightarrow x = \sqrt{y - 4} - 1$

[37] The parabola has an equation of the form $y = ax^2 + bx + c$. Substituting the x and y values of $P(2, 5)$, $Q(-2, -3)$, and $R(1, 6)$ into this equation yields:
$$\begin{cases} 4a + 2b + c = 5 & P \quad (E_1) \\ 4a - 2b + c = -3 & Q \quad (E_2) \\ a + b + c = 6 & R \quad (E_3) \end{cases}$$

Solving E_3 for c $\{c = 6 - a - b\}$ and substituting into E_1 and E_2 yields:
$$\begin{cases} 3a + b = -1 & (E_4) \\ 3a - 3b = -9 & (E_5) \end{cases}$$
$$E_4 - E_5 \Rightarrow 4b = 8 \Rightarrow b = 2; \ a = -1; \ c = 5. \text{ The equation is } y = -x^2 + 2x + 5.$$

[39] The parabola has an equation of the form $x = ay^2 + by + c$. Substituting the x and y values of $P(-1, 1)$, $Q(11, -2)$, and $R(5, -1)$ into this equation yields:
$$\begin{cases} a + b + c = -1 & P \quad (E_1) \\ 4a - 2b + c = 11 & Q \quad (E_2) \\ a - b + c = 5 & R \quad (E_3) \end{cases}$$

Solving E_3 for c $\{c = 5 - a + b\}$ and substituting into E_1 and E_2 yields:
$$\begin{cases} 2b = -6 & (E_4) \\ 3a - b = 6 & (E_5) \end{cases}$$
$$E_4 \Rightarrow b = -3; \ a = 1; \ c = 1. \text{ The equation is } x = y^2 - 3y + 1.$$

41 A cross section of the mirror is a parabola with $V(0, 0)$ and passing through $P(4, 1)$. The incoming light will collect at the focus F. A general equation of this form of a parabola is $y = ax^2$. Substituting $x = 4$ and $y = 1$ gives us $1 = a(4)^2 \Rightarrow a = \frac{1}{16}$. $p = 1/(4a) = 1/(\frac{1}{4}) = 4$. The light will collect 4 inches from the center of the mirror.

43 If we set up a coordinate system with a parabola opening upward and the vertex at the origin, then the phrase "3 feet across at the opening and 1 foot deep" implies that the points $(\pm\frac{3}{2}, 1)$ are on the parabola.

$$y = ax^2 \Rightarrow 1 = a(\tfrac{3}{2})^2 \Rightarrow a = \tfrac{4}{9}. \quad p = 1/(4a) = 1/(\tfrac{16}{9}) = \tfrac{9}{16} \text{ ft.}$$

45 $p = 5 \Rightarrow a = 1/(4p) = 1/(4 \cdot 5) = \frac{1}{20}$.

$y = ax^2 \{y = 2 \underline{\text{ft}} = 24 \text{ inches}\} \Rightarrow 24 = \frac{1}{20}x^2 \Rightarrow x^2 = 480 \Rightarrow x = \sqrt{480}$.

The width is twice the value of x. Width $= 2\sqrt{480} \approx 43.82$ in.

47 (a) Let the parabola have the equation $x^2 = 4py$. Since the point (r, h) is on the parabola, we can substitute r for x and h for y, giving us $r^2 = 4ph$. Solving for p we have $p = \dfrac{r^2}{4h}$.

(b) $p = 10$ and $h = 5 \Rightarrow r^2 = 4(10)(5) \Rightarrow r = 10\sqrt{2}$.

49 With $a = 125$ and $p = 50$, $S = \dfrac{8\pi p^2}{3}\left[\left(1 + \dfrac{a^2}{4p^2}\right)^{3/2} - 1\right] \approx 64{,}968 \text{ ft}^2$.

51 Depending on the type of calculator or software used, we may need to solve for y in terms of x. $x = -y^2 + 2y + 5 \Rightarrow y^2 - 2y + (x - 5) = 0$. This is a quadratic equation in y. Using the quadratic formula to solve for y yields

$$y = \frac{-(-2) \pm \sqrt{(-2)^2 - 4(1)(x-5)}}{2(1)} = 1 \pm \sqrt{6 - x}.$$

$[-11, 10]$ by $[-7, 7]$

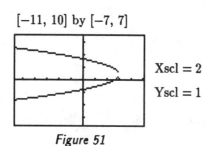

Xscl $= 2$

Yscl $= 1$

Figure 51

$[-2, 4]$ by $[-3, 3]$

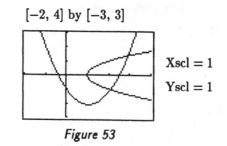

Xscl $= 1$

Yscl $= 1$

Figure 53

53 $x = y^2 + 1 \Rightarrow y = \pm\sqrt{x - 1}$. From the graph, we can see that there are 2 points of intersection. Their coordinates are approximately $(2.08, -1.04)$ and $(2.92, 1.38)$.

Note: Let C, V, F, and M denote the center, the vertices, the foci, and the endpoints of the minor axis, respectively. Let c denote the distance from the center of the ellipse to a focus.

1   $\dfrac{x^2}{9} + \dfrac{y^2}{4} = 1$ •   The x-intercepts are at $(\pm\sqrt{9},\ 0)$, or, equivalently, $(\pm 3,\ 0)$. The y-intercepts are at $(\pm\sqrt{4},\ 0) = (\pm 2,\ 0)$. The major axis {the longer of the two axes} is the horizontal axis and has length $2(3) = 6$. The minor axis {the shorter} is the vertical axis and has length $2(2) = 4$. To find the foci, it is helpful to remember the relationship
$$\left[\tfrac{1}{2}(\text{minor axis})\right]^2 + \left[c\right]^2 = \left[\tfrac{1}{2}(\text{major axis})\right]^2.$$
Using the values from above we have $\left[\tfrac{1}{2}(4)\right]^2 + c^2 = \left[\tfrac{1}{2}(6)\right]^2 \Rightarrow 4 + c^2 = 9 \Rightarrow$
$c = \pm\sqrt{5}$.

$$V(\pm 3,\ 0);\ F(\pm\sqrt{5},\ 0);\ M(0,\ \pm 2)$$

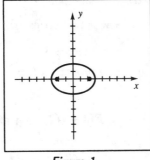

Figure 1

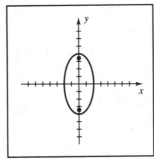

Figure 5

5   We first divide by 16 to obtain the "1" on the right side of the equation. $4x^2 + y^2 = 16 \Rightarrow \dfrac{x^2}{4} + \dfrac{y^2}{16} = 1$. Since the 16 in the denominator of the term with the variable y is larger than the 4 in the denominator of the term with the variable x, the vertices and foci are on the y-axis and the major axis is the vertical axis. $4 + c^2 = 16 \Rightarrow c^2 = 16 - 4 \Rightarrow c = \pm 2\sqrt{3}$.

$$V(0,\ \pm 4);\ F(0,\ \pm 2\sqrt{3});\ M(\pm 2,\ 0)$$

7 $4x^2 + 25y^2 = 1 \Rightarrow \frac{x^2}{\frac{1}{4}} + \frac{y^2}{\frac{1}{25}} = 1.$ $\frac{1}{25} + c^2 = \frac{1}{4} \Rightarrow c^2 = \frac{1}{4} - \frac{1}{25} = \frac{21}{100} \Rightarrow c = \pm\frac{1}{10}\sqrt{21}.$

$V(\pm\frac{1}{2}, 0); F(\pm\frac{1}{10}\sqrt{21}, 0); M(0, \pm\frac{1}{5})$

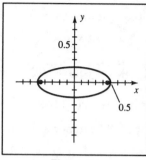

Figure 7

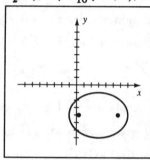

Figure 9

9 $\frac{(x-3)^2}{16} + \frac{(y+4)^2}{9} = 1$ • The effect of the factors $(x - 3)$ and $(y + 4)$ is to shift the center of the ellipse from $(0, 0)$ to $(3, -4)$. Since the larger denominator, 16, is in the term with x, the major axis will be horizontal. The endpoints are 4 units in either direction of the center. Their coordinates are the points $(3 \pm 4, -4)$, or, equivalently, $(7, -4)$ and $(-1, -4)$. The minor axis will be vertical with endpoints $(3, -4 \pm 3)$, or, equivalently, $(3, -1)$ and $(3, -7)$. $9 + c^2 = 16 \Rightarrow c^2 = 16 - 9 \Rightarrow c = \pm\sqrt{7}.$ Remember that c is the distance *from the center* to a focus. Hence, the coordinates of the foci are $(3 \pm \sqrt{7}, -4)$.

$C(3, -4); V(3 \pm 4, -4); F(3 \pm \sqrt{7}, -4); M(3, -4 \pm 3)$

11 $4x^2 + 9y^2 - 32x - 36y + 64 = 0 \Rightarrow$

$(4x^2 - 32x) + (9y^2 - 36y) = -64 \Rightarrow$

{ first group the x terms and the y terms }

$4(x^2 - 8x) + 9(y^2 - 4y) = -64 \Rightarrow$

{ factor out the coefficients of x^2 and y^2 }

$4(x^2 - 8x + \underline{16}) + 9(y^2 - 4y + \underline{4}) = -64 + \underline{64} + \underline{36} \Rightarrow$

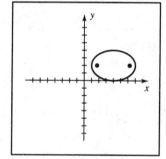

Figure 11

{ Complete the squares—remember that we have added $\underline{4}$ (16) and $\underline{9}$ (4), not 16 and 4. Thus,

we add 64 and 36 to the right side of the equation. }

$4(x-4)^2 + 9(y-2)^2 = 36 \Rightarrow \frac{(x-4)^2}{9} + \frac{(y-2)^2}{4} = 1.$

$c^2 = 9 - 4 \Rightarrow c = \pm\sqrt{5}.$

$C(4, 2); V(4 \pm 3, 2); F(4 \pm \sqrt{5}, 2); M(4, 2 \pm 2)$

Note: Let b denote the distance equal to $\frac{1}{2}$(minor axis length) and

let a denote the distance equal to $\frac{1}{2}$(major axis length).

15 $a = 2$ and $b = 6 \Rightarrow \dfrac{x^2}{a^2} + \dfrac{y^2}{b^2} = 1$ is $\dfrac{x^2}{4} + \dfrac{y^2}{36} = 1$.

17 The center of the ellipse is $(-2, 1)$. $a = 5$ and $b = 2$ give us $\dfrac{(x+2)^2}{25} + \dfrac{(y-1)^2}{4} = 1$.

19 Since the vertices are at $(\pm 8, 0)$, $\frac{1}{2}$(major axis) $= 8$. Since the foci are at $(\pm 5, 0)$,

$c = 5$. Using the relationship $\left[\frac{1}{2}(\text{minor axis})\right]^2 + \left[c\right]^2 = \left[\frac{1}{2}(\text{major axis})\right]^2$, we have

$\left[\frac{1}{2}(\text{minor axis})\right]^2 + 5^2 = 8^2 \Rightarrow \left[\frac{1}{2}(\text{minor axis})\right]^2 = 64 - 25 = 39$.

An equation is $\dfrac{x^2}{64} + \dfrac{y^2}{39} = 1$.

21 If the length of the minor axis is 3, then $b = \frac{3}{2}$.

An equation is $\dfrac{x^2}{\left(\frac{3}{2}\right)^2} + \dfrac{y^2}{5^2} = 1$, or, equivalently, $\dfrac{4x^2}{9} + \dfrac{y^2}{25} = 1$.

23 With the vertices at $(0, \pm 6)$, an equation of the ellipse is $\dfrac{x^2}{b^2} + \dfrac{y^2}{6^2} = 1$. Substituting

$x = 3$ and $y = 2$ and solving for b^2 yields $\dfrac{9}{b^2} + \dfrac{4}{36} = 1 \Rightarrow \dfrac{9}{b^2} = \dfrac{8}{9} \Rightarrow b^2 = \dfrac{81}{8}$.

An equation is $\dfrac{x^2}{\frac{81}{8}} + \dfrac{y^2}{36} = 1$, or, equivalently, $\dfrac{8x^2}{81} + \dfrac{y^2}{36} = 1$.

25 With vertices $V(0, \pm 4)$, an equation of the ellipse is $\dfrac{x^2}{b^2} + \dfrac{y^2}{16} = 1$.

Remember the formula for the eccentricity:

$$\boxed{e = \text{eccentricity} = \frac{c}{a} = \frac{\text{distance from center to a focus}}{\text{distance from center to a vertex}}}.$$

Hence, $e = \dfrac{c}{a} = \dfrac{3}{4}$ and $a = 4 \Rightarrow c = 3$. Thus, $b^2 + c^2 = a^2 \Rightarrow$

$b^2 = a^2 - c^2 = 4^2 - 3^2 = 16 - 9 = 7$. An equation is $\dfrac{x^2}{7} + \dfrac{y^2}{16} = 1$.

29 Remember to divide the lengths of the major and minor axes by 2.

$$\dfrac{x^2}{(\frac{1}{2} \cdot 8)^2} + \dfrac{y^2}{(\frac{1}{2} \cdot 5)^2} = 1 \Rightarrow \dfrac{x^2}{16} + \dfrac{y^2}{\frac{25}{4}} = 1 \Rightarrow \dfrac{x^2}{16} + \dfrac{4y^2}{25} = 1.$$

31 The graph of $x^2 + 4y^2 = 20$, or, equivalently, $\frac{x^2}{20} + \frac{y^2}{5} = 1$,

is that of an ellipse with x-intercepts at $(\pm \sqrt{20},\, 0)$ and

y-intercepts at $(0,\, \pm \sqrt{5})$. The graph of $x + 2y = 6$, or,

equivalently, $y = -\frac{1}{2}x + 3$, is that of a line with y-

intercept 3 and slope $-\frac{1}{2}$. Substituting $x = 6 - 2y$ into

$x^2 + 4y^2 = 20$ yields $(6 - 2y)^2 + 4y^2 = 20 \Rightarrow$

$8y^2 - 24y + 16 = 0 \Rightarrow 8(y^2 - 3y + 2) = 0 \Rightarrow$

$8(y - 1)(y - 2) = 0 \Rightarrow y = 1,\, 2;\ x = 4,\, 2.$

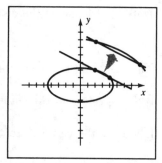

Figure 31

The two points of intersection are $(2,\, 2)$ and $(4,\, 1)$.

33 Refer to the definition of an ellipse. As in the discussion on page 385, we will let the

positive constant, k, equal $2a$. $k = 2a = 10 \Rightarrow a = 5$. $F(3,\, 0)$ and $F'(-3,\, 0) \Rightarrow c = 3$.

$$b^2 = a^2 - c^2 = 25 - 9 = 16.\ \text{An equation is } \frac{x^2}{25} + \frac{y^2}{16} = 1.$$

35 $k = 2a = 34 \Rightarrow a = 17$. $F(0,\, 15)$ and $F'(0,\, -15) \Rightarrow c = 15$.

$$b^2 = a^2 - c^2 = 289 - 225 = 64.\ \text{An equation is } \frac{x^2}{64} + \frac{y^2}{289} = 1.$$

37 $y = 11\sqrt{1 - \dfrac{x^2}{49}} \Rightarrow \dfrac{y}{11} = \sqrt{1 - \dfrac{x^2}{49}} \Rightarrow \dfrac{x^2}{49} + \dfrac{y^2}{121} = 1.$

Since $y \geq 0$ in the original equation, its graph is the upper half of the ellipse.

39 $x = -\frac{1}{3}\sqrt{9 - y^2} \Rightarrow -3x = \sqrt{9 - y^2} \Rightarrow 9x^2 = 9 - y^2 \Rightarrow x^2 + \dfrac{y^2}{9} = 1.$

Since $x \leq 0$ in the original equation, its graph is the left half of the ellipse.

41 $x = 1 + 2\sqrt{1 - \dfrac{(y + 2)^2}{9}} \Rightarrow \dfrac{x - 1}{2} = \sqrt{1 - \dfrac{(y + 2)^2}{9}} \Rightarrow \dfrac{(x - 1)^2}{4} + \dfrac{(y + 2)^2}{9} = 1.$

Since $x \geq 1$ in the original equation, its graph is the right half of the ellipse.

43 $y = 2 - 7\sqrt{1 - \dfrac{(x + 1)^2}{9}} \Rightarrow \dfrac{y - 2}{-7} = \sqrt{1 - \dfrac{(x + 1)^2}{9}} \Rightarrow \dfrac{(x + 1)^2}{9} + \dfrac{(y - 2)^2}{49} = 1.$

Since $y \leq 2$ in the original equation, its graph is the lower half of the ellipse.

45 Model this problem as an ellipse with $V(\pm 15,\, 0)$ and $M(0,\, \pm 10)$.

Substituting $x = 6$ into $\dfrac{x^2}{15^2} + \dfrac{y^2}{10^2} = 1$ yields $\dfrac{y^2}{100} = \dfrac{189}{225} \Rightarrow y^2 = 84$.

The desired height is $\sqrt{84} = 2\sqrt{21} \approx 9.165$ ft.

47 $e = \dfrac{c}{a} = 0.017 \Rightarrow c = 0.017a = 0.017(93{,}000{,}000) = 1{,}581{,}000$. The maximum and

minimum distances are $a + c = 94{,}581{,}000$ miles and $a - c = 91{,}419{,}000$ miles.

49 (a) Let c denote the distance from the center of the hemi-ellipsoid to F. Hence,

$$(\tfrac{1}{2}k)^2 + c^2 = h^2 \Rightarrow c^2 = h^2 - \tfrac{1}{4}k^2 \Rightarrow c = \sqrt{h^2 - \tfrac{1}{4}k^2}. \ \ d = d(V, F) = h - c \Rightarrow$$

$$d = h - \sqrt{h^2 - \tfrac{1}{4}k^2} \text{ and } d' = d(V, F') = h + c \Rightarrow d' = h + \sqrt{h^2 - \tfrac{1}{4}k^2}.$$

(b) From part (a), $d' = h + c \Rightarrow c = d' - h = 32 - 17 = 15.$ $c = \sqrt{h^2 - \tfrac{1}{4}k^2} \Rightarrow$

$$15 = \sqrt{17^2 - \tfrac{1}{4}k^2} \Rightarrow 225 = 289 - \tfrac{1}{4}k^2 \Rightarrow \tfrac{1}{4}k^2 = 64 \Rightarrow k^2 = 256 \Rightarrow k = 16 \text{ cm}.$$

$$d = h - c = 17 - 15 = 2 \Rightarrow F \text{ should be located 2 cm from } V.$$

51 $c^2 = (\tfrac{1}{2} \cdot 50)^2 - 15^2 = 625 - 225 = 400 \Rightarrow c = 20.$

Their feet should be $25 - 20 = 5$ ft from the vertices.

53 First determine an equation of the ellipse for the orbit of Earth. $e = \tfrac{c}{a} \Rightarrow c = ae =$

$0.093 \times 149.6 = 13.9128.$ $b^2 = a^2 - c^2 = 149.6^2 - 13.9128^2 \Rightarrow b \approx 148.95 \approx 149.0.$

An equation for the orbit of Earth is $\dfrac{x^2}{149.6^2} + \dfrac{y^2}{149.0^2} = 1.$

Graph $Y_1 = 149\sqrt{1 - (x^2/149.6^2)}$ and $Y_2 = -Y_1.$

The sun is at $(\pm 13.9128, 0)$. Plot the point $(13.9128, 0)$ for the sun.

$[-300, 300]$ by $[-200, 200]$ $[-3, 3]$ by $[-2, 2]$

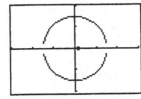

Xscl $= 100$
Yscl $= 100$

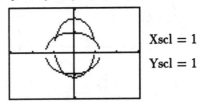

Xscl $= 1$
Yscl $= 1$

Figure 53   *Figure 57*

57 $\dfrac{(x + 0.1)^2}{1.7} + \dfrac{y^2}{0.9} = 1 \Rightarrow y = \pm\sqrt{0.9[1 - (x + 0.1)^2/1.7]}.$

$\dfrac{x^2}{0.9} + \dfrac{(y - 0.25)^2}{1.8} = 1 \Rightarrow y = 0.25 \pm \sqrt{1.8(1 - x^2/0.9)}.$

From the graph, the points of intersection are

approximately $(-0.88, 0.76)$, $(-0.48, -0.91)$, $(0.58, -0.81)$, and $(0.92, 0.59)$.

6.3 Exercises

Note: Let C, V, F, and W denote the center, the vertices, the foci, and the endpoints of the conjugate axis, respectively. Let c denote the distance from the center of the hyperbola to a focus.

$\boxed{1}$ $\dfrac{x^2}{9} - \dfrac{y^2}{4} = 1$ • See the note on abbreviations on the previous page. The hyperbola will have a right branch and a left branch since the term containing x is positive. The vertices will be on the horizontal transverse axis, $\pm\sqrt{9} = \pm 3$ units from the center. The endpoints of the vertical conjugate axis are $(0, \pm\sqrt{4}) = (0, \pm 2)$. The asymptotes have equations

$$y = \pm\left[\frac{\frac{1}{2}(\text{vertical axis length})}{\frac{1}{2}(\text{horizontal axis length})}\right](x).$$

Note that the terms are "vertical" and "horizontal" and not transverse and conjugate since the latter can be either vertical or horizontal. In this case, we have $y = \pm\dfrac{\frac{1}{2}(4)}{\frac{1}{2}(6)}(x) = \pm\frac{2}{3}x$. The positive sign corresponds to the asymptote with positive slope and the negative sign corresponds to the asymptote with negative slope. To find the foci, it is helpful to remember the relationship

$$\left[\tfrac{1}{2}(\text{transverse axis})\right]^2 + \left[\tfrac{1}{2}(\text{conjugate axis})\right]^2 = \left[c\right]^2.$$

Using the values from above we have $\left[\frac{1}{2}(6)\right]^2 + \left[\frac{1}{2}(4)\right]^2 = c^2 \Rightarrow 9 + 4 = c^2 \Rightarrow$ $c = \pm\sqrt{13}$.

$$V(\pm 3, 0); \ F(\pm\sqrt{13}, 0); \ W(0, \pm 2); \ y = \pm\tfrac{2}{3}x$$

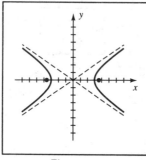

Figure 1

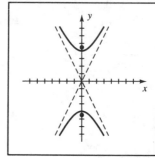

Figure 7

$\boxed{7}$ $y^2 - 4x^2 = 16 \Rightarrow \dfrac{y^2}{16} - \dfrac{x^2}{4} = 1$ • The hyperbola will have an upper branch and a lower branch since the term containing y is positive. The vertices will be $\pm\sqrt{16} = \pm 4$ units from the center on the vertical transverse axis. The endpoints of the horizontal conjugate axis are $(\pm\sqrt{4}, 0) = (\pm 2, 0)$. The asymptotes have equations $y = \pm\dfrac{\frac{1}{2}(8)}{\frac{1}{2}(4)}(x) = \pm 2x$. Using the foci relationship given in Exercise 1, we have $\left[\frac{1}{2}(8)\right]^2 + \left[\frac{1}{2}(4)\right]^2 = c^2 \Rightarrow 16 + 4 = c^2 \Rightarrow c = \pm 2\sqrt{5}$.

$$V(0, \pm 4); \ F(0, \pm 2\sqrt{5}); \ W(\pm 2, 0); \ y = \pm 2x$$

$\boxed{9}$ $16x^2 - 36y^2 = 1 \Rightarrow \dfrac{x^2}{\frac{1}{16}} - \dfrac{y^2}{\frac{1}{36}} = 1$. $c^2 = \frac{1}{16} + \frac{1}{36} \Rightarrow c = \pm\frac{1}{12}\sqrt{13}$.

$$V(\pm\tfrac{1}{4}, 0); \ F(\pm\tfrac{1}{12}\sqrt{13}, 0); \ W(0, \pm\tfrac{1}{6}); \ y = \pm\tfrac{2}{3}x$$

Note that the branches of the hyperbola almost coincide with the asymptotes.

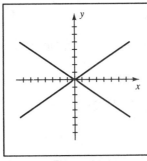

Figure 9

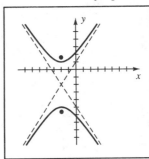

Figure 11

$\boxed{11}$ $\dfrac{(y+2)^2}{9} - \dfrac{(x+2)^2}{4} = 1$ •   The effect of the factors $(x+2)$ and $(y+2)$ is to shift the center of the hyperbola from $(0, 0)$ to $(-2, -2)$. Since the term involving y is positive, the transverse axis will be vertical. The vertices are 3 units in either direction of the center. Their coordinates are $(-2, -2\pm 3)$ or equivalently, $(-2, 1)$ and $(-2, -5)$. The conjugate axis will be horizontal with endpoints $(-2\pm 2, -2)$, or, equivalently, $(0, -2)$ and $(-4, -2)$. Remember that c is the distance *from the center* to a focus. $c^2 = 9 + 4 \Rightarrow c = \pm\sqrt{13}$. Hence, the coordinates of the foci are $(-2, -2\pm\sqrt{13})$. If the center of the hyperbola was at the origin, we would have asymptote equations $y = \pm\frac{3}{2}x$. Since the center of the hyperbola has been shifted by the factors $(y+2)$ and $(x+2)$, we can shift the asymptote equations using the same factors. Hence, these equations are $(y+2) = \pm\frac{3}{2}(x+2)$.

$$C(-2, -2); \ V(-2, -2\pm 3); \ F(-2, -2\pm\sqrt{13}); \ W(-2\pm 2, -2)$$

$\boxed{13}$ $144x^2 - 25y^2 + 864x - 100y - 2404 = 0 \Rightarrow$

$144(x^2 + 6x + \underline{\ 9\ }) - 25(y^2 + 4y + \underline{\ 4\ }) =$

$\qquad\qquad 2404 + \underline{\ 1296\ } - \underline{\ 100\ } \Rightarrow$

$144(x+3)^2 - 25(y+2)^2 = 3600 \Rightarrow \dfrac{(x+3)^2}{25} - \dfrac{(y+2)^2}{144} = 1$.

$c^2 = 25 + 144 \Rightarrow c = \pm 13$.

$C(-3, -2); \ V(-3\pm 5, -2); \ F(-3\pm 13, -2);$

$W(-3, -2\pm 12); \ (y+2) = \pm\frac{12}{5}(x+3)$

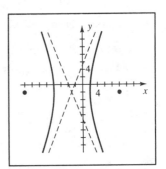

Figure 13

Note: Let b denote the distance equal to $\frac{1}{2}$(conjugate axis length) and

let a denote the distance equal to $\frac{1}{2}$(transverse axis length).

17 $a = 3$ and $c = 5 \Rightarrow b^2 = c^2 - a^2 = 16$. $\frac{x^2}{a^2} - \frac{y^2}{b^2} = 1$ is then $\frac{x^2}{9} - \frac{y^2}{16} = 1$.

19 The center of the hyperbola is $(-2, -3)$. $a = 1$ and $c = 2 \Rightarrow b^2 = 2^2 - 1^2 = 3$ and an

equation is $\frac{(y+3)^2}{1^2} - \frac{(x+2)^2}{3} = 1$, or, equivalently, $(y+3)^2 - \frac{(x+2)^2}{3} = 1$.

21 $F(0, \pm 4) \Rightarrow c = 4$. $V(0, \pm 1) \Rightarrow a = 1$. $a^2 + b^2 = c^2 \Rightarrow b^2 = 4^2 - 1^2 = 15$.

Since the vertices are on the y-axis, the "1^2" is associated with the y^2 term.

An equation is $\frac{y^2}{1} - \frac{x^2}{15} = 1$.

23 $F(\pm 5, 0)$ and $V(\pm 3, 0) \Rightarrow W(0, \pm 4)$. An equation is $\frac{x^2}{9} - \frac{y^2}{16} = 1$.

25 Conjugate axis of length 4 implies that $b = 2$. $F(0, \pm 5) \Rightarrow c = 5$.

$a^2 + b^2 = c^2 \Rightarrow a^2 = 5^2 - 2^2 = 21$. An equation is $\frac{y^2}{21} - \frac{x^2}{4} = 1$.

27 Since the asymptote equations are $y = \pm 2x$ and we know that the point $(3, 0)$ is on

the hyperbola, we conclude that the upper right corner of the rectangle formed by the

transverse and conjugate axes has coordinates $(3, 6)$ {substitute $x = 3$ in $y = 2x$ to

obtain the 6}. Thus we have endpoints of the conjugate axis at $(0, \pm 6)$.

An equation is $\frac{x^2}{9} - \frac{y^2}{36} = 1$.

29 $a = 5$, $b = 2(5) = 10$. $\frac{x^2}{5^2} - \frac{y^2}{10^2} = 1 \Rightarrow \frac{x^2}{25} - \frac{y^2}{100} = 1$.

31 Since the transverse axis is vertical, the y^2 term will be positive.

An equation is $\dfrac{y^2}{(\frac{1}{2} \cdot 10)^2} - \dfrac{x^2}{(\frac{1}{2} \cdot 14)^2} = 1$, or $\dfrac{y^2}{25} - \dfrac{x^2}{49} = 1$.

33 The graph of $y^2 - 4x^2 = 16$, or, equivalently, $\frac{y^2}{16} - \frac{x^2}{4} = 1$,

is that of a hyperbola with y-intercepts at $(0, \pm 4)$. The

graph of $y - x = 4$, or, equivalently, $y = x + 4$, is that of a

line with y-intercept 4 and slope 1. Substituting $y = x + 4$

into $y^2 - 4x^2 = 16$ yields $(x+4)^2 - 4x^2 = 16 \Rightarrow$

$3x^2 - 8x = 0 \Rightarrow x(3x - 8) = 0 \Rightarrow x = 0, \frac{8}{3}$; $y = 4, \frac{20}{3}$.

The two points of intersection are $(0, 4)$ and $(\frac{8}{3}, \frac{20}{3})$.

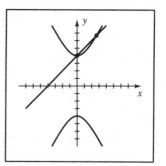

Figure 33

37 Refer to the definition of a hyperbola. As in the discussion on page 398,

we will let the positive constant, k, equal $2a$. $k = 2a = 16 \Rightarrow a = 8$.

$F(0, 10)$ and $F'(0, -10) \Rightarrow c = 10$. $b^2 = c^2 - a^2 = 100 - 64 = 36$.

An equation is $\frac{y^2}{64} - \frac{x^2}{36} = 1$.

39 $x = \frac{5}{4}\sqrt{y^2 + 16} \Rightarrow \frac{4}{5}x = \sqrt{y^2 + 16} \Rightarrow \frac{16}{25}x^2 = y^2 + 16 \Rightarrow \frac{16}{25}x^2 - y^2 = 16 \Rightarrow \frac{x^2}{25} - \frac{y^2}{16} = 1.$

Since $x > 0$ in the original equation, its graph is the right branch of the hyperbola.

41 $y = \frac{3}{7}\sqrt{x^2 + 49} \Rightarrow \frac{7}{3}y = \sqrt{x^2 + 49} \Rightarrow \frac{49}{9}y^2 = x^2 + 49 \Rightarrow \frac{y^2}{9} - \frac{x^2}{49} = 1.$

Since $y > 0$ in the original equation, its graph is the upper branch of the hyperbola.

43 $y = -\frac{9}{4}\sqrt{x^2 - 16} \Rightarrow -\frac{4}{9}y = \sqrt{x^2 - 16} \Rightarrow \frac{16}{81}y^2 = x^2 - 16 \Rightarrow \frac{x^2}{16} - \frac{y^2}{81} = 1.$

Since $y \le 0$ in the original equation,

its graph is the lower halves of the branches of the hyperbola.

45 $x = -\frac{2}{3}\sqrt{y^2 - 36} \Rightarrow -\frac{3}{2}x = \sqrt{y^2 - 36} \Rightarrow \frac{9}{4}x^2 = y^2 - 36 \Rightarrow \frac{y^2}{36} - \frac{x^2}{16} = 1.$

Since $x \le 0$ in the original equation,

its graph is the left halves of the branches of the hyperbola.

49 The path is a hyperbola with $V(\pm 3, 0)$ and $W(0, \pm\frac{3}{2})$.

An equation is $\dfrac{x^2}{(3)^2} - \dfrac{y^2}{(\frac{3}{2})^2} = 1$ or equivalently, $x^2 - 4y^2 = 9$.

If only the right branch is considered, then $x = \sqrt{9 + 4y^2}$ is an equation of the path.

51 Set up a coordinate system like the one in Example 6. Let the origin be located on the shoreline halfway between A and B and let $P(x, y)$ denote the coordinates of the ship. The coordinates of A and B (which can be thought of as the foci of the hyperbola) are $(-100, 0)$ and $(100, 0)$, respectively. Hence, $c = 100$. As in Exercise 37, the difference in distances is a constant—that is, $d(P, A) - d(P, B) = 2a = 160 \Rightarrow a = 80$. $b^2 = c^2 - a^2 = 100^2 - 80^2 \Rightarrow b = 60$. An equation of the hyperbola is

$\dfrac{x^2}{80^2} - \dfrac{y^2}{60^2} = 1$. Now, $y = 100 \Rightarrow \dfrac{x^2}{80^2} = 1 + \dfrac{100^2}{60^2} \Rightarrow x^2 = 80^2 \cdot \dfrac{13{,}600}{60^2} \Rightarrow$

$x = 80 \cdot \frac{10}{60}\sqrt{136} = \frac{80}{3}\sqrt{34}$. The ship's coordinates are $(\frac{80}{3}\sqrt{34}, 100) \approx (155.5, 100)$.

53 $\dfrac{(y - 0.1)^2}{1.6} - \dfrac{(x + 0.2)^2}{0.5} = 1 \Rightarrow 0.5(y - 0.1)^2 - 1.6(x + 0.2)^2 = 0.8 \Rightarrow$

$$y = 0.1 \pm \sqrt{1.6 + 3.2(x + 0.2)^2}.$$

$\dfrac{(y - 0.5)^2}{2.7} - \dfrac{(x - 0.1)^2}{5.3} = 1 \Rightarrow 5.3(y - 0.5)^2 - 2.7(x - 0.1)^2 = 14.31 \Rightarrow$

$$y = 0.5 \pm \sqrt{\tfrac{1}{5.3}\left[14.31 + 2.7(x - 0.1)^2\right]}.$$

From the graph of *Figure 53* on the next page,

the point of intersection in the first quadrant is approximately $(0.741, 2.206)$.

[−15, 15] by [−10, 10]

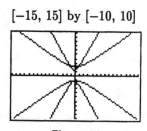

Xscl = 1
Yscl = 1

[−15, 15] by [−10, 10]

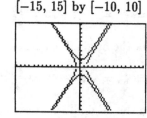

Xscl = 2
Yscl = 2

Figure 53   *Figure 55*

[55] $\dfrac{(x-0.3)^2}{1.3} - \dfrac{y^2}{2.7} = 1 \Rightarrow y = \pm\sqrt{2.7[-1+(x-0.3)^2/1.3]}.$

$\dfrac{y^2}{2.8} - \dfrac{(x-0.2)^2}{1.2} = 1 \Rightarrow y = \pm\sqrt{2.8[1+(x-0.2)^2/1.2]}.$

The two graphs nearly intersect in the second and fourth quadrants,

but there are no points of intersection.

[57] (a) The comet's path is hyperbolic with $a^2 = 26 \times 10^{14}$ and $b^2 = 18 \times 10^{14}$.

$c^2 = a^2 + b^2 = 26 \times 10^{14} + 18 \times 10^{14} = 44 \times 10^{14} \Rightarrow c \approx 6.63 \times 10^7.$

The coordinates of the sun are approximately $(6.63 \times 10^7, 0)$.

(b) The minimum distance between the comet and the sun will be

$c - a = \sqrt{44 \times 10^{14}} - \sqrt{26 \times 10^{14}} = 1.53 \times 10^7 \text{ mi}.$

Since r must be in meters, $1.53 \times 10^7 \text{ mi} \times 1610 \text{ m/mi} \approx 2.47 \times 10^{10} \text{ m}.$ At this

distance, v must be greater than $\sqrt{\dfrac{2k}{r}} \approx \sqrt{\dfrac{2(1.325 \times 10^{20})}{2.47 \times 10^{10}}} \approx 103{,}600 \text{ m/sec}.$

6.4 Exercises

---

[1] For this exercise (and others), we solve for t in terms of x, and then substitute that

expression for t in the equation that relates y and t.

$x = t - 2 \Rightarrow t = x + 2.$ $y = 2t + 3 = 2(x + 2) + 3 = 2x + 7.$

As t varies from 0 to 5, (x, y) varies from $(-2, 3)$ to $(3, 13)$.

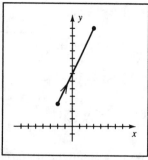

Figure 1

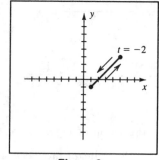

Figure 3

[3] $x = t^2 + 1 \Rightarrow t^2 = x - 1.$ $y = t^2 - 1 = x - 2.$ As t varies from -2 to 2,

(x, y) varies from $(5, 3)$ to $(1, -1)$ { when $t = 0$ } and back to $(5, 3)$.

⑤ Since y is linear in t, it is easier to solve the second equation for t than it is to solve the first equation for t. Hence, we solve for t in terms of y, and then substitute that expression for t in the equation that relates x and t. $y = 2t + 3 \Rightarrow t = \frac{1}{2}(y - 3)$. $x = 4\left[\frac{1}{2}(y - 3)\right]^2 - 5 \Rightarrow (y - 3)^2 = x + 5$. This is a parabola with vertex at $(-5, 3)$. Since t takes on all real values, so does y, and the curve C is the entire parabola.

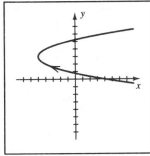

Figure 5

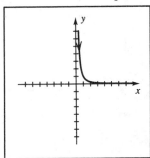

Figure 7

⑦ $y = e^{-2t} = (e^t)^{-2} = x^{-2} = 1/x^2$. This is a rational function. Only the first quadrant portion is used and as t varies from $-\infty$ to ∞, x varies from 0 to ∞, excluding 0.

⑨ $x = 2\sin t$ and $y = 3\cos t \Rightarrow \frac{x}{2} = \sin t$ and $\frac{y}{3} = \cos t \Rightarrow$

$\frac{x^2}{4} + \frac{y^2}{9} = \sin^2 t + \cos^2 t = 1$. As t varies from 0 to 2π,

(x, y) traces the ellipse from $(0, 3)$ in a clockwise direction back to $(0, 3)$.

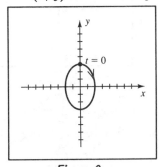

Figure 9

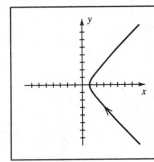

Figure 11

⑪ $x = \sec t$ and $y = \tan t \Rightarrow x^2 - y^2 = \sec^2 t - \tan^2 t = 1$.

As t varies from $-\frac{\pi}{2}$ to $\frac{\pi}{2}$, (x, y) traces the right branch of the hyperbola along the asymptote $y = -x$ to $(1, 0)$ and then along the asymptote $y = x$.

13 $y = 2\ln t = \ln t^2 \{\text{since } t > 0\} = \ln x.$

As t varies from 0 to ∞, so does x, and y varies from $-\infty$ to ∞.

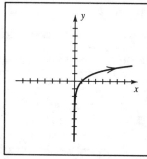

Figure 13

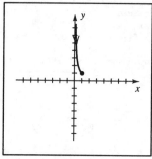

Figure 15

15 $y = \csc t = \dfrac{1}{\sin t} = \dfrac{1}{x}.$

As t varies from 0 to $\frac{\pi}{2}$, (x, y) varies asymptotically from the positive y-axis to $(1, 1)$.

17 $x = t$ and $y = \sqrt{t^2 - 1} \Rightarrow y = \sqrt{x^2 - 1} \Rightarrow x^2 - y^2 = 1.$

Since y is nonnegative, the graph is the top half of both branches of the hyperbola.

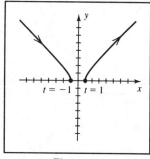

Figure 17

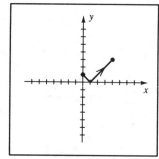

Figure 19

19 $x = t$ and $y = \sqrt{t^2 - 2t + 1} \Rightarrow y = \sqrt{x^2 - 2x + 1} = \sqrt{(x-1)^2} = |x - 1|.$

As t varies from 0 to 4, (x, y) traces $y = |x - 1|$ from $(0, 1)$ to $(4, 3)$.

21 $x = (t+1)^3 \Rightarrow t = x^{1/3} - 1.$ $y = (t+2)^2 = (x^{1/3} + 1)^2.$

This is probably an unfamiliar graph. The graph of $y = x^{1/3}$ is similar to the graph of $y = x^{1/2}$ $\{y = \sqrt{x}\}$ but is symmetric with respect to the origin. The "+1" shifts $y = x^{1/3}$ up 1 unit and the "squaring" makes all y values nonnegative. Since we have restrictions on the variable t, we only have a portion of this graph. As t varies from 0 to 2, (x, y) varies from $(1, 4)$ to $(27, 16)$.

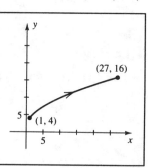

Figure 21

23 All of the curves are a portion of the parabola $x = y^2$.

C_1: $x = t^2 = y^2$. y takes on all real values and we have the entire parabola.

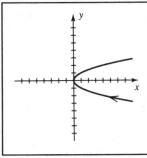

Figure 23 (C_1)

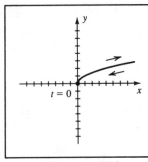

Figure 23 (C_2)

C_2: $x = t^4 = (t^2)^2 = y^2$. C_2 is only the top half since $y = t^2$ is nonnegative.

As t varies from $-\infty$ to ∞, the top portion is traced twice.

C_3: $x = \sin^2 t = (\sin t)^2 = y^2$. C_3 is the portion of the curve from $(1, -1)$ to $(1, 1)$.

The point $(1, 1)$ is reached at $t = \frac{\pi}{2} + 2\pi n$ and the point $(1, -1)$ when

$$t = \frac{3\pi}{2} + 2\pi n.$$

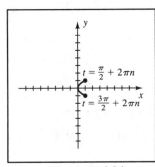

Figure 23 (C_3)

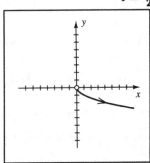

Figure 23 (C_4)

C_4: $x = e^{2t} = (e^t)^2 = (-e^t)^2 = y^2$. C_4 is the bottom half of the parabola since y

is negative. As t approaches $-\infty$, the parabola approaches the origin.

25 In each part, the motion is on the unit circle since $x^2 + y^2 = 1$.

(a) For $0 \le t \le \pi$, x {$\cos t$} varies from 1 to -1 and y {$\sin t$} is nonnegative.

P(x, y) moves from (1, 0) counterclockwise to $(-1, 0)$.

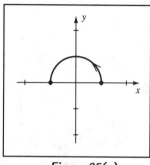 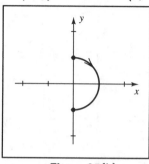 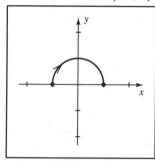

Figure 25(a)   *Figure 25(b)*   *Figure 25(c)*

(b) For $0 \le t \le \pi$, y {$\cos t$} varies from 1 to -1 and x {$\sin t$} is nonnegative.

P(x, y) moves from (0, 1) clockwise to $(0, -1)$.

(c) For $-1 \le t \le 1$, x {t} varies from -1 to 1 and y {$\sqrt{1-t^2}$} is nonnegative.

P(x, y) moves from $(-1, 0)$ clockwise to (1, 0).

27 $x = a \cos t + h$ and $y = b \sin t + k \Rightarrow \dfrac{x - h}{a} = \cos t$ and $\dfrac{y - k}{b} = \sin t \Rightarrow$

$$\frac{(x - h)^2}{a^2} + \frac{(y - k)^2}{b^2} = \cos^2 t + \sin^2 t = 1.$$ This is the equation of an ellipse with

center (h, k) and semiaxes of lengths a and b (axes of lengths $2a$ and $2b$).

29 Some choices for parts (a) and (b) are given—there are an infinite number of choices.

(a) (1) $x = t$, $y = t^2$; $t \in \mathbb{R}$

Letting x equal t is usually the simplest choice we can make.

(2) $x = \tan t$, $y = \tan^2 t$; $-\frac{\pi}{2} < t < \frac{\pi}{2}$

(3) $x = t^3$, $y = t^6$; $t \in \mathbb{R}$

(b) (1) $x = e^t$, $y = e^{2t}$; $t \in \mathbb{R}$

This choice only gives $x > 0$ since e^t is always positive for $t \in \mathbb{R}$.

(2) $x = \sin t$, $y = \sin^2 t$; $t \in \mathbb{R}$

This choice only gives $-1 \le x \le 1$ since $-1 \le \sin t \le 1$ for $t \in \mathbb{R}$.

(3) $x = \tan^{-1} t$, $y = (\tan^{-1} t)^2$; $t \in \mathbb{R}$

This choice only gives $-\frac{\pi}{2} < x < \frac{\pi}{2}$ since $-\frac{\pi}{2} < \tan^{-1} t < \frac{\pi}{2}$ for $t \in \mathbb{R}$.

$\boxed{31}$ (a) $x = a \sin \omega t$ and $y = b \cos \omega t \Rightarrow \frac{x}{a} = \sin \omega t$ and $\frac{y}{b} = \cos \omega t \Rightarrow \frac{x^2}{a^2} + \frac{y^2}{b^2} = 1$.

The figure is an ellipse with center $(0, 0)$ and axes of lengths $2a$ and $2b$.

(b) $f(t + p) = a \sin[\omega_1(t + p)] = a \sin[\omega_1 t + \omega_1 p] = a \sin[\omega_1 t + 2\pi n] =$

$$a \sin \omega_1 t = f(t).$$

$$g(t + p) = b \cos[\omega_2(t + p)] = b \cos\left[\omega_2 t + \frac{\omega_2}{\omega_1} 2\pi n\right] = b \cos\left[\omega_2 t + \frac{m}{n} 2\pi n\right] =$$

$$b \cos[\omega_2 t + 2\pi m] = b \cos \omega_2 t = g(t).$$

Since f and g are periodic with period p,

the curve retraces itself every p units of time.

$\boxed{33}$ (a) Let $x = 3 \sin(240\pi t)$ and $y = 4 \sin(240\pi t)$ for $0 \le t \le 0.01$.

(b) From the graph, $y_{\text{int}} = 0$ and $y_{\text{max}} = 4$.

Thus, the phase difference is $\phi = \sin^{-1} \frac{y_{\text{int}}}{y_{\text{max}}} = \sin^{-1} \frac{0}{4} = 0°$.

$[-9, 9]$ by $[-6, 6]$

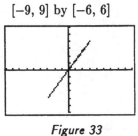

Xscl $= 1$

Yscl $= 1$

Figure 33

$[-120, 120]$ by $[-80, 80]$

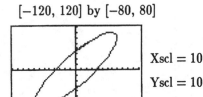

Xscl $= 10$

Yscl $= 10$

Figure 35

$\boxed{35}$ (a) Let $x = 80 \sin(60\pi t)$ and $y = 70 \cos(60\pi t - \pi/3)$ for $0 \le t \le 0.035$.

(b) See *Figure 35*. From the graph, $y_{\text{int}} = 35$ and $y_{\text{max}} = 70$.

Thus, the phase difference is $\phi = \sin^{-1} \frac{y_{\text{int}}}{y_{\text{max}}} = \sin^{-1} \frac{35}{70} = 30°$.

Note: Tstep must be sufficiently small to obtain the maximum of 70 on the

graph. $70 \cos(60\pi t - \pi/3) = 70 \sin(60\pi t + \pi/6)$

$\boxed{37}$ $x(t) = \sin(6\pi t)$, $y(t) = \cos(5\pi t)$ for $0 \le t \le 2$

$[-1, 1]$ by $[-1, 1]$

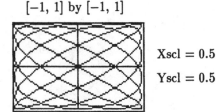

Xscl $= 0.5$

Yscl $= 0.5$

Figure 37

[39] Let $\theta = \angle FDP$ and $\alpha = \angle GDP = \angle EDP$. Then $\angle ODG = \left(\frac{\pi}{2} - t\right)$ and

$\alpha = \theta - \left(\frac{\pi}{2} - t\right) = \theta + t - \frac{\pi}{2}$. Arcs AF and PF are equal in length since each is

the distance rolled. Thus, $at = b\theta$, or $\theta = \left(\frac{a}{b}\right)t$ and $\alpha = \frac{a+b}{b}t - \frac{\pi}{2}$.

Note that $\cos\alpha = \sin\left(\frac{a+b}{b}t\right)$ and $\sin\alpha = -\cos\left(\frac{a+b}{b}t\right)$.

For the location of the points as illustrated, the coordinates of P are:

$x = d(O, G) + d(G, B) = d(O, G) + d(E, P) = (a+b)\cos t + b\sin\alpha$

$$= (a+b)\cos t - b\cos\left(\frac{a+b}{b}t\right).$$

$y = d(B, P) = d(G, D) - d(D, E) = (a+b)\sin t - b\cos\alpha$

$$= (a+b)\sin t - b\sin\left(\frac{a+b}{b}t\right).$$

It can be verified that these equations are valid for all locations of the points.

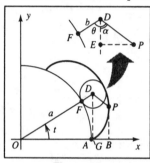

Figure 39

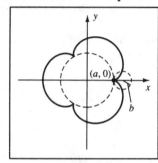

Figure 41

[41] $b = \frac{1}{3}a \Rightarrow a = 3b$. Substituting into the equations from Exercise 39 yields:

$$x = (3b+b)\cos t - b\cos\left(\frac{3b+b}{b}t\right) = 4b\cos t - b\cos 4t$$

$$y = (3b+b)\sin t - b\sin\left(\frac{3b+b}{b}t\right) = 4b\sin t - b\sin 4t$$

As an aid in graphing, to determine where the path of the smaller circle will intersect

the path of the larger circle (for the original starting point of intersection at $A(a, 0)$),

we can solve $x^2 + y^2 = a^2$ for t.

$x^2 + y^2 = 16b^2\cos^2 t - 8b^2\cos t\cos 4t + b^2\cos^2 4t +$

$$16b^2\sin^2 t - 8b^2\sin t\sin 4t + b^2\sin^2 4t$$

$$= 17b^2 - 8b^2(\cos t\cos 4t + \sin t\sin 4t)$$

$$= 17b^2 - 8b^2[\cos(t - 4t)] = 17b^2 - 8b^2\cos 3t.$$

Thus, $x^2 + y^2 = a^2 \Rightarrow 17b^2 - 8b^2\cos 3t = a^2 = 9b^2 \Rightarrow 8b^2 = 8b^2\cos 3t \Rightarrow$

$1 = \cos 3t \Rightarrow 3t = 2\pi n \Rightarrow t = \frac{2\pi}{3}n$.

It follows that the intersection points are at $t = \frac{2\pi}{3}, \frac{4\pi}{3}$, and 2π.

43 Change "Param" from "Function" under ⎡MODE⎤. Make the assignments $3(\sin T)\hat{} 5$ to X_{1T}, $3(\cos T)\hat{} 5$ to Y_{1T}, 0 to Tmin, 6.28 to Tmax, and 0.105 to Tstep. Algebraically, we have $x = 3\sin^5 t$ and $y = 3\cos^5 t \Rightarrow \frac{x}{3} = \sin^5 t$ and $\frac{y}{3} = \cos^5 t \Rightarrow (\frac{x}{3})^{2/5} + (\frac{y}{3})^{2/5} = \sin^2 t + \cos^2 t = 1$. The graph traces an astroid.

[−6, 6] by [−4, 4]

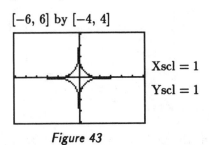

Xscl = 1
Yscl = 1

Figure 43

[−30, 30] by [−20, 20]

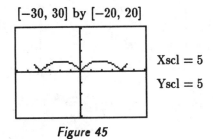

Xscl = 5
Yscl = 5

Figure 45

45 The graph traces a curtate cycloid.

47 The figure is a mask with a mouth, nose, and eyes. This graph may be obtained with a graphing utility that has the capability to graph 5 sets of parametric equations.

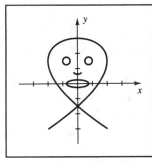

Figure 47

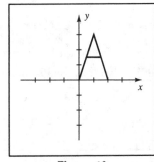

Figure 49

49 C_1 is the line $y = 3x$ from $(0, 0)$ to $(1, 3)$. For C_2, $x − 1 = \tan t$ and $1 − \frac{1}{3}y = \tan t \Rightarrow x − 1 = 1 − \frac{1}{3}y \Rightarrow y = −3x + 6$. This line is sketched from $(1, 3)$ to $(2, 0)$. C_3 is the horizontal line $y = \frac{3}{2}$ from $(\frac{1}{2}, \frac{3}{2})$ to $(\frac{3}{2}, \frac{3}{2})$. The figure is the letter A.

6.5 Exercises

Note: For the following exercises, the substitutions $y = r\sin\theta$, $x = r\cos\theta$, $r^2 = x^2 + y^2$, and $\tan\theta = \frac{y}{x}$ are used without mention. We have found it helpful to find the "pole" values { when the graph intersects the pole } to determine which values of θ should be used in the construction of an r-θ chart. The numbers listed on each line of the r-θ chart correspond to the numbers labeled on the figures.

$\boxed{1}$ $r = 5 \Rightarrow r^2 = 25 \Rightarrow x^2 + y^2 = 25$, a circle centered at the origin with radius 5.

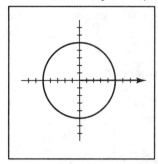

Figure 1

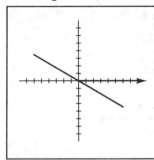

Figure 3

$\boxed{3}$ $\theta = -\frac{\pi}{6} \Rightarrow \tan\theta = \tan\left(-\frac{\pi}{6}\right) \Rightarrow \frac{y}{x} = -\frac{1}{\sqrt{3}} \Rightarrow y = -\frac{1}{3}\sqrt{3}\,x.$

This is a line through the origin with slope $-\frac{1}{3}\sqrt{3}$.

$\boxed{5}$ $r = 3\cos\theta \Rightarrow \{\text{multiply by } r \text{ to obtain } r^2\}$

$r^2 = 3r\,\cos\theta \Rightarrow x^2 + y^2 = 3x \Rightarrow$

$\{\text{recognize this as an equation of a circle and complete the square}\}$

$(x^2 - 3x + \underline{\frac{9}{4}}) + y^2 = \underline{\frac{9}{4}} \Rightarrow (x - \frac{3}{2})^2 + y^2 = \frac{9}{4}.$

This is a circle with center $(\frac{3}{2}, 0)$ and radius $\sqrt{\frac{9}{4}} = \frac{3}{2}$. From the table, we see that as θ varies from 0 to $\frac{\pi}{2}$, r will vary from 3 to 0. This corresponds to the portion of the circle in the first quadrant. As θ varies from $\frac{\pi}{2}$ to π, r varies from 0 to -3. Remember that -3 in the π direction is the same as 3 in the 0 direction. This corresponds to the portion of the circle in the fourth quadrant.

Variation of θ			Variation of r		
1)	0	$\rightarrow$ $\frac{\pi}{2}$	3	$\rightarrow$	0
2)	$\frac{\pi}{2}$	$\rightarrow$ π	0	$\rightarrow$	-3

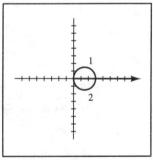

Figure 5

$\boxed{7}$ $r = 4\cos\theta + 2\sin\theta \Rightarrow r^2 = 4r\,\cos\theta + 2r\,\sin\theta \Rightarrow$

$x^2 + y^2 = 4x + 2y \Rightarrow$

$x^2 - 4x + \underline{4} + y^2 - 2y + \underline{1} = \underline{4} + \underline{1} \Rightarrow$

$(x - 2)^2 + (y - 1)^2 = 5.$

Variation of θ			Variation of r		
1)	0	$\rightarrow$ $\frac{\pi}{2}$	4	$\rightarrow$	2
2)	$\frac{\pi}{2}$	$\rightarrow$ π	2	$\rightarrow$	-4

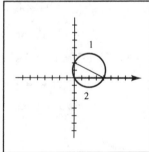

Figure 7

9 $r = 4(1 - \sin\theta)$ is a cardioid since the coefficient of $\sin\theta$ has the same magnitude as the constant term.

$0 = 4(1 - \sin\theta) \Rightarrow \sin\theta = 1 \Rightarrow \theta = \frac{\pi}{2} + 2\pi n$. The "v" in the heart-shaped curve corresponds to the pole value $\frac{\pi}{2}$.

Variation of θ			Variation of r		
1)	0	$\rightarrow$ $\frac{\pi}{2}$	4	$\rightarrow$	0
2)	$\frac{\pi}{2}$	$\rightarrow$ π	0	$\rightarrow$	4
3)	π	$\rightarrow$ $\frac{3\pi}{2}$	4	$\rightarrow$	8
4)	$\frac{3\pi}{2}$	$\rightarrow$ 2π	8	$\rightarrow$	4

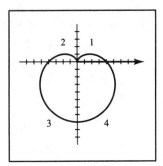

Figure 9

11 $r = -6(1 + \cos\theta)$ is a cardioid.

$0 = -6(1 + \cos\theta) \Rightarrow \cos\theta = -1 \Rightarrow \theta = \pi + 2\pi n$. The "v" in the heart-shaped curve corresponds to the pole value π.

Variation of θ			Variation of r		
1)	0	$\rightarrow$ $\frac{\pi}{2}$	-12	$\rightarrow$	-6
2)	$\frac{\pi}{2}$	$\rightarrow$ π	-6	$\rightarrow$	0
3)	π	$\rightarrow$ $\frac{3\pi}{2}$	0	$\rightarrow$	-6
4)	$\frac{3\pi}{2}$	$\rightarrow$ 2π	-6	$\rightarrow$	-12

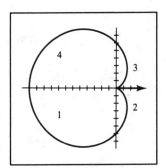

Figure 11

13 $r = 2 + 4\sin\theta$ is a limaçon with a loop since the constant term has a smaller magnitude than the coefficient of $\sin\theta$. $0 = 2 + 4\sin\theta \Rightarrow \sin\theta = -\frac{1}{2} \Rightarrow \theta = \frac{7\pi}{6} + 2\pi n$, $\frac{11\pi}{6} + 2\pi n$. Trace through the table and the figure to make sure you understand what values of θ form the loop.

Variation of θ			Variation of r		
1)	0	$\rightarrow$ $\frac{\pi}{2}$	2	$\rightarrow$	6
2)	$\frac{\pi}{2}$	$\rightarrow$ π	6	$\rightarrow$	2
3)	π	$\rightarrow$ $\frac{7\pi}{6}$	2	$\rightarrow$	0
4)	$\frac{7\pi}{6}$	$\rightarrow$ $\frac{3\pi}{2}$	0	$\rightarrow$	-2
5)	$\frac{3\pi}{2}$	$\rightarrow$ $\frac{11\pi}{6}$	-2	$\rightarrow$	0
6)	$\frac{11\pi}{6}$	$\rightarrow$ 2π	0	$\rightarrow$	2

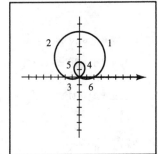

Figure 13

$\boxed{15}$ $r = \sqrt{3} - 2\sin\theta$ is a limaçon with a loop. $0 = \sqrt{3} - 2\sin\theta \Rightarrow \sin\theta = \sqrt{3}/2 \Rightarrow$

$\theta = \frac{\pi}{3} + 2\pi n, \frac{2\pi}{3} + 2\pi n$. Let $a = \sqrt{3} - 2 \approx -0.27$ and $b = \sqrt{3} + 2 \approx 3.73$.

	Variation of θ		Variation of r	
1)	0	$\rightarrow$ $\frac{\pi}{3}$	$\sqrt{3} \rightarrow$	0
2)	$\frac{\pi}{3}$	$\rightarrow$ $\frac{\pi}{2}$	$0 \rightarrow$	a
3)	$\frac{\pi}{2}$	$\rightarrow$ $\frac{2\pi}{3}$	$a \rightarrow$	0
4)	$\frac{2\pi}{3}$	$\rightarrow$ π	$0 \rightarrow$	$\sqrt{3}$
5)	π	$\rightarrow$ $\frac{3\pi}{2}$	$\sqrt{3} \rightarrow$	b
6)	$\frac{3\pi}{2}$	$\rightarrow$ 2π	$b \rightarrow$	$\sqrt{3}$

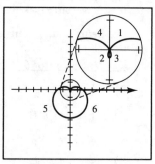

Figure 15

$\boxed{17}$ $r = 2 - \cos\theta$ •

$0 = 2 - \cos\theta \Rightarrow \cos\theta = 2 \Rightarrow$ no pole values.

	Variation of θ		Variation of r	
1)	0	$\rightarrow$ $\frac{\pi}{2}$	$1 \rightarrow$	2
2)	$\frac{\pi}{2}$	$\rightarrow$ π	$2 \rightarrow$	3
3)	π	$\rightarrow$ $\frac{3\pi}{2}$	$3 \rightarrow$	2
4)	$\frac{3\pi}{2}$	$\rightarrow$ 2π	$2 \rightarrow$	1

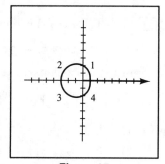

Figure 17

$\boxed{19}$ $r = 4\csc\theta \Rightarrow r\sin\theta = 4 \Rightarrow y = 4$.

r is undefined at $\theta = \pi n$.

This is a horizontal line with y-intercept $(0, 4)$.

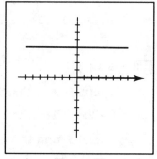

Figure 19

$\boxed{21}$ $r = 8\cos 3\theta$ is a 3-leafed rose since 3 is odd.

$0 = 8\cos 3\theta \Rightarrow \cos 3\theta = 0 \Rightarrow 3\theta = \frac{\pi}{2} + \pi n \Rightarrow \theta = \frac{\pi}{6} + \frac{\pi}{3}n$.

	Variation of θ		Variation of r	
1)	0	$\rightarrow$ $\frac{\pi}{6}$	$8 \rightarrow$	0
2)	$\frac{\pi}{6}$	$\rightarrow$ $\frac{\pi}{3}$	$0 \rightarrow$	-8
3)	$\frac{\pi}{3}$	$\rightarrow$ $\frac{\pi}{2}$	$-8 \rightarrow$	0
4)	$\frac{\pi}{2}$	$\rightarrow$ $\frac{2\pi}{3}$	$0 \rightarrow$	8
5)	$\frac{2\pi}{3}$	$\rightarrow$ $\frac{5\pi}{6}$	$8 \rightarrow$	0
6)	$\frac{5\pi}{6}$	$\rightarrow$ π	$0 \rightarrow$	-8

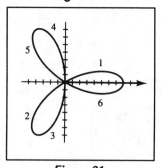

Figure 21

[23] $r = 3\sin 2\theta$ is a 4-leafed rose. $0 = 3\sin 2\theta \Rightarrow \sin 2\theta = 0 \Rightarrow 2\theta = \pi n \Rightarrow \theta = \frac{\pi}{2}n$.

	Variation of θ			Variation of r	
1)	0	$\rightarrow$	$\frac{\pi}{4}$	$0 \rightarrow$	3
2)	$\frac{\pi}{4}$	$\rightarrow$	$\frac{\pi}{2}$	$3 \rightarrow$	0
3)	$\frac{\pi}{2}$	$\rightarrow$	$\frac{3\pi}{4}$	$0 \rightarrow$	-3
4)	$\frac{3\pi}{4}$	$\rightarrow$	π	$-3 \rightarrow$	0
5)	π	$\rightarrow$	$\frac{5\pi}{4}$	$0 \rightarrow$	3
6)	$\frac{5\pi}{4}$	$\rightarrow$	$\frac{3\pi}{2}$	$3 \rightarrow$	0
7)	$\frac{3\pi}{2}$	$\rightarrow$	$\frac{7\pi}{4}$	$0 \rightarrow$	-3
8)	$\frac{7\pi}{4}$	$\rightarrow$	2π	$-3 \rightarrow$	0

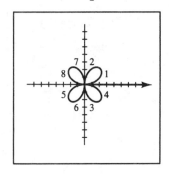

Figure 23

[25] $r^2 = 4\cos 2\theta$ (lemniscate) • $0 = 4\cos 2\theta \Rightarrow \cos 2\theta = 0 \Rightarrow 2\theta = \frac{\pi}{2} + \pi n \Rightarrow \theta = \frac{\pi}{4} + \frac{\pi}{2}n$.
Note that as θ varies from 0 to $\frac{\pi}{4}$, we have $r^2 = 4$ or $r = \pm 2$ to $r^2 = 0$—both parts labeled "1" are traced with this range of θ. When θ varies from $\frac{\pi}{4}$ to $\frac{3\pi}{4}$, 2θ varies from $\frac{\pi}{2}$ to $\frac{3\pi}{2}$, and $\cos 2\theta$ is negative. Since r^2 can't equal a negative value, no portion of the graph is traced for these values of θ.

	Variation of θ			Variation of r	
1)	0	$\rightarrow$	$\frac{\pi}{4}$	$\pm 2 \rightarrow$	0
2)	$\frac{\pi}{4}$	$\rightarrow$	$\frac{\pi}{2}$	undefined	
3)	$\frac{\pi}{2}$	$\rightarrow$	$\frac{3\pi}{4}$	undefined	
4)	$\frac{3\pi}{4}$	$\rightarrow$	π	$0 \rightarrow$	± 2

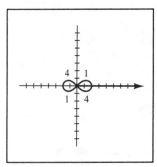

Figure 25

[27] $r = 2^{\theta}$, $\theta \geq 0$ (spiral) •

	Variation of θ			Variation of r		
1)	0	$\rightarrow$	$\frac{\pi}{2}$	1	$\rightarrow$	2.97
2)	$\frac{\pi}{2}$	$\rightarrow$	π	2.97	$\rightarrow$	8.82
3)	π	$\rightarrow$	$\frac{3\pi}{2}$	8.82	$\rightarrow$	26.22
4)	$\frac{3\pi}{2}$	$\rightarrow$	2π	22.62	$\rightarrow$	77.88

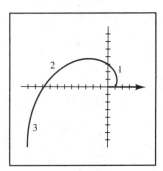

Figure 27

29 $r = 2\theta$, $\theta \geq 0$ •

Variation of θ			Variation of r		
1)	0	$\to$	$\frac{\pi}{2}$	0	$\to$	π
2)	$\frac{\pi}{2}$	$\to$	π	π	$\to$	2π
3)	π	$\to$	$\frac{3\pi}{2}$	2π	$\to$	3π

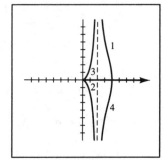

Figure 29

31 To simplify $\sin^2(\frac{\theta}{2})$, recall the half-angle identity for the sine. $r = 6\sin^2\left(\frac{\theta}{2}\right) = 6\left(\frac{1 - \cos\theta}{2}\right) = 3(1 - \cos\theta)$ is a cardioid. $0 = 3(1 - \cos\theta) \Rightarrow \cos\theta = 1 \Rightarrow \theta = 2\pi n$.

Variation of θ			Variation of r	
1)	0	$\to$	$\frac{\pi}{2}$	$0 \to$	3
2)	$\frac{\pi}{2}$	$\to$	π	$3 \to$	6
3)	π	$\to$	$\frac{3\pi}{2}$	$6 \to$	3
4)	$\frac{3\pi}{2}$	$\to$	2π	$3 \to$	0

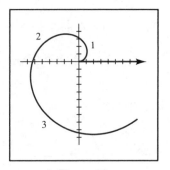

Figure 31

33 Note that $r = 2\sec\theta$ is equivalent to $x = 2$. If $0 < \theta < \frac{\pi}{2}$ or $\frac{3\pi}{2} < \theta < 2\pi$, then $\sec\theta > 0$ and the graph of $r = 2 + 2\sec\theta$ is to the right of $x = 2$. If $\frac{\pi}{2} < \theta < \frac{3\pi}{2}$, $\sec\theta < 0$ and $r = 2 + 2\sec\theta$ is to the left of $x = 2$. r is undefined at $\theta = \frac{\pi}{2} + \pi n$.

$0 = 2 + 2\sec\theta \Rightarrow \sec\theta = -1 \Rightarrow \theta = \pi + 2\pi n.$

Variation of θ			Variation of r	
1)	0	$\to$	$\frac{\pi}{2}$	$4 \to$	∞
2)	$\frac{\pi}{2}$	$\to$	π	$-\infty \to$	0
3)	π	$\to$	$\frac{3\pi}{2}$	$0 \to$	$-\infty$
4)	$\frac{3\pi}{2}$	$\to$	2π	$\infty \to$	4

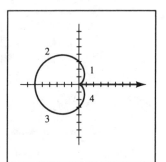

Figure 33

35 Use (1) in the "relationship between rectangular and polar coordinates."

(a) $x = r\cos\theta = 3\cos\frac{\pi}{4} = 3\left(\frac{\sqrt{2}}{2}\right) = \frac{3}{2}\sqrt{2}.$ $y = r\sin\theta = 3\sin\frac{\pi}{4} = 3\left(\frac{\sqrt{2}}{2}\right) = \frac{3}{2}\sqrt{2}.$

Hence, rectangular coordinates for $(3, \frac{\pi}{4})$ are $(\frac{3}{2}\sqrt{2}, \frac{3}{2}\sqrt{2})$.

(b) $x = -1\cos\frac{2\pi}{3} = -1(-\frac{1}{2}) = \frac{1}{2}.$ $y = -1\sin\frac{2\pi}{3} = -1\left(\frac{\sqrt{3}}{2}\right) = -\frac{1}{2}\sqrt{3}.$

37 (a) $x = 8\cos\left(-\frac{2\pi}{3}\right) = 8(-\frac{1}{2}) = -4.$ $y = 8\sin\left(-\frac{2\pi}{3}\right) = 8\left(-\frac{\sqrt{3}}{2}\right) = -4\sqrt{3}.$

(b) $x = -3\cos\frac{5\pi}{3} = -3(\frac{1}{2}) = -\frac{3}{2}.$ $y = -3\sin\frac{5\pi}{3} = -3\left(-\frac{\sqrt{3}}{2}\right) = \frac{3}{2}\sqrt{3}.$

39 Let $\theta = \arctan\frac{3}{4}.$ Then we have $\cos\theta = \frac{4}{5}$ and $\sin\theta = \frac{3}{5}.$

$$x = 6\cos\theta = 6(\frac{4}{5}) = \frac{24}{5}.\quad y = 6\sin\theta = 6(\frac{3}{5}) = \frac{18}{5}.$$

41 Use (2) in the "relationship between rectangular and polar coordinates."

(a) $r^2 = x^2 + y^2 = (-1)^2 + (1)^2 = 2 \Rightarrow r = \sqrt{2}$ { since we want $r > 0$ }.

$$\tan\theta = \frac{y}{x} = \frac{1}{-1} = -1 \Rightarrow \theta = \frac{3\pi}{4}\ \{\theta\text{ in QII}\}.$$

(b) $r^2 = (-2\sqrt{3})^2 + (-2)^2 = 16 \Rightarrow r = \sqrt{16} = 4.$

$$\tan\theta = \frac{-2}{-2\sqrt{3}} = \frac{1}{\sqrt{3}} \Rightarrow \theta = \frac{7\pi}{6}\ \{\theta\text{ in QIII}\}.$$

43 (a) $r^2 = 7^2 + (-7\sqrt{3})^2 = 196 \Rightarrow r = \sqrt{196} = 14.$

$$\tan\theta = \frac{-7\sqrt{3}}{7} = -\sqrt{3} \Rightarrow \theta = \frac{5\pi}{3}\ \{\theta\text{ in QIV}\}.$$

(b) $r^2 = 5^2 + 5^2 = 50 \Rightarrow r = \sqrt{50} = 5\sqrt{2}.$ $\tan\theta = \frac{5}{5} = 1 \Rightarrow \theta = \frac{\pi}{4}\ \{\theta\text{ in QI}\}.$

45 (a) Since $\frac{7\pi}{3}$ is coterminal with $\frac{\pi}{3}$, $(3, \frac{7\pi}{3})$ represents the same point as $(3, \frac{\pi}{3})$.

(b) The point $(3, -\frac{\pi}{3})$ is in QIV, not QI, as is $(3, \frac{\pi}{3})$.

(c) The angle $\frac{4\pi}{3}$ is π radians larger than $\frac{\pi}{3}$, so its terminal side is on the same line as the terminal side of the angle $\frac{\pi}{3}$. $r = -3$ in the $\frac{4\pi}{3}$ direction is equivalent to $r = 3$ in the $\frac{\pi}{3}$ direction, so $(-3, \frac{4\pi}{3})$ represents the same point as $(3, \frac{\pi}{3})$.

(d) The point $(3, -\frac{2\pi}{3})$ is diametrically opposite the point $(3, \frac{\pi}{3})$.

(e) From part (d), we deduce that the point $(-3, -\frac{2\pi}{3})$ would represent the same point as $(3, \frac{\pi}{3})$.

(f) The point $(-3, -\frac{\pi}{3})$ is in QII, not QI.

Thus, choices (a), (c), and (e) represent the same point as $(3, \pi/3)$.

47 $x = -3 \Rightarrow r\cos\theta = -3 \Rightarrow r = \frac{-3}{\cos\theta} \Rightarrow r = -3\sec\theta$

49 $x^2 + y^2 = 16 \Rightarrow r^2 = 16 \Rightarrow r = \pm 4$ { both are circles with radius 4 }.

51 $2y = -x \Rightarrow \frac{y}{x} = -\frac{1}{2} \Rightarrow \tan\theta = -\frac{1}{2} \Rightarrow \theta = \tan^{-1}\left(-\frac{1}{2}\right)$

53 $y^2 - x^2 = 4 \Rightarrow r^2\sin^2\theta - r^2\cos^2\theta = 4 \Rightarrow -r^2(\cos^2\theta - \sin^2\theta) = 4 \Rightarrow$

$$-r^2\cos 2\theta = 4 \Rightarrow r^2 = \frac{-4}{\cos 2\theta} \Rightarrow r^2 = -4\sec 2\theta$$

55 $(x-1)^2 + y^2 = 1 \Rightarrow x^2 - 2x + 1 + y^2 = 1 \Rightarrow x^2 + y^2 = 2x \Rightarrow$

$$r^2 = 2r\cos\theta \Rightarrow r = 2\cos\theta$$

$\boxed{57}$ $r\cos\theta = 5 \Rightarrow x = 5$. This is a vertical line with x-intercept $(5, 0)$.

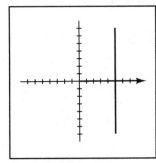

Figure 57

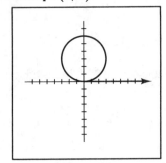

Figure 59

$\boxed{59}$ $r - 6\sin\theta = 0 \Rightarrow r^2 = 6r\sin\theta \Rightarrow x^2 + y^2 = 6y \Rightarrow x^2 + y^2 - 6y + \underline{9} = \underline{9} \Rightarrow$

$$x^2 + (y-3)^2 = 9.$$

$\boxed{61}$ $\theta = \frac{\pi}{4} \Rightarrow \tan\theta = \tan\frac{\pi}{4} \Rightarrow \frac{y}{x} = 1 \Rightarrow y = x$.

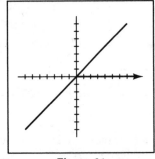

Figure 61

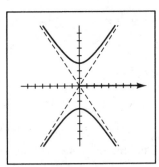

Figure 63

$\boxed{63}$ $r^2(4\sin^2\theta - 9\cos^2\theta) = 36 \Rightarrow 4r^2\sin^2\theta - 9r^2\cos^2\theta = 36 \Rightarrow$

$$4y^2 - 9x^2 = 36 \Rightarrow \frac{y^2}{9} - \frac{x^2}{4} = 1.$$

$\boxed{65}$ $r^2\cos 2\theta = 1 \Rightarrow r^2(\cos^2\theta - \sin^2\theta) = 1 \Rightarrow r^2\cos^2\theta - r^2\sin^2\theta = 1 \Rightarrow x^2 - y^2 = 1$.

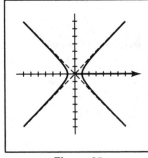

Figure 65

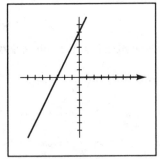

Figure 67

$\boxed{67}$ $r(\sin\theta - 2\cos\theta) = 6 \Rightarrow r\sin\theta - 2r\cos\theta = 6 \Rightarrow y - 2x = 6$.

$\boxed{69}$ $r(\sin\theta + r\cos^2\theta) = 1 \Rightarrow r\sin\theta + r^2\cos^2\theta = 1 \Rightarrow y + x^2 = 1$, or $y = -x^2 + 1$.

See *Figure 69*.

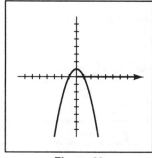

Figure 69

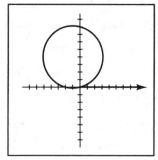

Figure 71

$\boxed{71}$ $r = 8\sin\theta - 2\cos\theta \Rightarrow r^2 = 8r\sin\theta - 2r\cos\theta \Rightarrow x^2 + y^2 = 8y - 2x \Rightarrow$

$$x^2 + 2x + \underline{\ 1\ } + y^2 - 8y + \underline{\ 16\ } = \underline{\ 1\ } + \underline{\ 16\ } \Rightarrow (x+1)^2 + (y-4)^2 = 17.$$

$\boxed{73}$ $r = \tan\theta \Rightarrow r^2 = \tan^2\theta \Rightarrow x^2 + y^2 = \dfrac{y^2}{x^2} \Rightarrow x^4 + x^2y^2 = y^2 \Rightarrow y^2 - x^2y^2 = x^4 \Rightarrow$

$y^2(1 - x^2) = x^4 \Rightarrow y^2 = \dfrac{x^4}{1 - x^2}.$ This is a tough one even after getting the equation in

x and y. Since we have solved for y^2, the right side must be positive. Since x^4 is

always nonnegative, we must have $1 - x^2 > 0$, or equivalently, $|x| < 1$. We have

vertical asymptotes at $x = \pm 1$, i.e., when the denominator is 0.

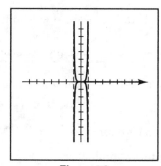

Figure 73

$[-9, 9]$ by $[-6, 6]$

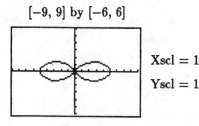

Xscl = 1

Yscl = 1

Figure 77

$\boxed{77}$ (a) $I = \frac{1}{2}I_0[1 + \cos(\pi\sin\theta)] \Rightarrow r = 2.5[1 + \cos(\pi\sin\theta)]$ for $\theta \in [0, 2\pi]$.

(b) The signal is maximum in an east–west direction and minimum in a north–south

direction.

$\boxed{79}$ Change to "Pol" mode under $\boxed{\text{MODE}}$, assign $2(\sin\theta)^2(\tan\theta)^2$ to r1 under $\boxed{\text{Y} =}$,

and $-\pi/3$ to θmin, $\pi/3$ to θmax, and 0.04 to θstep under $\boxed{\text{WINDOW}}$. The graph is

symmetric with respect to the polar axis.

$[-9, 9]$ by $[-6, 6]$

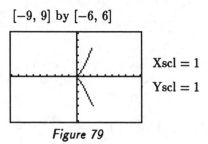

Xscl = 1

Yscl = 1

Figure 79

$[-12, 12]$ by $[-9, 9]$

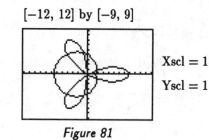

Xscl = 1

Yscl = 1

Figure 81

[81] Assign $8\cos(3\theta)$ to r1, $4 - 2.5\cos\theta$ to r2, 0 to θmin, 2π to θmax, and $\pi/30$ to θstep. From the graph, there are six points of intersection. The approximate polar coordinates are $(1.75, \pm 0.45)$, $(4.49, \pm 1.77)$, and $(5.76, \pm 2.35)$. See *Figure 81* on the preceding page.

6.6 Exercises

Note: For the ellipse, the major axis is vertical if the denominator contains $\sin\theta$, horizontal if the denominator contains $\cos\theta$. For the hyperbola, the transverse axis is vertical if the denominator contains $\sin\theta$, horizontal if the denominator contains $\cos\theta$. The focus at the pole is called F and V is the vertex associated with (or closest to) F. $d(V, F)$ denotes the distance from the vertex to the focus. The foci are not asked for in the directions, but are listed. For the parabola, the directrix is on the right, on the left, above, or below the focus depending on the term "+cos", "−cos", "+sin", or "−sin", respectively, appearing in the denominator.

[1] Divide the numerator and denominator by the constant term in the denominator, i.e.,

6. $r = \dfrac{12}{6 + 2\sin\theta} = \dfrac{2}{1 + \frac{1}{3}\sin\theta} \Rightarrow e = \frac{1}{3} < 1$, ellipse. From the previous note, we see that the denominator has "+sin θ" and we have vertices when $\theta = \frac{\pi}{2}$ and $\frac{3\pi}{2}$. They are $V(\frac{3}{2}, \frac{\pi}{2})$ and $V'(3, \frac{3\pi}{2})$. The distance from the focus at the pole to the vertex V is $\frac{3}{2}$. The distance from V' to F' must also be $\frac{3}{2}$ and we see that $F' = (\frac{3}{2}, \frac{3\pi}{2})$. We will use the following notation to summarize this in future problems:

$$d(V, F) = \tfrac{3}{2} \Rightarrow F' = (\tfrac{3}{2}, \tfrac{3\pi}{2}).$$

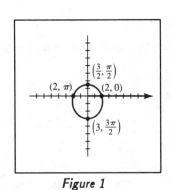

Figure 1

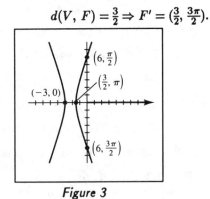

Figure 3

[3] Divide both the numerator and the denominator by 2 to obtain the "1" in the standard form. $r = \dfrac{12}{2 - 6\cos\theta} = \dfrac{6}{1 - 3\cos\theta} \Rightarrow e = 3 > 1$, hyperbola.

$$V(\tfrac{3}{2}, \pi) \text{ and } V'(-3, 0). \quad d(V, F) = \tfrac{3}{2} \Rightarrow F' = (-\tfrac{9}{2}, 0).$$

$\boxed{5}$ $r = \dfrac{3}{2 + 2\cos\theta} = \dfrac{\frac{3}{2}}{1 + 1\cos\theta} \Rightarrow e = 1$, parabola. Note that the expression is

undefined in the $\theta = \pi$ direction. The vertex is in the $\theta = 0$ direction, $V(\frac{3}{4},\ 0)$.

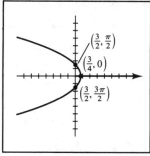

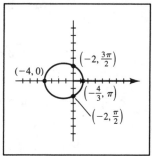

Figure 5   Figure 7

$\boxed{7}$ $r = \dfrac{4}{\cos\theta - 2} = \dfrac{-2}{1 - \frac{1}{2}\cos\theta} \Rightarrow e = \frac{1}{2} < 1$, ellipse.

$$V(-\tfrac{4}{3},\ \pi) \text{ and } V'(-4,\ 0).\quad d(V,\ F) = \tfrac{4}{3} \Rightarrow F' = (-\tfrac{8}{3},\ 0)$$

$\boxed{9}$ We multiply by $\dfrac{\sin\theta}{\sin\theta}$ to obtain the standard form of a conic.

$$r = \dfrac{6\csc\theta}{2\csc\theta + 3} \cdot \dfrac{\sin\theta}{\sin\theta} = \dfrac{6}{2 + 3\sin\theta} = \dfrac{3}{1 + \frac{3}{2}\sin\theta} \Rightarrow e = \tfrac{3}{2} > 1, \text{ hyperbola.}$$

$V(\frac{6}{5},\ \frac{\pi}{2})$ and $V'(-6,\ \frac{3\pi}{2})$. $d(V,\ F) = \frac{6}{5} \Rightarrow F' = (-\frac{36}{5},\ \frac{3\pi}{2})$.

Since the original equation is undefined when $\csc\theta$ is undefined { which is

when $\theta = \pi n$ }, the points $(3,\ 0)$ and $(3,\ \pi)$ are excluded from the graph.

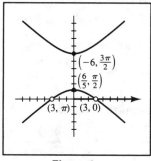

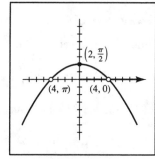

Figure 9   Figure 11

$\boxed{11}$ $r = \dfrac{4\csc\theta}{1 + \csc\theta} \cdot \dfrac{\sin\theta}{\sin\theta} = \dfrac{4}{1 + 1\sin\theta} \Rightarrow e = 1$, parabola. The vertex is in the $\theta = \frac{\pi}{2}$

direction, $V(2,\ \frac{\pi}{2})$. Since the original equation is undefined when $\csc\theta$ is undefined,

the points $(4,\ 0)$ and $(4,\ \pi)$ are excluded from the graph.

$\boxed{13}$ $r = \dfrac{12}{6 + 2\sin\theta} \Rightarrow 6r + 2r\sin\theta = 12 \Rightarrow 6r + 2y = 12 \Rightarrow$

{ Isolate the term with the "r" so that by squaring both sides of the equation,

we can make the substitution $r^2 = x^2 + y^2$. }

$$3r = 6 - y \Rightarrow 9r^2 = 36 - 12y + y^2 \Rightarrow 9(x^2 + y^2) = 36 - 12y + y^2 \Rightarrow$$

$$9x^2 + 8y^2 + 12y - 36 = 0$$

Note: For the following exercises, the substitutions

$$x = r\cos\theta, \ y = r\sin\theta, \text{ and } r^2 = x^2 + y^2 \text{ are made without mention.}$$

$\boxed{15}$ $r = \dfrac{12}{2 - 6\cos\theta} \Rightarrow 2r - 6x = 12 \Rightarrow r = 3x + 6 \Rightarrow r^2 = 9x^2 + 36x + 36 \Rightarrow$

$$8x^2 - y^2 + 36x + 36 = 0$$

$\boxed{17}$ $r = \dfrac{3}{2 + 2\cos\theta} \Rightarrow 2r + 2x = 3 \Rightarrow 2r = 3 - 2x \Rightarrow 4r^2 = 4x^2 - 12x + 9 \Rightarrow$

$$4y^2 + 12x - 9 = 0$$

$\boxed{19}$ $r = \dfrac{4}{\cos\theta - 2} \Rightarrow x - 2r = 4 \Rightarrow x - 4 = 2r \Rightarrow x^2 - 8x + 16 = 4r^2 \Rightarrow$

$$3x^2 + 4y^2 + 8x - 16 = 0$$

$\boxed{21}$ $r = \dfrac{6\csc\theta}{2\csc\theta + 3} \cdot \dfrac{\sin\theta}{\sin\theta} = \dfrac{6}{2 + 3\sin\theta} \Rightarrow 2r + 3y = 6 \Rightarrow 2r = 6 - 3y \Rightarrow$

$4r^2 = 36 - 36y + 9y^2 \Rightarrow 4x^2 - 5y^2 + 36y - 36 = 0.$

r is undefined when $\theta = 0$ or π. For the rectangular equation, these points

correspond to $y = 0$ (or $r\sin\theta = 0$). Substituting $y = 0$ into the above rectangular

equation yields $4x^2 = 36$, or $x = \pm 3$. $\therefore$ exclude $(\pm 3, 0)$

$\boxed{23}$ $r = \dfrac{4\csc\theta}{1 + \csc\theta} \cdot \dfrac{\sin\theta}{\sin\theta} = \dfrac{4}{1 + 1\sin\theta} \Rightarrow r + y = 4 \Rightarrow r = 4 - y \Rightarrow$

$r^2 = y^2 - 8y + 16 \Rightarrow x^2 + 8y - 16 = 0.$ r is undefined when $\theta = 0$ or π.

For the rectangular equation, these points correspond to $y = 0$ (or $r\sin\theta = 0$).

Substituting $y = 0$ into the above rectangular equation yields $x^2 = 16$, or $x = \pm 4$.

$\therefore$ exclude $(\pm 4, 0)$

$\boxed{25}$ $r = 2\sec\theta \Rightarrow r\cos\theta = 2 \Rightarrow x = 2.$ Remember that d is the distance from the focus at

the pole to the directrix. Thus, $d = 2$ and since the directrix is on the

right of the focus at the pole, we use "$+\cos\theta$". $r = \dfrac{2(\frac{1}{3})}{1 + \frac{1}{3}\cos\theta} \cdot \dfrac{3}{3} = \dfrac{2}{3 + \cos\theta}.$

$\boxed{27}$ $r\cos\theta = -3 \Rightarrow x = -3.$ Thus, $d = 3$ and since the directrix is on the left of the

focus at the pole, we use "$-\cos\theta$". $r = \dfrac{3(\frac{4}{3})}{1 - \frac{4}{3}\cos\theta} \cdot \dfrac{3}{3} = \dfrac{12}{3 - 4\cos\theta}.$

$\boxed{29}$ $r\sin\theta = -2 \Rightarrow y = -2.$ Thus, $d = 2$ and since the directrix is below the focus at

the pole, we use "$-\sin\theta$". $r = \dfrac{2(1)}{1 - 1\sin\theta} = \dfrac{2}{1 - \sin\theta}.$

$\boxed{31}$ $r = 4\csc\theta \Rightarrow r\sin\theta = 4 \Rightarrow y = 4.$ Thus, $d = 4$ and since the directrix is above

the focus at the pole, we use "$+\sin\theta$". $r = \dfrac{4(\frac{2}{5})}{1 + \frac{2}{5}\sin\theta} \cdot \dfrac{5}{5} = \dfrac{8}{5 + 2\sin\theta}.$

$\boxed{33}$ For a parabola, $e = 1$. The vertex is 4 units above the focus at the pole,

so $d = 2(4)$ and we should use "$+\sin\theta$" in the denominator. $r = \dfrac{8}{1 + \sin\theta}$

35 (a) See *Figure 35.* $e = \frac{c}{a} = \frac{d(C, F)}{d(C, V)} = \frac{3}{4}$.

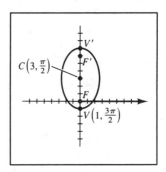

(b) Since the vertex is below the focus at the pole,

use "$-\sin\theta$". $r = \dfrac{d(\frac{3}{4})}{1 - \frac{3}{4}\sin\theta}$ and $r = 1$ when

$\theta = \frac{3\pi}{2} \Rightarrow 1 = \dfrac{d(\frac{3}{4})}{1 - \frac{3}{4}(-1)} \Rightarrow 1 = \dfrac{\frac{3}{4}d}{\frac{7}{4}} \Rightarrow$

$d = \frac{7}{3}$. Thus, $r = \dfrac{(\frac{7}{3})(\frac{3}{4})}{1 - \frac{3}{4}\sin\theta} \cdot \dfrac{4}{4} = \dfrac{7}{4 - 3\sin\theta}$.

Figure 35

An equivalent rectangular equation is $\dfrac{x^2}{7} + \dfrac{(y - 3)^2}{16} = 1$.

37 (a) Let V and C denote the vertex closest to the sun and the center of the ellipse,

respectively. Let s denote the distance from V to the directrix to the left of V.

$d(O, V) = d(C, V) - d(C, O) = a - c = a - ea = a(1 - e)$.

Also, by the first theorem in §6.6, $\dfrac{d(O, V)}{s} = e \Rightarrow s = \dfrac{d(O, V)}{e} = \dfrac{a(1 - e)}{e}$.

Now, $d = s + d(O, V) = \dfrac{a(1 - e)}{e} + a(1 - e) = \dfrac{a(1 - e^2)}{e}$ and $de = a(1 - e^2)$.

Thus, the equation of the orbit is $r = \dfrac{(1 - e^2)a}{1 - e\cos\theta}$.

(b) The minimum distance occurs when $\theta = \pi$. $r_{\text{per}} = \dfrac{(1 - e^2)a}{1 - e(-1)} = a(1 - e)$.

The maximum distance occurs when $\theta = 0$. $r_{\text{aph}} = \dfrac{(1 - e^2)a}{1 - e(1)} = a(1 + e)$.

39 (a) Since $e = 0.9673 < 1$, the orbit of Halley's Comet is elliptical.

(b) The polar equation for the orbit of Saturn is $r = \dfrac{9.006(1 + 0.056)}{1 - 0.056\cos\theta}$.

The polar equation for Halley's comet is $r = \dfrac{0.5871(1 + 0.9673)}{1 - 0.9673\cos\theta}$.

[−36, 36] by [−24, 24]

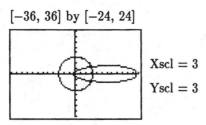

Xscl = 3

Yscl = 3

[−18, 18] by [−12, 12]

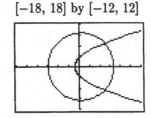

Xscl = 3

Yscl = 3

Figure 39

Figure 41

41 (a) Since $e = 1.003 > 1$, the orbit of Comet 1959 III is hyperbolic.

(b) The polar equation for Comet 1959 III is $r = \dfrac{1.251(1 + 1.003)}{1 - 1.003\cos\theta}$.

Note: Let the notation be the same as in §6.1–6.3.

$\boxed{1}$ $y^2 = 64x \Rightarrow x = \frac{1}{64}y^2 \Rightarrow a = \frac{1}{64}.$ $p = \frac{1}{4(\frac{1}{64})} = 16.$ $V(0, 0)$; $F(16, 0)$; $l: x = -16$

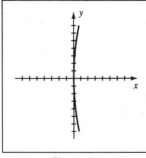

Figure 1

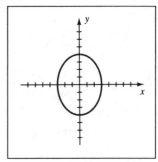

Figure 3

$\boxed{3}$ $9y^2 = 144 - 16x^2 \Rightarrow \frac{x^2}{9} + \frac{y^2}{16} = 1.$ $c^2 = 16 - 9 \Rightarrow c = \pm\sqrt{7}.$

$$V(0, \pm 4); \ F(0, \pm\sqrt{7}); \ M(\pm 3, 0)$$

$\boxed{5}$ $x^2 - y^2 - 4 = 0 \Rightarrow \frac{x^2}{4} - \frac{y^2}{4} = 1.$ $c^2 = 4 + 4 \Rightarrow c = \pm 2\sqrt{2}.$

$$V(\pm 2, 0); \ F(\pm 2\sqrt{2}, 0); \ W(0, \pm 2); \ y = \pm x$$

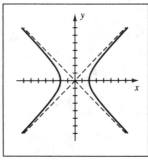

Figure 5

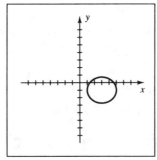

Figure 8

$\boxed{8}$ $3x^2 + 4y^2 - 18x + 8y + 19 = 0 \Rightarrow$

$3(x^2 - 6x + \underline{\ 9\ }) + 4(y^2 + 2y + \underline{\ 1\ }) = -19 + \underline{\ 27\ } + \underline{\ 4\ } \Rightarrow$

$3(x - 3)^2 + 4(y + 1)^2 = 12 \Rightarrow \frac{(x - 3)^2}{4} + \frac{(y + 1)^2}{3} = 1.$ $c^2 = 4 - 3 \Rightarrow c = \pm 1.$

$$C(3, -1); \ V(3 \pm 2, -1); \ F(3 \pm 1, -1); \ M(3, -1 \pm \sqrt{3})$$

$\boxed{12}$ $4x^2 - y^2 - 40x - 8y + 88 = 0 \Rightarrow$

$4(x^2 - 10x + \underline{\;25\;}) - (y^2 + 8y + \underline{\;16\;}) = -88 + \underline{\;100\;} - \underline{\;16\;} \Rightarrow$

$4(x-5)^2 - (y+4)^2 = -4 \Rightarrow \dfrac{(y+4)^2}{4} - \dfrac{(x-5)^2}{1} = 1.$ $c^2 = 4 + 1 \Rightarrow c = \pm\sqrt{5}.$

$C(5, -4);\ V(5, -4 \pm 2);\ F(5, -4 \pm \sqrt{5});\ W(5 \pm 1, -4);\ (y+4) = \pm 2(x-5)$

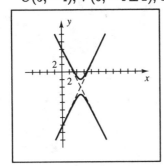 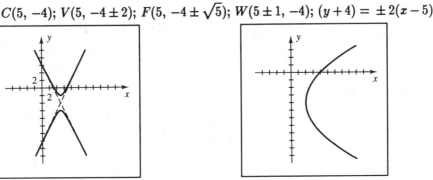

Figure 12 *Figure 13*

$\boxed{13}$ $y^2 - 8x + 8y + 32 = 0 \Rightarrow x = \frac{1}{8}y^2 + y + 4 \Rightarrow a = \frac{1}{8}.$ $p = \dfrac{1}{4(\frac{1}{8})} = 2.$

$V(2, -4);\ F(4, -4);\ l{:}\,x = 0$

$\boxed{17}$ The vertex is halfway between the x-intercepts, so it is of the form $V(-7,\,k)$.

$y = a(x + 10)(x + 4)$ and $x = 0,\ y = 80 \Rightarrow 80 = a(10)(4) \Rightarrow a = 2.$

$x = -7 \Rightarrow y = 2(3)(-3) = -18.$ Hence, $y = 2(x + 7)^2 - 18.$

$\boxed{18}$ The vertex is $V(-4,\,k)$. $y = a(x + 11)(x - 3)$ and $x = 2,\ y = 39 \Rightarrow$

$39 = a(13)(-1) \Rightarrow a = -3.$ $x = -4 \Rightarrow y = -3(7)(-7) = 147.$

Hence, $y = -3(x + 4)^2 + 147.$

$\boxed{20}$ $F(-4, 0)$ and $l{:}\,x = 4 \Rightarrow p = -4$ and $V(0, 0)$.

An equation is $(y - 0)^2 = \big[4(-4)\big](x - 0)$, or $y^2 = -16x.$

$\boxed{22}$ The general equation of a parabola that is symmetric to the x-axis and has its vertex

at the origin is $x = ay^2$. Substituting $x = 5$ and $y = -1$ into that equation yields

$a = 5$. An equation is $x = 5y^2.$

$\boxed{24}$ $F(\pm 10, 0)$ and $V(\pm 5, 0) \Rightarrow b^2 = 10^2 - 5^2 = 75.$

An equation is $\dfrac{x^2}{5^2} - \dfrac{y^2}{75} = 1$ or $\dfrac{x^2}{25} - \dfrac{y^2}{75} = 1.$

$\boxed{26}$ $F(\pm 2, 0) \Rightarrow c^2 = 4.$ Now $\dfrac{x^2}{a^2} + \dfrac{y^2}{b^2} = 1$ can be written as $\dfrac{x^2}{a^2} + \dfrac{y^2}{a^2 - 4} = 1$ since

$b^2 = a^2 - c^2$. Substituting $x = 2$ and $y = \sqrt{2}$ into that equation yields

$\dfrac{4}{a^2} + \dfrac{2}{a^2 - 4} = 1 \Rightarrow 4a^2 - 16 + 2a^2 = a^4 - 4a^2 \Rightarrow a^4 - 10a^2 + 16 = 0 \Rightarrow$

$(a^2 - 2)(a^2 - 8) = 0 \Rightarrow a^2 = 2, 8.$ Since $a > c$, a^2 must be 8 and b^2 is equal to 4.

An equation is $\dfrac{x^2}{8} + \dfrac{y^2}{4} = 1.$

$\boxed{27}$ $M(\pm 5, 0) \Rightarrow b = 5.$ $e = \dfrac{c}{a} = \dfrac{\sqrt{a^2 - b^2}}{a} = \dfrac{\sqrt{a^2 - 25}}{a} = \dfrac{2}{3} \Rightarrow \dfrac{2}{3}a = \sqrt{a^2 - 25} \Rightarrow$

$\dfrac{4}{9}a^2 = a^2 - 25 \Rightarrow \dfrac{5}{9}a^2 = 25 \Rightarrow a^2 = 45.$ An equation is $\dfrac{x^2}{25} + \dfrac{y^2}{45} = 1.$

$\boxed{30}$ The vertex of the square in the first quadrant has coordinates $(x, x).$

Since it is on the ellipse, $\dfrac{x^2}{a^2} + \dfrac{y^2}{b^2} = 1 \Rightarrow b^2 x^2 + a^2 x^2 = a^2 b^2 \Rightarrow x^2 = \dfrac{a^2 b^2}{a^2 + b^2}.$

x^2 is $\frac{1}{4}$ of the area of the square, hence $A = \dfrac{4a^2 b^2}{a^2 + b^2}.$

$\boxed{31}$ The focus is a distance of $p = 1/(4a) = 1/(4 \cdot \frac{1}{8}) = 2$ units from the origin.

An equation of the circle is $x^2 + (y - 2)^2 = 2^2 = 4.$

$\boxed{32}$ $y = \dfrac{1}{64}\omega^2 x^2 + k \Rightarrow x^2 = \dfrac{64}{\omega^2}(y - k) \Rightarrow 4p = \dfrac{64}{\omega^2} \Rightarrow p = \dfrac{16}{\omega^2} = 2 \Rightarrow$

$\omega = 2\sqrt{2}$ rad/sec ≈ 0.45 rev/sec.

$\boxed{33}$ $y = t - 1 \Rightarrow t = y + 1.$ $x = 3 + 4t = 3 + 4(y + 1) = 4y + 7.$

As t varies from -2 to 2, (x, y) varies from $(-5, -3)$ to $(11, 1).$

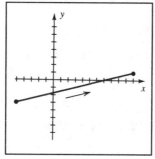

Figure 33

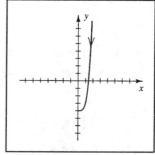

Figure 34

$\boxed{34}$ $x = \sqrt{-t} \Rightarrow t = -x^2.$ $y = t^2 - 4 = x^4 - 4.$ As t varies from $-\infty$ to 0,

x varies from ∞ to 0 and the graph is the right half of the quartic.

$\boxed{35}$ $x = \cos^2 t - 2 \Rightarrow x + 2 = \cos^2 t;$ $y = \sin t + 1 \Rightarrow (y - 1)^2 = \sin^2 t.$

$\sin^2 t + \cos^2 t = 1 = x + 2 + (y - 1)^2 \Rightarrow (y - 1)^2 = -(x + 1).$ This is a parabola with

vertex at $(-1, 1)$ and opening to the left. $t = 0$ corresponds to the vertex and as t

varies from 0 to 2π, the point (x, y) moves to $(-2, 2)$ at $t = \frac{\pi}{2}$, back to the vertex at

$t = \pi$, down to $(-2, 0)$ at $t = \frac{3\pi}{2}$, and finishes at the vertex at $t = 2\pi.$

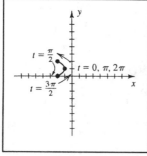

Figure 35

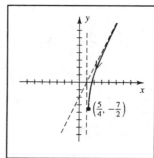

Figure 37

[37] $x = \frac{1}{t} + 1 \Rightarrow x - 1 = \frac{1}{t} \Rightarrow t = \frac{1}{x-1}$ and

$$y = \frac{2}{t} - t = 2(x-1) - \left(\frac{1}{x-1}\right) = \frac{2(x^2 - 2x + 1) - 1}{x-1} = \frac{2x^2 - 4x + 1}{x-1}.$$

This is a rational function with a vertical asymptote at $x = 1$ and an oblique asymptote of $y = 2x - 2$. The graph has a minimum point at $\left(\frac{5}{4}, -\frac{7}{2}\right)$ when $t = 4$ and then approaches the oblique asymptote as t approaches 0. See *Figure 37* on the preceding page.

[38] All of the curves are a portion of the circle $x^2 + y^2 = 16$.

C_1: $y = \sqrt{16 - t^2} = \sqrt{16 - x^2}$.

Since $y = \sqrt{16 - t^2}$, y must be nonnegative and we have the top half of the circle.

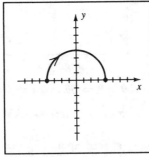

Figure 38 (C_1)

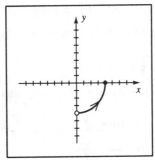

Figure 38 (C_2)

C_2: $x = -\sqrt{16 - t} = -\sqrt{16 - (-\sqrt{t})^2} = -\sqrt{16 - y^2}$.

This is the left half of the circle. Since $y = -\sqrt{t}$, y can only be nonpositive.

Hence we have only the third quadrant portion of the circle.

C_3: $x = 4\cos t$, $y = 4\sin t \Rightarrow \frac{x}{4} = \cos t, \frac{y}{4} = \sin t \Rightarrow \frac{x^2}{16} = \cos^2 t, \frac{y^2}{16} = \sin^2 t \Rightarrow$

$\frac{x^2}{16} + \frac{y^2}{16} = \cos^2 t + \sin^2 t = 1 \Rightarrow x^2 + y^2 = 16$. This is the entire circle.

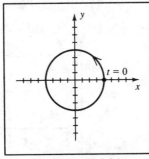

Figure 38 (C_3)

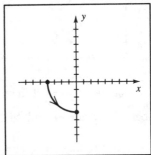

Figure 38 (C_4)

C_4: $y = -\sqrt{16 - e^{2t}} = -\sqrt{16 - (e^t)^2} = -\sqrt{16 - x^2}$.

This is the bottom half of the circle. Since e^t is positive, x takes on all positive real values. Note that $(0, -4)$ is *not* included on the graph since $x \neq 0$.

39 $x = r\cos\theta = 5\cos\frac{7\pi}{4} = 5\left(\frac{\sqrt{2}}{2}\right) = \frac{5}{2}\sqrt{2}$. $y = r\sin\theta = 5\sin\frac{7\pi}{4} = 5\left(-\frac{\sqrt{2}}{2}\right) = -\frac{5}{2}\sqrt{2}$.

40 $r^2 = x^2 + y^2 = (2\sqrt{3})^2 + (-2)^2 = 16 \Rightarrow r = 4$.

$$\tan\theta = \frac{y}{x} = \frac{-2}{2\sqrt{3}} = -\frac{1}{\sqrt{3}} \Rightarrow \theta = \frac{11\pi}{6} \{\theta \text{ in QIV}\}.$$

41 $r = -4\sin\theta \Rightarrow r^2 = -4r\sin\theta \Rightarrow x^2 + y^2 = -4y \Rightarrow$

$$x^2 + y^2 + 4y + \underline{4} = \underline{4} \Rightarrow x^2 + (y+2)^2 = 4.$$

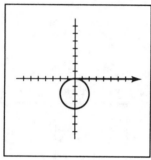

Figure 41

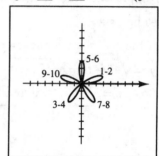

Figure 43

43 $r = 3\sin 5\theta$ is a 5-leafed rose. $0 = 3\sin 5\theta \Rightarrow \sin 5\theta = 0 \Rightarrow 5\theta = \pi n \Rightarrow \theta = \frac{\pi}{5}n$. The numbers 1–10 correspond to θ ranging from 0 to π in $\frac{\pi}{10}$ increments. One leaf is centered on the line $\theta = \frac{\pi}{2}$ and the others are equally spaced apart $\left(\frac{360°}{5} = 72°\right)$.

45 $r = 3 - 3\sin\theta$ is a cardioid since the coefficient of $\sin\theta$ has the same magnitude as the constant term.

$0 = 3 - 3\sin\theta \Rightarrow \sin\theta = 1 \Rightarrow \theta = \frac{\pi}{2} + 2\pi n$.

	Variation of θ		Variation of r	
1)	0	$\rightarrow \frac{\pi}{2}$	3 $\rightarrow$	0
2)	$\frac{\pi}{2}$	$\rightarrow \pi$	0 $\rightarrow$	3
3)	π	$\rightarrow \frac{3\pi}{2}$	3 $\rightarrow$	6
4)	$\frac{3\pi}{2}$	$\rightarrow 2\pi$	6 $\rightarrow$	3

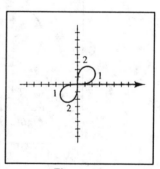

Figure 45

47 $r^2 = 9\sin 2\theta$ •

$0 = 9\sin 2\theta \Rightarrow \sin 2\theta = 0 \Rightarrow 2\theta = \pi n \Rightarrow \theta = \frac{\pi}{2}n$.

	Variation of θ		Variation of r	
1)	0	$\rightarrow \frac{\pi}{4}$	0 $\rightarrow$	± 3
2)	$\frac{\pi}{4}$	$\rightarrow \frac{\pi}{2}$	$\pm 3 \rightarrow$	0
3)	$\frac{\pi}{2}$	$\rightarrow \frac{3\pi}{4}$	undefined	
4)	$\frac{3\pi}{4}$	$\rightarrow \pi$	undefined	

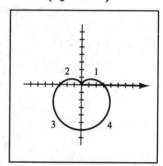

Figure 47

48 $2r = \theta \Rightarrow r = \frac{1}{2}\theta$. Positive values of θ yield the "counterclockwise spiral" while the "clockwise spiral" is obtained from the negative values of θ.

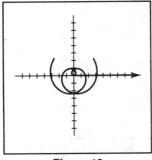

Figure 48

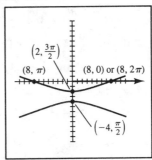

Figure 49

49 $r = \frac{8}{1 - 3\sin\theta} \Rightarrow e = 3 > 1$, hyperbola. See §6.6 for more details on this problem.

$$V(2, \tfrac{3\pi}{2}) \text{ and } V'(-4, \tfrac{\pi}{2}). \quad d(V, F) = 2 \Rightarrow F'(-6, \tfrac{\pi}{2}).$$

50 $r = 6 - r\cos\theta \Rightarrow r + r\cos\theta = 6 \Rightarrow r(1 + \cos\theta) = 6 \Rightarrow r = \frac{6}{1 + 1\cos\theta} \Rightarrow$

$e = 1$, parabola. The vertex is in the $\theta = 0$ direction, $V(3, 0)$.

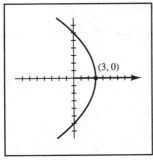

Figure 50

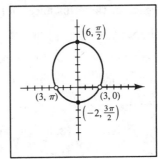

Figure 52

52 $r = \frac{-6\csc\theta}{1 - 2\csc\theta} \cdot \frac{-\sin\theta}{-\sin\theta} = \frac{6}{2 - \sin\theta} = \frac{3}{1 - \frac{1}{2}\sin\theta} \Rightarrow e = \frac{1}{2} < 1$, ellipse.

$V(2, \tfrac{3\pi}{2})$ and $V'(6, \tfrac{\pi}{2})$. $d(V, F) = 2 \Rightarrow F' = (4, \tfrac{\pi}{2})$.

Since the original equation is undefined when $\csc\theta$ is undefined,

the points $(3, 0)$ and $(3, \pi)$ are excluded from the graph.

53 $y^2 = 4x \Rightarrow r^2\sin^2\theta = 4r\cos\theta \Rightarrow r = \frac{4r\cos\theta}{r\sin^2\theta} = 4 \cdot \frac{\cos\theta}{\sin\theta} \cdot \frac{1}{\sin\theta} \Rightarrow r = 4\cot\theta\,\csc\theta$.

56 $x^2 + y^2 = 2xy \Rightarrow r^2 = 2r^2\cos\theta\sin\theta \Rightarrow 1 = 2\sin\theta\cos\theta \Rightarrow \sin 2\theta = 1 \Rightarrow$

$2\theta = \frac{\pi}{2} + 2\pi\text{n} \Rightarrow \theta = \frac{\pi}{4}, \frac{5\pi}{4}$ on $[0, 2\pi)$, which are the same lines.

In rectangular coordinates: $x^2 + y^2 = 2xy \Rightarrow x^2 - 2xy + y^2 = 0 \Rightarrow$

$(x - y)^2 = 0 \Rightarrow x - y = 0$, or $y = x$.

57 $r^2 = \tan\theta \Rightarrow x^2 + y^2 = \frac{y}{x} \Rightarrow x^3 + xy^2 = y$.

[59] $r^2 = 4 \sin 2\theta \Rightarrow r^2 = 4(2\sin\theta \cos\theta) \Rightarrow r^2 = 8\sin\theta \cos\theta \Rightarrow$

$$r^2 \cdot r^2 = 8(r \sin\theta)(r \cos\theta) \Rightarrow (x^2 + y^2)^2 = 8xy.$$

[60] $\theta = \sqrt{3} \Rightarrow \tan^{-1}\left(\frac{y}{x}\right) = \sqrt{3} \Rightarrow \frac{y}{x} = \tan\sqrt{3} \Rightarrow y = (\tan\sqrt{3})\,x.$

Note that $\tan\sqrt{3} \approx -6.15$. This is a line through the origin making an angle of

approximately $99.24°$ with the positive x-axis. The line is *not* $y = \frac{\pi}{3}x$.

[62] $r^2 \sin\theta = 6\csc\theta + r\cot\theta \Rightarrow$

$$r^2 \sin^2\theta = 6 + r\cos\theta \;\{\text{multiply by } \sin\theta \text{ to get } r^2 \sin^2\theta\} \Rightarrow y^2 = 6 + x$$

Chapter 6 Discussion Exercises

[1] For $y = ax^2$, the horizontal line through the focus is $y = p$. Since $a = 1/(4p)$, we have $p = (1/(4p))x^2 \Rightarrow x^2 = 4p^2 \Rightarrow x = 2|p|$. Doubling this value for the width gives us $w = 4|p|$.

[3] Refer to Figure 14 and the derivation on text page 385.

[5] $P(x, y)$ is a distance of $(2 + d)$ from $(0, 0)$ and a distance of d from $(4, 0)$. The difference of these distances is $(2 + d) - d = 2$, a <u>positive constant</u>. By the definition of a hyperbola, $P(x, y)$ lies on the right branch of the hyperbola with foci $(0, 0)$ and $(4, 0)$. The center of the hyperbola is halfway between the foci, i.e., $(2, 0)$. The vertex is halfway from $(2, 0)$ to $(4, 0)$ since the distance from the circle to P equals the distance from P to $(4, 0)$.

Thus, the vertex is $(3, 0)$ and $a = 1$.

$b^2 = c^2 - a^2 = 2^2 - 1^2 = 3$ and

an equation of the right branch of the hyperbola is

$\dfrac{(x-2)^2}{1} - \dfrac{y^2}{3} = 1,\; x \geq 3 \quad \text{or} \quad x = 2 + \sqrt{1 + \dfrac{y^2}{3}}.$

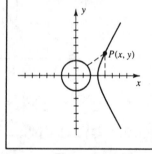

Figure 5

[7] **n even:** There are $2n$ leaves, each having a leaf angle of $(180/n)°$. There is no open space between the leaves.

n odd: There are n leaves, each having a leaf angle of $(180/n)°$. There is $180°$ of open space—each space is $(180/n)°$, equispaced between the leaves. If $n = 4k - 1$, where k is a natural number, there is a leaf centered on the $\theta = 3\pi/2$ axis; and if $n = 4k + 1$, there is a leaf centered on the $\theta = \pi/2$ axis.

For $r = \sin n\theta$, the pole values start at $0°$ and occur every $(180/n)°$. For $r = \cos n\theta$, the pole values start at $(90/n)°$ and occur every $(180/n)°$.

Appendix V: Complex Numbers

5 $(3+5i)(2-7i) = (3+5i)2 + (3+5i)(-7i) = 6 + 10i - 21i - 35i^2$ { group the real

parts and the imaginary parts } $= (6 - 35i^2) + (10 - 21)i = (6 + 35) - 11i = 41 - 11i$

9 $(5-2i)^2 = 5^2 - 2(5)(2i) + (2i)^2 = (25 - 4) - 20i = 21 - 20i$

11 $i(3+4i)^2 = i[(9-16) + 2(3)(4i)] = i(-7 + 24i) = -24 - 7i$

13 $(3+4i)(3-4i)$ { note that this difference of squares ... } $=$

$$3^2 - (4i)^2 = 9 - (-16) = \{ \dots \text{ becomes a "sum of squares" } \} \ 9 + 16 = 25$$

17 Since $i^k = 1$ if k is a multiple of 4, we will write i^{73} as $i^{72}i^1$,

knowing that i^{72} will reduce to 1. $i^{73} = i^{72}i = (i^4)^{18}i = 1^{18}i = i$

21 Multiply by the conjugate of the denominator to eliminate all i's in the denominator.

The new denominator is the sum of the squares of the coefficients—in this case, 6^2

and 2^2. $\quad \dfrac{1-7i}{6-2i} \cdot \dfrac{6+2i}{6+2i} = \dfrac{(6+14) + (2-42)i}{36 - (-4)} = \dfrac{20 - 40i}{40} = \dfrac{1}{2} - i$

25 Multiplying the denominator by i will eliminate the i's in the denominator.

$$\dfrac{4-2i}{-5i} = \dfrac{4-2i}{-5i} \cdot \dfrac{i}{i} = \dfrac{4i - 2i^2}{-5i^2} = \dfrac{2+4i}{5} = \dfrac{2}{5} + \dfrac{4}{5}i$$

27 $(2+5i)^3 = (2)^3 + 3(2)^2(5i) + 3(2)(5i)^2 + (5i)^3 = (8 + 150i^2) + (60i + 125i^3) =$

$$(8 - 150) + (60 - 125)i = -142 - 65i$$

29 A common mistake is to multiply $\sqrt{-4}\sqrt{-16}$ and obtain $\sqrt{64}$, or 8.

The correct procedure is $\sqrt{-4}\sqrt{-16} = \sqrt{4}\,i \cdot \sqrt{16}\,i = (2i)(4i) = 8i^2 = -8$.

$(2 - \sqrt{-4})(3 - \sqrt{-16}) = (2 - 2i)(3 - 4i) = (6 - 8) + (-6i - 8i) = -2 - 14i$

31 $\dfrac{4 + \sqrt{-81}}{7 - \sqrt{-64}} = \dfrac{4 + 9i}{7 - 8i} \cdot \dfrac{7 + 8i}{7 + 8i} = \dfrac{(28 - 72) + (32 + 63)i}{49 - (-64)} = \dfrac{-44 + 95i}{113} = -\dfrac{44}{113} + \dfrac{95}{113}i$

33 $\dfrac{\sqrt{-36}\,\sqrt{-49}}{\sqrt{-16}} = \dfrac{(6i)(7i)}{4i} \cdot \dfrac{-i}{-i} = \dfrac{(-42)(-i)}{-4i^2} = \dfrac{42i}{4} = \dfrac{21}{2}i$

35 We need to equate the real parts and the imaginary parts on each side of " $=$ ".

$8 + (3x + y)i = 2x - 4i \Rightarrow 2x = 8$ and $3x + y = -4 \Rightarrow x = 4, \ y = -16$

37 $(3x + 2y) - y^3i = 9 - 27i \Rightarrow y^3 = 27 \ \{ y = 3 \}$ and $3x + 2y = 9 \Rightarrow x = 1, \ y = 3$

39 $x^2 - 6x + 13 = 0 \Rightarrow$

$$x = \dfrac{-(-6) \pm \sqrt{(-6)^2 - 4(1)(13)}}{2(1)} = \dfrac{6 \pm \sqrt{36 - 52}}{2} = \dfrac{6 \pm \sqrt{-16}}{2} = \dfrac{6 \pm 4i}{2} = 3 \pm 2i$$

41 $x^2 + 4x + 13 = 0 \Rightarrow x = \dfrac{-4 \pm \sqrt{16 - 52}}{2} = \dfrac{-4 \pm 6i}{2} = -2 \pm 3i$

47 Solving $x^3 = -125$ would only give us the solution $x = -5$. So first we need to factor $x^3 + 125$ as the sum of cubes. $x^3 + 125 = 0 \Rightarrow (x+5)(x^2 - 5x + 25) = 0 \Rightarrow$

$$x = -5 \text{ or } x = \frac{5 \pm \sqrt{25 - 100}}{2} = \frac{5 \pm 5\sqrt{3}\,i}{2}. \text{ The three solutions are } -5, \tfrac{5}{2} \pm \tfrac{5}{2}\sqrt{3}\,i.$$

49 $x^4 = 256 \Rightarrow x^4 - 256 = 0 \Rightarrow (x^2 - 16)(x^2 + 16) = 0 \Rightarrow x = \pm 4, \ \pm 4i$

51 $4x^4 + 25x^2 + 36 = 0 \Rightarrow (x^2 + 4)(4x^2 + 9) = 0 \Rightarrow x = \pm 2i, \ \pm \tfrac{3}{2}i$

53 $x^3 + 3x^2 + 4x = 0 \Rightarrow x(x^2 + 3x + 4) = 0 \Rightarrow x = 0, \ -\tfrac{3}{2} \pm \tfrac{1}{2}\sqrt{7}\,i$

55 If $z = a + bi$ and $w = c + di$, then

$$\overline{z + w} = \overline{(a + bi) + (c + di)} \qquad \text{definition of } z \text{ and } w$$
$$= \overline{(a + c) + (b + d)i} \qquad \text{write in complex number form}$$
$$= (a + c) - (b + d)i \qquad \text{definition of conjugate}$$
$$= (a - bi) + (c - di) \qquad \text{rearrange terms}$$
$$= \bar{z} + \bar{w}. \qquad \text{definition of conjugates of } z \text{ and } w$$

In the step described by "rearrange terms",

we are really looking ahead to the terms we want to obtain, $\bar{z}$ and $\bar{w}$.

59 If $\bar{z} = z$, then $a - bi = a + bi$ and hence $-bi = bi$, or $2bi = 0$.

Thus, $b = 0$ and $z = a$ is real. Conversely, if z is real, then $b = 0$ and hence

$$\bar{z} = \overline{a + 0i} = a - 0i = a + 0i = z.$$

Appendix VI: Lines

1 $A(-3, 2)$, $B(5, -4) \Rightarrow m_{AB} = \dfrac{(-4) - 2}{5 - (-3)} = \dfrac{-6}{8} = -\dfrac{3}{4}$

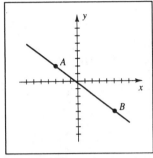

Figure 1

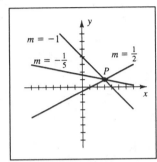

Figure 15

7 Show that the slopes of opposite sides are equal.

$A(-3, 1)$, $B(5, 3)$, $C(3, 0)$, $D(-5, -2) \Rightarrow m_{AB} = \frac{1}{4} = m_{DC}$ and $m_{DA} = \frac{3}{2} = m_{CB}$.

9 Show that the slopes of opposite sides are equal (parallel lines) and the slopes of two adjacent sides are negative reciprocals (perpendicular lines). $A(6, 15)$,

$B(11, 12)$, $C(-1, -8)$, $D(-6, -5) \Rightarrow m_{DA} = \frac{5}{3} = m_{CB}$ and $m_{AB} = -\frac{3}{5} = m_{DC}$.

11 $A(-1, -3)$ is 5 units to the left and 5 units down from $B(4, 2)$. D will have the same relative position from $C(-7, 5)$, that is, $(-7 - 5, 5 - 5) = (-12, 0)$.

15 $P(3, 1)$; $\quad m = \frac{1}{2}, -1, -\frac{1}{5}$ • See *Figure 15* above.

19 (a) Parallel to the y-axis implies the equation is of the form $x = k$.

The x-value of $A(5, -2)$ is 5, hence $x = 5$ is the equation.

(b) Perpendicular to the y-axis implies the equation is of the form $y = k$.

The y-value of $A(5, -2)$ is -2, hence $y = -2$ is the equation.

21 Using the point-slope form, the equation of the line through $A(5, -3)$ with slope -4

is $y + 3 = -4(x - 5) \Rightarrow y + 3 = -4x + 20 \Rightarrow 4x + y = 17$.

23 $A(4, 0)$; slope -3 { use the point-slope form of a line } $\Rightarrow$

$y - 0 = -3(x - 4) \Rightarrow y = -3x + 12 \Rightarrow 3x + y = 12$.

25 $A(4, -5)$, $B(-3, 6) \Rightarrow m_{AB} = -\frac{11}{7}$.

$y + 5 = -\frac{11}{7}(x - 4) \Rightarrow 7(y + 5) = -11(x - 4) \Rightarrow 7y + 35 = -11x + 44 \Rightarrow 11x + 7y = 9$.

27 $5x - 2y = 4 \Leftrightarrow y = \frac{5}{2}x - 2$. Using the same slope, $\frac{5}{2}$, with $A(2, -4)$, gives us

$y + 4 = \frac{5}{2}(x - 2) \Rightarrow 2(y + 4) = 5(x - 2) \Rightarrow 2y + 8 = 5x - 10 \Rightarrow 5x - 2y = 18$.

29 $2x - 5y = 8 \Leftrightarrow y = \frac{2}{5}x - \frac{8}{5}$. Using the negative reciprocal of $\frac{2}{5}$ for the slope,

$y + 3 = -\frac{5}{2}(x - 7) \Rightarrow 2(y + 3) = -5(x - 7) \Rightarrow 2y + 6 = -5x + 35 \Rightarrow 5x + 2y = 29$.

[31] $A(4, 0)$, $B(0, -3) \Rightarrow m = \frac{3}{4}$.

Using the slope-intercept form with $b = -3$ gives us $y = \frac{3}{4}x - 3$.

[33] $A(5, 2)$, $B(-1, 4) \Rightarrow m = -\frac{1}{3}$.

$$y - 2 = -\frac{1}{3}(x - 5) \Rightarrow y = -\frac{1}{3}x + \frac{5}{3} + 2 \Rightarrow y = -\frac{1}{3}x + \frac{11}{3}.$$

[35] We need the line through the midpoint of segment AB that is perpendicular to

segment AB. $A(3, -1)$, $B(-2, 6) \Rightarrow M_{AB} = (\frac{1}{2}, \frac{5}{2})$ and $m_{AB} = -\frac{7}{5}$.

$$y - \frac{5}{2} = \frac{5}{7}(x - \frac{1}{2}) \Rightarrow 7(y - \frac{5}{2}) = 5(x - \frac{1}{2}) \Rightarrow 7y - \frac{35}{2} = 5x - \frac{5}{2} \Rightarrow 5x - 7y = -15.$$

[37] An equation of the line with slope -1 through the origin is

$$y - 0 = -1(x - 0), \text{ or } y = -x.$$

[39] We can solve the given equation for y to obtain the slope-intercept form, $y = mx + b$.

$$2x = 15 - 3y \Rightarrow 3y = -2x + 15 \Rightarrow y = -\frac{2}{3}x + 5; \ m = -\frac{2}{3}, \ b = 5$$

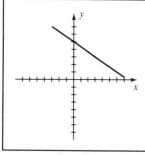

Figure 39

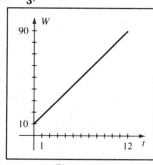

Figure 55

[43] (a) An equation of the horizontal line with y-intercept 3 is $y = 3$.

(b) An equation of the line through the origin with slope $-\frac{1}{2}$ is $y = -\frac{1}{2}x$.

(c) An equation of the line with slope $-\frac{3}{2}$ and y-intercept 1 is $y = -\frac{3}{2}x + 1$.

(d) An equation of the line through $(3, -2)$ with slope -1 is $y + 2 = -(x - 3)$.

Alternatively, we have a slope of -1 and a y-intercept of 1, i.e., $y = -x + 1$.

[45] Since we want to obtain a "1" on the right side of the equation, we will divide by 6.

$$[4x - 2y = 6] \cdot \frac{1}{6} \Rightarrow \frac{4x}{6} - \frac{2y}{6} = \frac{6}{6} \Rightarrow \frac{2x}{3} - \frac{y}{3} = 1 \Rightarrow \frac{x}{\frac{3}{2}} + \frac{y}{-3} = 1$$

The x-intercept is $(\frac{3}{2}, 0)$ and the y-intercept is $(0, -3)$.

[47] The radius of the circle is the vertical distance from the center of the circle to the line

$y = 5$, that is, $r = 5 - (-2) = 7$. An equation is $(x - 3)^2 + (y + 2)^2 = 49$.

[49] $L = 28 \Rightarrow 1.53t - 6.7 = 28 \Rightarrow t = \dfrac{28 + 6.7}{1.53} \approx 22.68$, or approximately 23 weeks.

[51] (a) $L = 40 \Rightarrow W = 1.70(40) - 42.8 = 25.2$ tons

(b) Error in $L = \pm 2 \Rightarrow$ Error in $W = 1.70(\pm 2) = \pm 3.4$ tons

53 (a) $y = mx = \dfrac{\text{change in } y \text{ from the beginning of the season}}{\text{change in } x \text{ from the beginning of the season}}(x) = \dfrac{5-0}{14-0}x = \dfrac{5}{14}x.$

   (b) $x = 162 \Rightarrow y = \frac{5}{14}(162) \approx 58.$

55 (a) Using the slope-intercept form, $W = mt + b = mt + 10.$

   $$W = 30 \text{ when } t = 3 \Rightarrow 30 = 3m + 10 \Rightarrow m = \frac{20}{3} \text{ and } W = \frac{20}{3}t + 10.$$

   (b) $t = 6 \Rightarrow W = \frac{20}{3}(6) + 10 \Rightarrow W = 50 \text{ lb}$

   (c) $W = 70 \Rightarrow 70 = \frac{20}{3}t + 10 \Rightarrow 60 = \frac{20}{3}t \Rightarrow t = 9 \text{ years old}$

   (d) The graph has endpoints at $(0, 10)$ and $(12, 90)$. See *Figure 55* (preceding page).

57 Using $(10, 2480)$ and $(25, 2440)$,

   $$\text{we have } H - 2440 = \frac{2440 - 2480}{25 - 10}(T - 25), \text{ or } H = -\frac{8}{3}T + \frac{7520}{3}.$$

59 (a) Using the slope-intercept form with $m = 0.032$ and $b = 13.5$,

   $$\text{we have } T = 0.032t + 13.5.$$

   (b) $t = 2000 - 1915 = 85 \Rightarrow T = 0.032(85) + 13.5 = 16.22°\text{C}.$

61 (a) Expenses $(E) = (\$1000) + (5\% \text{ of } R) + (\$2600) + (50\% \text{ of } R) \Rightarrow$

   $$E = 1000 + 0.05R + 2600 + 0.50R = 0.55R + 3600.$$

   (b) Profit $(P) = $ Revenue $(R) - $ Expenses $(E) \Rightarrow$

   $$P = R - (0.55R + 3600) = R - 0.55R - 3600 = 0.45R - 3600.$$

   (c) *Break even* means P would be 0. $P = 0 \Rightarrow 0 = 0.45R - 3600 \Rightarrow 0.45R = 3600 \Rightarrow$

   $$R = 3600(\tfrac{100}{45}) = \$8000/\text{month}$$

63 The targets are on the x-axis {which is the line $y = 0$}.

   To determine if a target is hit, set $y = 0$ and solve for x.

   (a) $y - 2 = -1(x - 1) \Rightarrow x + y = 3.$ $y = 0 \Rightarrow x = 3$ and a creature is hit.

   (b) $y - \frac{5}{3} = -\frac{4}{9}(x - \frac{3}{2}) \Rightarrow 4x + 9y = 21.$ $y = 0 \Rightarrow x = 5.25$ and no creature is hit.

65 $s = \dfrac{v_2 - v_1}{h_2 - h_1} \Rightarrow 0.07 = \dfrac{v_2 - 22}{185 - 0} \Rightarrow v_2 = 22 + 0.07(185) = 34.95 \text{ mi/hr}.$

67 The slope of AB is $\dfrac{-1.11905 - (-1.3598)}{-0.55 - (-1.3)} = 0.321.$ Similarly, the slopes of BC and

   CD are also 0.321. Therefore, the points all lie on the same line. Since the common

   slope is 0.321, let $a = 0.321.$ $y = 0.321x + b \Rightarrow -1.3598 = 0.321(-1.3) + b \Rightarrow$

   $b = -0.9425.$ Thus, the points are linearly related by the equation

   $$y = 0.321x - 0.9425.$$

69 $x - 3y = -58 \Leftrightarrow y = (x + 58)/3$ and $3x - y = -70 \Leftrightarrow y = 3x + 70$. Assign $(x + 58)/3$ to Y_1 and $3x + 70$ to Y_2. Using a standard viewing rectangle, we don't see the lines. Zooming out gives us an indication where the lines intersect and by tracing and zooming in or using an intersect feature, we find that the lines intersect at $(-19, 13)$.

$[-30, 3]$ by $[-2, 20]$

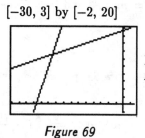

Xscl = 2
Yscl = 2

Figure 69

$[-15, 15]$ by $[-10, 10]$

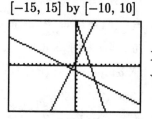

Xscl = 1
Yscl = 1

Figure 71

71 From the graph, we can see that the points of intersection are $A(-0.8, -0.6)$, $B(4.8, -3.4)$, and $C(2, 5)$. The lines intersecting at A are perpendicular since they have slopes of 2 and $-\frac{1}{2}$. Since $d(A, B) = \sqrt{39.2}$ and $d(A, C) = \sqrt{39.2}$, the triangle is isosceles. Thus, the polygon is a right isosceles triangle.

73 The data appear to be linear. Using the two arbitrary points $(-7, -25)$ and $(4.6, 12.2)$, the slope of the line is $\dfrac{12.2 - (-25)}{4.6 - (-7)} \approx 3.2$. An equation of the line is $y + 25 = 3.2(x + 7) \Rightarrow y = 3.2x - 2.6$.

$[-8, 5]$ by $[-27, 15]$

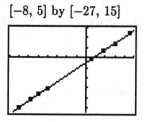

Xscl = 1
Yscl = 5

Figure 73

$[1980, 1988]$ by $[300, 625]$

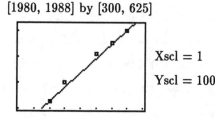

Xscl = 1
Yscl = 100

Figure 75

75 (a) See *Figure 75*.

(b) To find a first approximation for the line use the arbitrary points $(1982, 325)$ and $(1987, 600)$. The resulting line is $y = 55x - 108,685$. Adjustments may be made to this equation.

(c) Let $y = 55x - 108,685$. When $x = 1984$, $y = 435$ and when $x = 1995$, $y = 1040$.